Fundamentals of Rock Mechanics

Fundamentals of Rock Mechanics

Contributors

Lixin Wu and Shanjun Liu et al.

AURIS
Reference

www.aurisreference.com

Fundamentals of Rock Mechanics

Contributors: Lixin Wu and Shanjun Liu et al.

Published by Auris Reference Limited

www.aurisreference.com

United Kingdom

Fundamentals of Rock Mechanics

ISBN: 978-1-78154-904-9

British Library Cataloguing in Publication Data
A CIP record for this book is available from the British Library

Printed in the United Kingdom

Exclusively distributed by CBS Publishers & Distributors Pvt. Ltd.

Sales & Distribution Rights only for India, Pakistan, Bangladesh, Sri Lanka, Nepal and Bhutan.This book is not to be sold outside these territories.

Contents

List of Abbreviations .... *vii*

List of Contributors..*ix*

Preface.. ....*xi*

Chapter 1 **Remote Sensing Rock Mechanics and Earthquake Thermal Infrared Anomalies**... 1

Lixin Wu and Shanjun Liu

Chapter 2 **Performance of EVA-Based Membranes for SCL in Hard Rock** 47

Karl Gunnar Holter

Chapter 3 **An Experimental Investigation of Shale Mechanical Properties through Drained and Undrained Test Mechanisms** 111

Md. Aminul Islam and Paal Skalle

Chapter 4 **Rock Magnetic Properties of Sedimentary Rocks in Central Hokkaido — Insights into Sedimentary and Tectonic Processes on an Active Margin** 173

Yasuto Itoh, Machiko Tamaki, and Osamu Takano

Chapter 5 **Theories on Rock Cutting, Grinding and Polishing Mechanisms** 199

Irfan Celal Engin

Chapter 6 **Simulation of Asymmetric Destabilization of Mine-void Rock Masses Using a Large 3D Physical Model** 227

X. P. Lai, P. F. Shan, J. T. Cao, F. Cui and H. Sun

Chapter 7 **Water Ingress Assessment for Rock Tunnels: A Tool for Risk Planning** 261

Wing Kei Kong

Citations 281

Index.. 283

List of Abbreviations

AE	Acoustic Emission
AMS	Anisotropy of Magnetic Susceptibility
ASTM	American Society for Testing and Materials
BPS	Buried Pressure Sensor
CFP	Close Field Photogrammetry
COA	Crack Optical Acquirement
DCS	Degree of Capillary Saturation
DRM	Detrital Remanent Magnetization
EDZ	Excavation Disturbed Zone
ESTCS	Exploiting Extremely Steep and Thick Coal Seams
HDD	Horizontal Directional Drillholes
HPHT	High Pressure/High Temperature
IRR	Infrared Radiation
ISRM	International Society for Rock Mechanics
LTCC	Longwall Top Coal Caving
NRM	Natural Remanent Magnetization
PAFD	Progressive Alternating Field Demagnetization
RQD	Rock Quality Designation
RSO	Roof Separation Observation
RSRM	Remote Sensing Rock Mechanics
SCL	Sprayed Concrete Tunnel Linings
SEM	Scanning Electron Microscope
SHS	Simulated Hydraulic Support
SPATE	Stress Pattern Analysis by Thermal Emission
SSTCC	Sub Horizontal Section Top Coal Caving
TIR	Thermal Infrared
TSA	Thermo Elastic Stress Analysis

List of Contributors

Lixin Wu
Academy of Disaster Reduction & Emergency Management, Beijing Normal University, Beijing, China
Institute for Geo-informatics & Digital Mine Research, Northeastern University, Shenyang, China

Shanjun Liu
Institute for Geo-informatics & Digital Mine Research, Northeastern University, Shenyang, China

Karl Gunnar Holter
Department of Geology and Mineral Resources Engineering, Norwegian University of Science and Technology, Sem Sælands vei 1, 7491 Trondheim, Norway

Md. Aminul Islam
Department of Petroleum Engineering and Applied Geophysics, Norwegian University of Science and Technology, Trondheim, Norway

Paal Skalle
Department of Petroleum Engineering and Applied Geophysics, Norwegian University of Science and Technology, Trondheim, Norway

Yasuto Itoh
Graduate School of Science, Osaka Prefecture University, Osaka, Japan

Machiko Tamaki
Japan Oil Engineering Co. Ltd., Tokyo, Japan

Osamu Takano
JAPEX Research Center, Japan Petroleum Exploration Co. Ltd., Chiba, Japan

Irfan Celal Engin
Afyon Kocatepe University, Engineering Faculty, Department of Mining Engineering, Afyonkarahisar, Turkey

X. P. Lai
School of Energy and Mining Engineering, Xi'an University of Science and Technology, Xi'an 710054, China

Key Laboratory of Western Mines and Hazard Prevention, Ministry of Education of China, Xi'an 710054, China

P. F. Shan
School of Energy and Mining Engineering, Xi'an University of Science and Technology, Xi'an 710054, China
Key Laboratory of Western Mines and Hazard Prevention, Ministry of Education of China, Xi'an 710054, China

J. T. Cao
School of Energy and Mining Engineering, Xi'an University of Science and Technology, Xi'an 710054, China
Key Laboratory of Western Mines and Hazard Prevention, Ministry of Education of China, Xi'an 710054, China

F. Cui
School of Energy and Mining Engineering, Xi'an University of Science and Technology, Xi'an 710054, China
Key Laboratory of Western Mines and Hazard Prevention, Ministry of Education of China, Xi'an 710054, China

H. Sun
School of Energy and Mining Engineering, Xi'an University of Science and Technology, Xi'an 710054, China
Key Laboratory of Western Mines and Hazard Prevention, Ministry of Education of China, Xi'an 710054, China

Wing Kei Kong
A-P Design, MWH Australia Pty Ltd, Level 3, 35 Boundary Street, South Brisbane, QLD 4101, Australia

Preface

Rock mechanics is a theoretical and applied science of the mechanical behavior of rock and rock masses; compared to geology, it is that branch of mechanics concerned with the response of rock and rock masses to the force fields of their physical environment. The text Fundamentals of Rock Mechanics focuses on mechanical behavior of rock and rock masses, and presents the essentials of rock mechanics. First chapter focuses on remote sensing rock mechanics and earthquake thermal infrared anomalies. The performance of EVA-based membranes for SCL in hard rock has been measured in second chapter. The goal of third chapter is to provide additional insight regarding the organization of the non-linear model input parameters in borehole simulations and to assist other researchers involved in the rock physics-related research fields. In fourth chapter, we present preliminary results of rock magnetic analyses of the Cretaceous Yezo Supergroup, the Eocene Ishikari Group and the Miocene Kawabata formation in order to detect tectonic movements around the basin and to describe the microfabric of sedimentary rocks related to the tectonic regime and sedimentation processes in the mobile zone. Theories on rock cutting, grinding and polishing mechanisms have been proposed in fifth chapter. In sixth chapter, a methodology has been developed for assessing destabilization potential of the host rock mass from mine voids. Last chapter deals with water ingress assessment for rock tunnels.

Chapter 1

REMOTE SENSING ROCK MECHANICS AND EARTHQUAKE THERMAL INFRARED ANOMALIES

Lixin Wu[1,2] and Shanjun Liu[2]

[1]Academy of Disaster Reduction & Emergency Management, Beijing Normal University, Beijing, China

[2]Institute for Geo-informatics & Digital Mine Research, Northeastern University, Shenyang, China

INTRODUCTION

Rock fracturing is the cause of many geo-hazards including tectonic earthquake (EQ), rock burst, rock sloping and rock pillar failure. Radiation signals such as acoustic emission, radio frequency emission and electromagnetic (EM) radiation from loaded deforming rock, are able to provide useful information for monitoring, interpreting and predicting rock fracturing (Renata, 1977, Brady and Rowell, 1986, Yamada et al., 1989, Martelli et al., 1989). Based on thermo-elastic theory, thermo-elastic stress analysis (TSA) and stress pattern analysis by thermal emission (SPATE) were developed for the stress measurement of solid materials, including homogeneous metal, macromolecular and composite materials, respectively in 1960's and 1970's (Mounatin and Webber, 1978). Luong applied thermovision to study experimentally the damage processes of concrete and rock (Luong, 1990), but no reach to the remote sensing on geo-hazards.

In the experiments for investigating the mechanism of satellite thermal infrared (TIR) anomaly before tectonic EQ (Gorny et al., 1988, Qiang et al., 1990), it was discovered that there do exist TIR anomaly before rock fracturing (Geng et al., 1992). Later, it was furthermore discovered that there are obvious TIR features as precursors of rock fracturing, and that the loaded stress around

0.79 σ_c can be taken as a precaution index for the stability monitoring of loaded rocks (Wu and Wang, 1998). To explore the laws of infrared radiation (IRR) variation in the process of rock loading, deforming and fracturing, and to reveal the possible mechanism of satellite TIR anomaly before EQ, a large amount of IRR imaging experiments on rock loaded to fracturing were conducted in China (Wu et al., 2000, 2001, 2002, 2003,2004a, 2004b, 2004c, 2004d, 2006a, 2006b; Deng et al., 2001, Liu et al., 2002). Hence, a new intersection discipline, Remote Sensing Rock Mechanics (RSRM), which takes Remote Sensing, Rock Mechanics, Rock Physics and Informatics as its foundations and serves for remote sensing on geo-hazards, was originated (Geng et al., 1992; Wu et al., 2000).

Based on retrospection to past experiments on RSRM, it was pointed out that there are two IRR anomalies, being IRR image anomaly and IRR temperature curve anomaly respectively, can act as rock fracturing precursors. The average IRR temperature (AIRT), being the integral reflection of surface IRR energy, is applied as a quantitative index to study the temporal evolution of IRR from loaded rock and to seek for the potential precursors of rock fracturing. The temporal evolution of AIRT are the comprehensive effect of a series of physical-mechanical processes inside a loaded rock, such as rock thermo-elastic acting, pore gas desorbing & escaping, fractures producing & extending, rock frictionating, heat transferring and environment radiation. The thermo-elastic effect and the frictional thermal are two of the main mechanisms of increased IRR from loaded rock. RSRM experiments had revealed the laws of changed IRR from loaded rock and provided scientific interpretations for the mechanisms of satellite TIR anomaly before tectonic EQs of Ms>5.5.

REMOTE SENSING ROCK MECHANICS EXPERIMENTS

Experiment Methods and Tools

The typical RSRM experiment is comprised of a loader (uni-axial or bi-axial), an infrared imager and rock samples. As in Figure 1, a bi-axial loader was applied for loading along two directions, and an infrared imager was applied to detect the surface IRR from loaded rock. The maximum imaging rate of the imager is 60f/s, and the recording rate was usually set as 1f/s to record the IRR images continuously. Usually, tectonic EQ might be resulted from the suddenly fracturing of compressively-sheared crust rock, the suddenly breaking of faults at disjointed zones, the suddenly sliding of compressively-sheared faults or the stability losing of compressively loaded intersected faults. To simulate the different mechanisms of rock fracturing and EQ, several typical loading schemes were applied as in Figure 1.

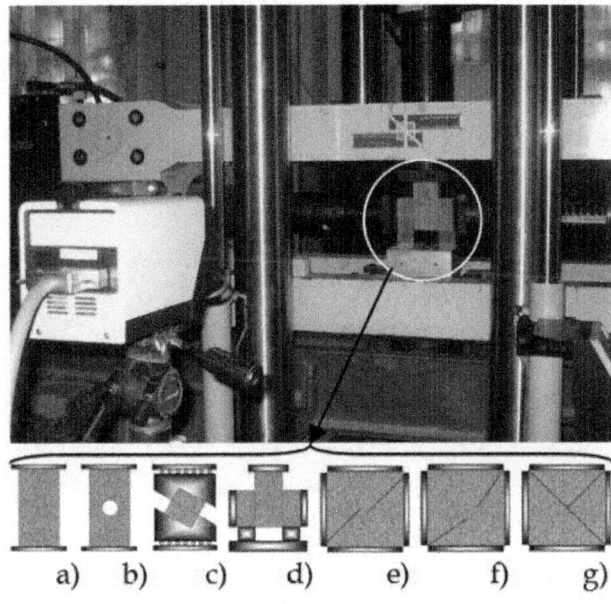

Figure 1. RSRM experiment schemes to simulate different mechanisms of rock fracturing or tectonic EQ: a) uni-axially load on a standard cylinder rock sample; b) uniaxially load on a cylinder rock sample with a central hole; c) compressively-sheared load on a hexahedral rock sample; d) bi-axially load on three jointed rock samples to frictional sliding; e) bi-axially load on a damage rock sample with en echelon faults; f) bi-axially load on a damage rock sample with disjointed faults; and g) bi-axially load on three jointed rock samples simulating intersected faults.

Rock Fracturing Precursor: IRR Image Anomaly

Uni-axially Loaded Rock

Lots of rock samples made from coal, ironstone, sandstone, marble, limestone, granite, granodiorite, gabbro and gneiss were uni-axially loaded and thermal imaging detected. The sample size was standard of diameter and length, respectively, 50 and 100mm. It was discovered that the IRR images of the uni-axially loaded rock have different features for different fracturing pattern (Wu et al., 2006a). As in Figures 2~4, there are three fracturing patterns, "X"-shaped, "//"-shaped and "|"-shaped respectively, occurred in our experiments. The "X"-shaped and "//"-shaped positive IRR abnormal strips foretell the coming of "X"-shaped shearing fracturing and the coming of "//"-shaped shearing fracturing respectively, while the "|"-shaped negative IRR abnormal strip foretells the coming of tensile fracturing.

The "X"-shaped positive IRR abnormal strips generated with loading along the "X"-shaped shearing zone before peak stress, and got distinguished after peak stress, as in Figure 2. The rock sample got finally fractured along the "X"-shape shearing zone. The evolution of IRR abnormal strip had also reflected the fracturing being not symmetrical upper-and lower, in that the upper part was clear with higher temperature, while the lower part is fuzzy with lower temperature.

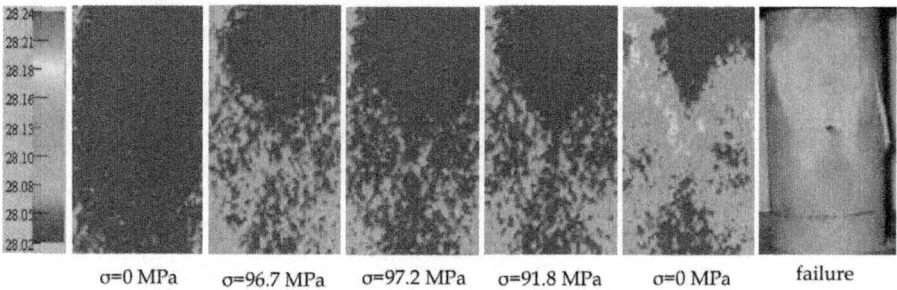

σ=0 MPa   σ=96.7 MPa   σ=97.2 MPa   σ=91.8 MPa   σ=0 MPa   failure

Figure 2. The IRR image positive anomaly of "X"-shaped shearing fracturing of an uni-axially loaded marble sample.

The "//"-shaped positive IRR abnormal strips generated with loading along the "//"-shaped shearing zone at the upper part of sample before peak stress, and got distinguished after peak stress, as in Fig 3. The evolution of the positive IRR image anomaly had also reflected the fracturing being not symmetrical in that the upper part of the IRR anomaly strip was clear with higher temperature, while the lower part was fuzzy excepting for the final fracturing near the bottom of the sample. Besides, there was strong IRR anomaly spot at the fracturing center for the intensive accumulation of mechanical energy and for the intensive generation of frictional thermal at the local central place.

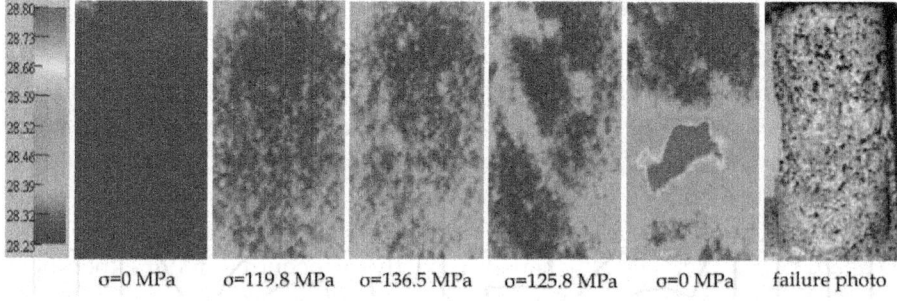

σ=0 MPa   σ=119.8 MPa   σ=136.5 MPa   σ=125.8 MPa   σ=0 MPa   failure photo

Figure 3. The IRR image positive anomaly of "//"-shaped shearing-fracturing of an uni-axially loaded granite sample.

The "|"-shaped negative IRR abnormal strip generated with loading along the tensile fracturing zone of a rock sample before the peak stress, and got distinguished gradually at the peak stress and after fracturing, as the approximately vertical dark strip in Figure 4. The same phenomenon for a sandstone sample with a calcite vein was also reported (Wu, et al., 2000).

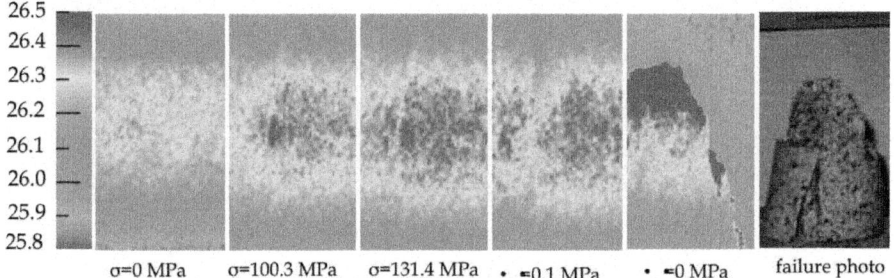

σ=0 MPa   σ=100.3 MPa   σ=131.4 MPa   • ≈0.1 MPa   • ≈0 MPa   failure photo

Figure 4. The IRR image negative anomaly of "|"-shaped tensile fracturing of an uni-axially loaded granite sample.

Uni-Axially Loaded Rock with a Central Hole

More than 10 samples with a central hole, modeling the structure stability of loaded rock tunnels, made from marble and granite were infrared imaging detected. The rock samples had two kinds of shapes respectively being cylinder with diameter and length, respectively, 50 and 100mm, and regular block with thickness, width and length, respectively, 70, 35 and 100 mm. It was discovered that there were distinguished positive IRR image anomalies before rock fracturing, and the place of anomaly were exactly the coming fracturing place. As in Figure 5, the positive IRR image anomalies had reflected the two kinds of fracturing, respectively being diagonal fracturing (sample 1~5) and fork fracturing (sample 6). The IRR anomalies, along the fracturing planes and shaped as spots or strips, generated not only on rock surface but also on the hole's surface (lateral sample 4 and 5). The temperature increment is 1~3°C and 4~8°C respectively for marble and granite samples.

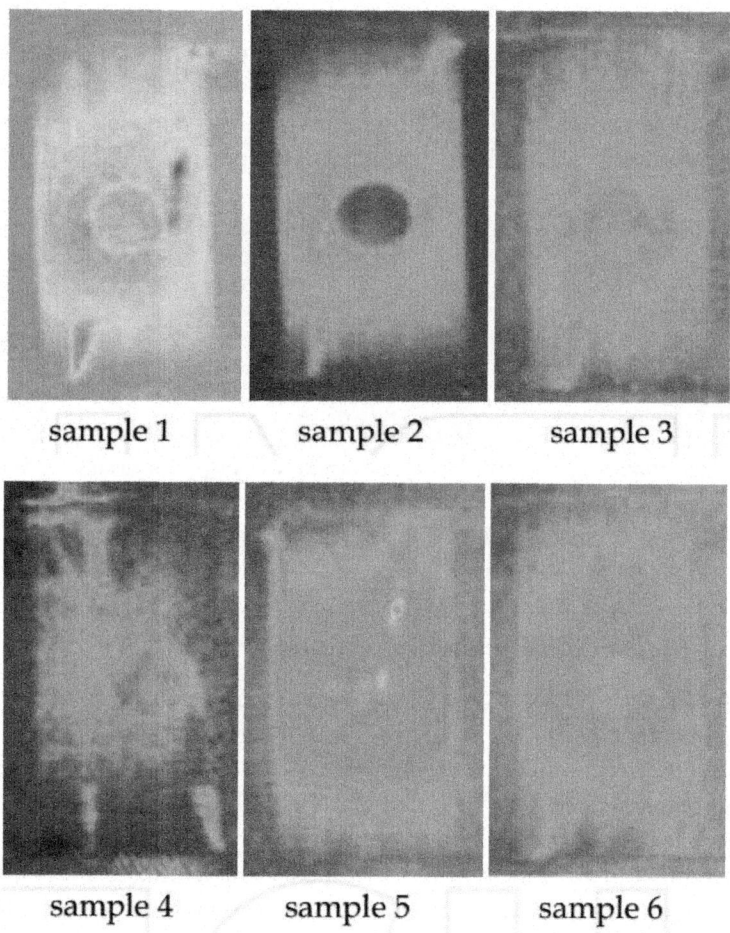

sample 1   sample 2   sample 3

sample 4   sample 5   sample 6

Figure 5. The IRR image positive anomalies of a group of uni-axially loaded marble samples with a central hole.

Compressively Sheared Rock

More than 20 samples, size $7 \times 7 \times 7 \times cm^3$, made from sandstone, marble, limestone, granite and gneiss, were compressively sheared and infrared imaging detected. Three pairs of steel platens with shearing angle being 45°, 60°and 70° respectively were applied. The loading rate was controlled as 2~5kN/s. It was discovered that the IRR temperature of rock surface changed with loading, and a strip-shaped positive IRR image anomaly generated along the central shearing plane before fracturing. With loading, the positive abnormal strip got more and more distinguished and migrated gradually from the upper end to the lower end of the sample, which foretold that the compressive-shearing fracturing was developing gradually from the upper end to the lower end of the sample along the central shear plane. Figure 6shows the typical IRR image series of a compressively sheared limestone sample.

As a special geological phenomenon occurring with the formation of great fault, penniform-shaped fractures are a group of secondary fractures produced with the formation of great primary fracture (Nicolas et al., 1977). It happened to occur in our experiments that there were penniform-shaped fractures produced with a primary fracture in the compressive loaded rock samples, as in Figure 7. The IRR positive anomaly strips generated aside the primary IRR strip, passing through the central shearing plane, had reflected the penniform-shaped fracturing events.

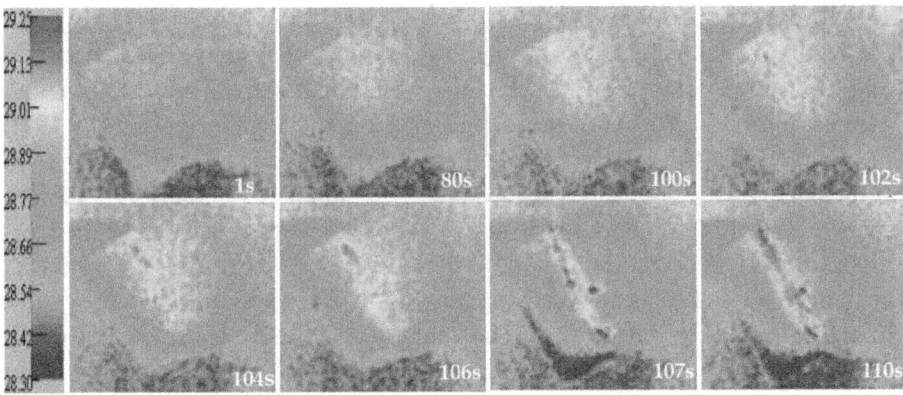

Figure 6. The IRR image positive anomaly of the fracturing of a compressively sheared limestone sample (time in second).

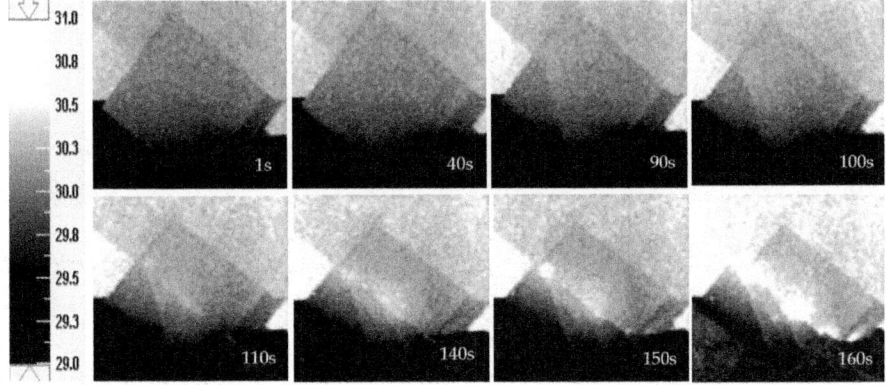

Figure 7. The IRR image anomaly of the penniform-shaped fracturing of a compressively sheared marble sample.

Bi-Sheared Frictional Sliding Rock Blocks

Ten groups of rock samples made from gabbro, granodiorite, limestone and marble were IRR detected in the process of bi-sheared frictional sliding or viscosity sliding. Each group, as in Figure 8 and 9, was comprised of three jointed rock blocks whose size respectively be $50\times50\times100mm^3$, $50\times70\times150mm^3$ and $50\times50\times100mm^3$ from left to right, and its friction area was constant, $50\times100mm^2$. Four contact conditions, symmetrical (yes for rock property and for its smooth friction surface, as in Figure 8), uncertain symmetrical (yes for rock property but not for its coarse friction surface, as in Figure 9), unstable asymmetrical (yes for rock property but not for its staged friction surface) and stable asymmetrical (not for rock property but yes for its smooth friction surface), were designed and tested respectively (Wu et al., 2004b).

It is revealed that the evolution of rock surface IRR temperature field is not only correlated with rock stress, but also correlated with the features of friction surface and rock properties at both sides. General law lies in that the IRR at the place of stress concentration and strong friction zone is stronger than that at the place of stress relaxation and weak friction zone. In condition of friction surface be symmetrical, the IRR image is double butterfly-wings shaped, as in Figure 8. However, in condition of friction surface be uncertain symmetrical, unstable asymmetrical or stable asymmetrical, the temporal-spatial evolution of IRR anomaly is uncertain or unstable, as in Figure 9. The positive IRR anomaly spots, foretelling the evolution of stress, energy and viscosity-sliding process, may be beads-shaped, needle-shaped, suspended needle-shaped, strip-shaped, single butterfly-wings shaped or its evolution in order (Wu et al., 2004b).

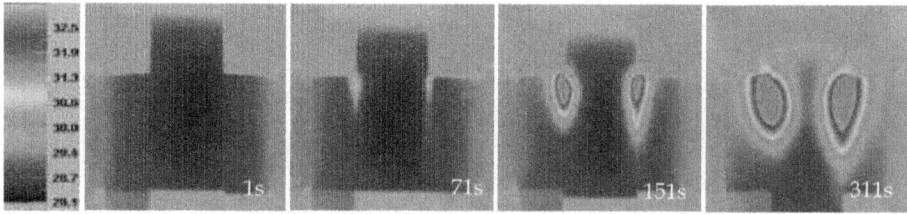

Figure 8. The IRR image positive anomaly of the stick-slipping of symmetrical rock samples (time in second).

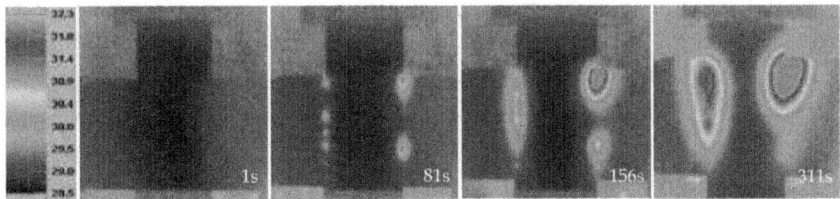

Figure 9. The IRR image positive anomaly of the stick-slipping of asymmetrical rock samples (time in second).

Bi-Axially Loaded Rock

Infrared imaging detection on the rupturing of en echelon and collinearly disjointed jointed faults were done in the process of bi-axial loading. It was revealed that the IRR from loaded rock surface is correlated with loading stress, which could be divided into five stages as loading beginning, linear elastic, stress locking, stress unlocking and fracturing(Wu et al., 2004a). During the stress-unlocking stage, positive IRR anomaly strip generated at the disjointed zone, as in Figure 10 and 11. The positive IRR anomaly strip around the disjointed zone has general evolution features as: firstly, the strip gets enhancing; then, gets weakening (or 'silence'); and finally, gets enhancing again. The re-enhancing of IRR anomaly strip after the weakening stage is a meaningful precursor foretelling the place of primary fracturing of faults or the coming epicenter of an EQ.

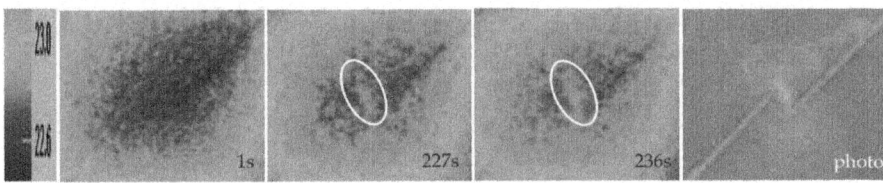

Figure 10. The local IRR positive anomaly foretell the fracturing of en echelon disjointed faults (marble, time in second).

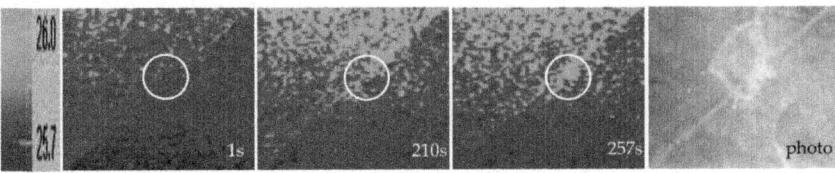

Figure 11. The local IRR positive anomaly foretell the fracturing of collinearly disjointed faults (marble, time in second).

Rock Fracturing Precursor: IRR Temperature Anomaly

Quantitative Index: AIRT

The evolution of surface IRR from loaded rock is the comprehensive effect of rock thermo-elastic acting, pore gas desorbing & escaping, fractures producing & extending, rock frictionating, heat transferring and environment radiation. Being the integral reflection of surface IRR energy, the average IRR temperature (AIRT) is selected as a quantitative index to study the evolution of IRR from loaded rock and to seek for rock fracturing precursors (Wu et al., 2006b).

The infrared imager detects and records the thermal images of loaded rock surface. The thermogram is comprised of a matrix of color pixels which representing the IRR brightness temperature of each pixel of the rock surface. For example, the imaging matrix of TVS-8100MKII infrared imager is 160×120. The IRR temperature of each pixel tends to fluctuate with time due to the instability of the detector unit and the influence of environmental radiation, and the IRR temperature of each pixel will not always be the same for the local difference of rock stress and rock strain. The maximum, minimum and average value of loaded rock surface IRR temperature, respectively being IRRT_{max}, IRRT_{min} and IRRT_{ave}, could be quantitatively obtained from thermogram. The analysis revealed that IRRT_{max} and IRRT_{min} will not change obviously except that IRRT_{max} might rise suddenly just before rock fracturing, while IRRT_{ave} is to change stably with loading, as in Figure 12. The physical interpretation lies in that the surface IRRT_{ave} is a general reflection of the energy balance inside the loaded rock.

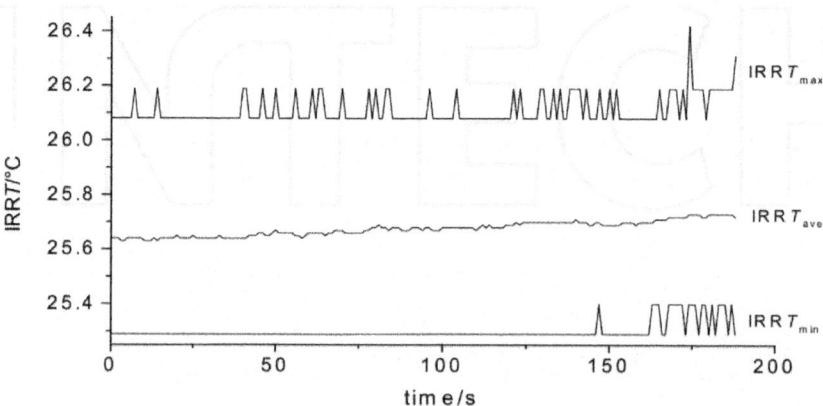

Figure 12. The evolution of three indexes of IRR brightness temperature of loaded rock surface.

Hence, the IRRT_{ave} of rock surface, denominated as AIRT, is selected as a quantitative index to study the precursors of rock fracturing and geo-hazards. The procedures for AIRT-based precursor analysis includes:

- to define a unified boundary of the analyzed region (resampling region) for all thermogrames;

- to resample the IRR temperature value from the data file of each thermograme in time order; 3) to calculate the AIRT of the resampling region of each thermograme;

- to draw the AIRT–time curve of the rock sample;

- to analyze the evaluation features of the AIRT–time curve and to identify the messages as a precursor of rock fracturing and hazard;

- to compare with the qualitative image anomaly so as to analyze and to confirm the AIRT abnormal precursor.

Influence Factors of AIRT Curves

Loading stages and rock deformation

The stress-strain curve is a basic method for describing rock deformation and for interpreting rock mechanical behaviors (Hudson and Harrison, 1997). Generally, the deformation process of loaded rock is divided into four stages respectively being stage-I of defects compaction, stage-II of linear elastic deformation, stage-III of plastic deformation and stage-IV of fracturing failure, as in Figure 13. The four characteristic points, E, Y, P and F are called as elastic-starting point, yield-starting point, peak-stress point and failure-impending point respectively.

Stage-I: the downward-concave curve section tells that there are some defects such as pores, fissures and joints inside the rock body, and that the defects is under compaction, which cause the stress to rise slowly. The more the defects, the severe the curve downward concave.

Stage-II: the curve section linearly developed tells that the compaction of defects has finished and the rock is undergoing elastic deformation. The higher the angle of the section line inclined, the stronger the rock.

Stage-III: the upward-concave curve section tells that there are new fractures developing inside the rock. The plastic deformation starts, and the new generated fractures together with the initial defects are possible to cause friction between its two side-faces.

Stage-IV: the curve section turning to drop tells that the fractures are getting wider, longer and to connect with each other. The rock is losing its strength and stability, and the final fracturing failure or rock hazard is impending.

For the difference between rock compositions, the details of stress-strain curve of different rock will be different. As to brittle rock, its stage-II is close to point P and its stage-IV will be cliff-shaped. Usually, most of the crust rocks are brittle. Five kinds of typical crust rock, granodiorite, gabbro, gneiss, limestone and marble had been tested in our experiments. The typical load-displacement curves of the tested rock samples are shown in Figure 14. It tells that all the tested rocks are brittle.

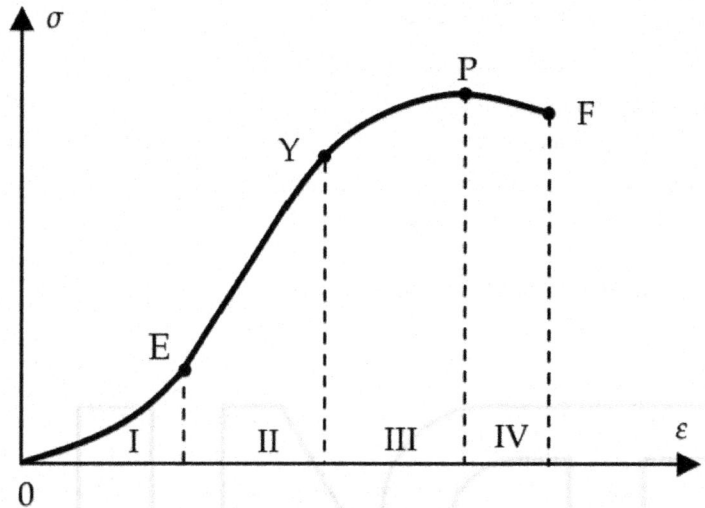

Figure 13. Typical AIRT curve of uni-axially loaded rock.

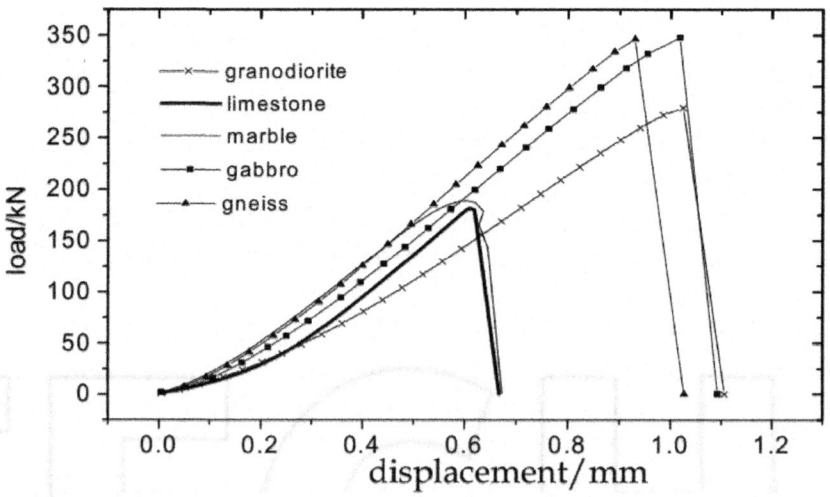

Figure 14. The load-displacement curves of five tested rocks.

The influence of loading condition

The experiments discovered that the evolutions of AIRT have different laws in different loading condition, such as uni-axial loading (constant displacement controlled), compressively-sheared loading (approximately constant load controlled) and bi-axial loading (constant displacement controlled).

- Uni-axial loading

As shown in the left-hand side of Figure 15, the surface facing to the infrared imager is to be detected, and a rectangle region close to the boundary of the rock sample is defined for data resampling and analyzing (Liu et al., 2002; Wu et al., 2002). Multiple experiments revealed that there was slight variation of AIRT at different deformation stage although the AIRT linearly increased with load and deformation. At stage-I, the AIRT will rise slowly or drop a little; at stage-II, the AIRT will rise stably; at stage-III, the AIRT will rise quickly than that in stage-I and stage-II. The right-hand side of Figure 15 shows the comparison of the evolution of AIRT and the load with rock deformation, which is rock displacement in generally, of a marble sample.

- Compressive-shear loading

In condition of compressively shear, the rock sample will always get fracturing along the shearing plane, which locates near to the central plane of the loaded sample. To minimize data resampling work and to focus on the key region, a narrow rectangle along the shearing plane is defined as the resampling and analyzing region (Wu et al., 2004c). Multiple experiments revealed that the temporal evolution of the AIRT is different with the shearing angle. Three shearing angles (γ) being 45°, 60° and 70° respectively, are applied. As the shearing angle changes from 45° to 70°, the temporal evolution changes from monotonic rise, to drop-to-rise and to monotonic drop in order, as in Figure 16.

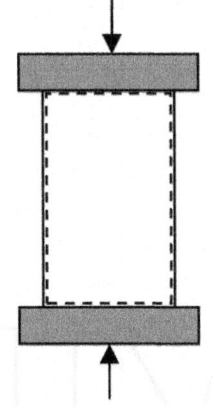

resampling region (in broken line)

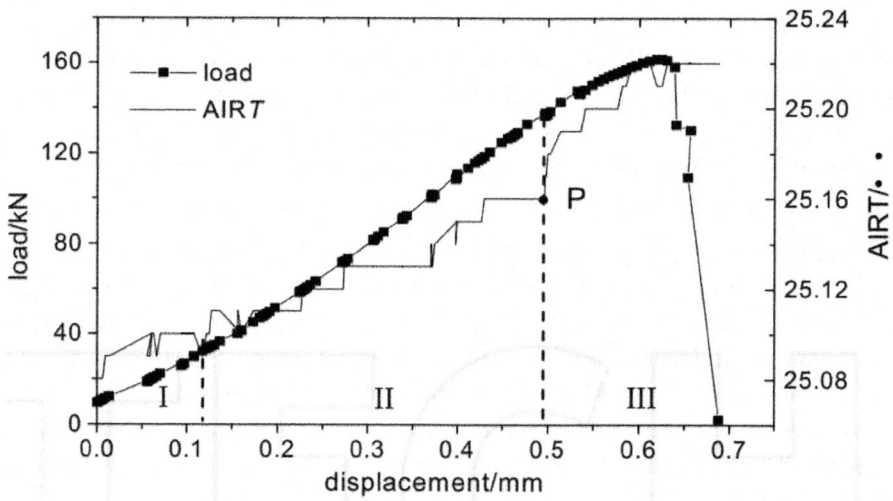

Figure 15. The typical AIRT curve of uni-axially loaded rock sample (marble).

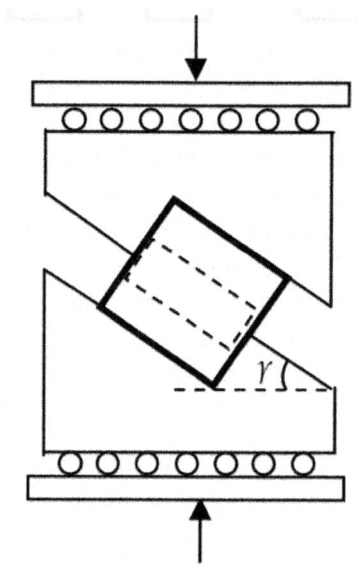

resampling region(in broken line)

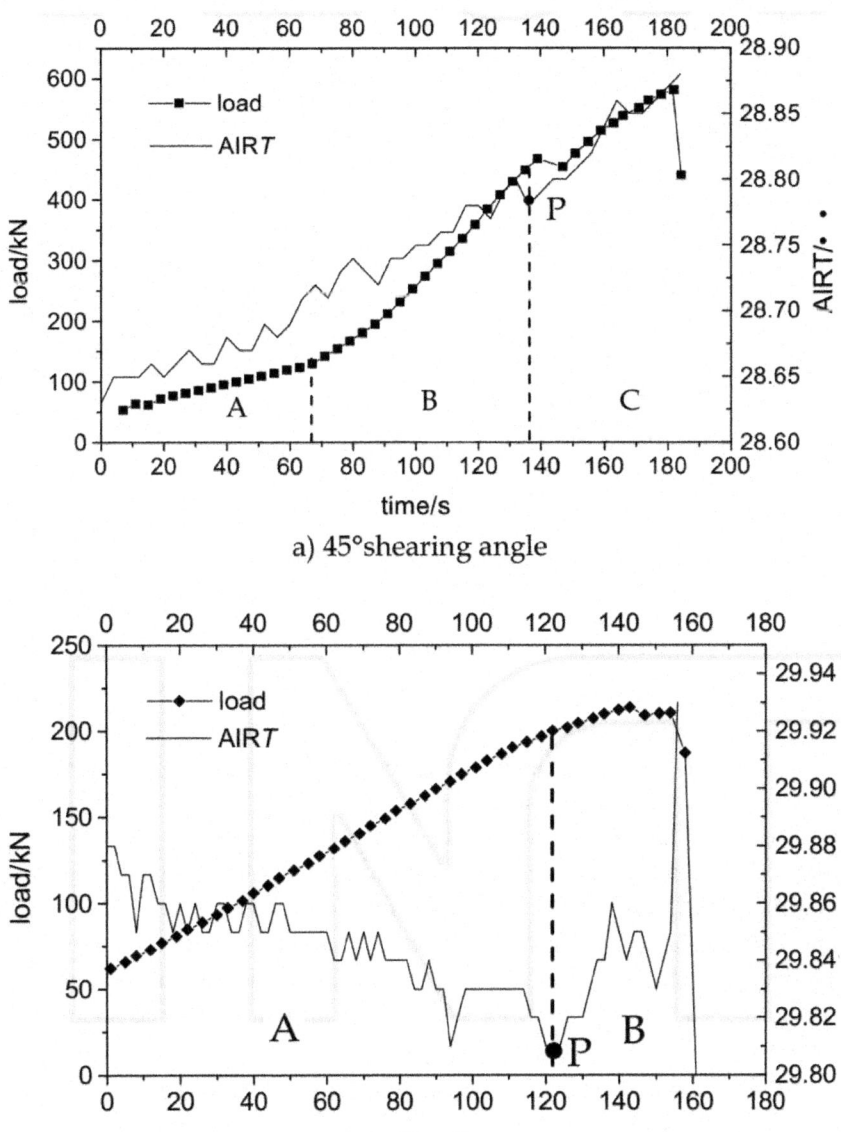

a) 45°shearing angle

b) 60°shearing angle

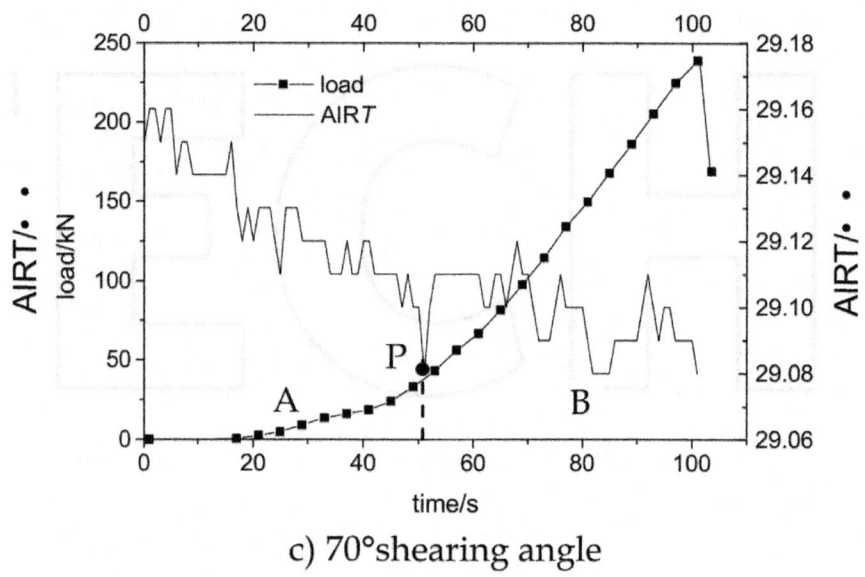

Figure 16. The typical AIR*T* curve of compressively-sheared rock samples at different shearing angles.

The mechanism lays in that the ratio of compressive-stress to shear-stress along the shearing plane decrease with the rise of the shearing angle. The smaller is the shearing angle, the higher is the compression-shear ratio. In condition of 45°, the load-time curve developed in three stages, as stage A, B and C in Figure 16a, with load increasing from slow to rapid, and to slow again. In condition of 60°the load-time curve developed in two stages, as stage A and B in Figure 16b, with load speed changing from approximate constant to be decrease slightly. In condition of 70°, the load-time curve developed in two stages, as stage A and B in Figure 16c, with loading speed changing from slow to rapid.

The compressive action on loaded rock is to cause surface IRR temperature rise, while the tensile action on loaded rock is to cause surface IRR temperature drop. Actually, both compressive action and tensile action are to occur along the compressively sheared plane, and the detected surface IRR is the comprehensive effect of the two actions. It was reached that (Wu et al., 2004c): 1) in condition of shearing angle being 45°, the surface AIR*T* will rise monotonically with loading in that the temperature increment from compressive action and friction is stronger than the temperature decrement from tensile action in the whole loading process; 2) in condition of shearing angle being 60°, the surface AIR*T* will drop monotonically with loading in that the temperature increment from compressive action and friction is weaker than the temperature decrement from tensile action before stage-III (point Y in Figure 13, and point P in Figure 16b); with the friction effect getting strong in stage-III, the surface AIRT will get to rise in that the temperature from compression and friction get stronger than the temperature decrement from tensile action; 3) in condition of 70°, the surface AIRT will drop monotonically with loading in that the temperature increment from both compression and friction are weaker than the temperature decrement from tensile action.

- Biaxial loading

By using of bi-axial loading system and infrared imaging system, the IRR features of two kinds of disjointed jointed faults, respectively be collinearly and non-collinearly disjointed faults, were experimentally studied (Wu et al., 2004a). Since all the faults got fractured finally at the disjointed zone, a circle covering the disjointed region is defined as the resampling region, as in the left-hand side of Figure 17. It could be known from the right-hand side of Figure 17 that the IRR from loaded samples is related with load stress, and the evolution stage could be classified into five stages (I~V) relating with initial compacting, elastic deforming, stress blocking, stress deblocking and rock fracturing respectively. From stage-II to stage-IV, the evolution of AIRT has the features of rising to dropping, and to rising again.

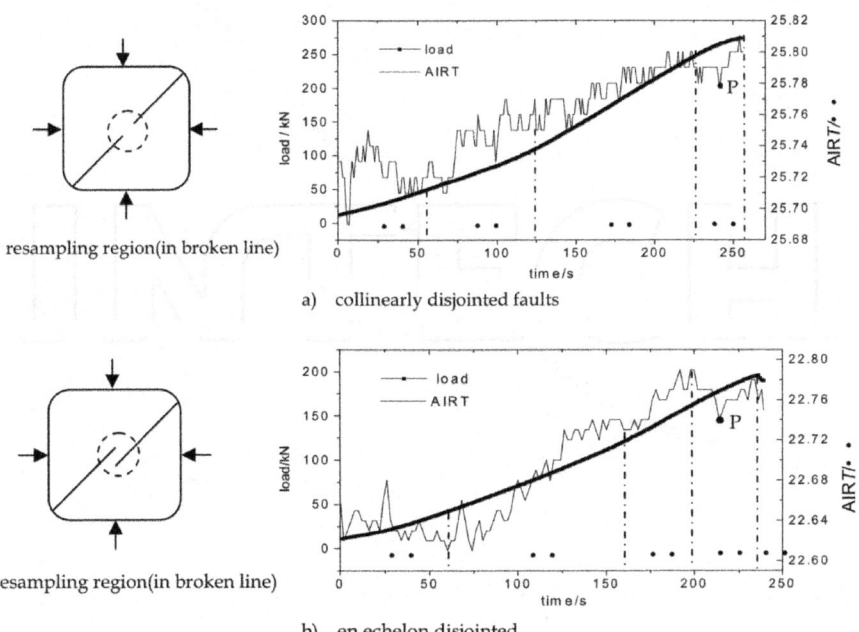

a) collinearly disjointed faults

b) en echelon disjointed

Figure 17. The typical AIRT evolution of two kinds of disjointed faults bi-axially loaded.

The influence of rock characteristics

It was discovered in our experiments that for the most of rock samples in condition of uni-axial loading, its AIRT approximately rose with loading. But there were a few abnormal samples made from limestone had shown AIRT features of dropping with loading, as in Figure 18. The cause lies in that limestone has much more pores than the other rocks. Usually, there are many gases, such as CH_4, CO_2, CO and O_2etc., enclosed inside the pores of rock body (Wang, 2003). With the decrement of pore volume due to the loading compaction and with the increment of fractures produced inside the limestone sample, the pore gases will get escaping. The escaping behavior of pore gas needs to absorb thermal energy from the rock sample. If the heat from compression and friction is lower than that absorbed by pore gases, the surface AIRT is to drop with loading. If look carefully at the load-displacement curves in Figure 14, it could be founded that the curve of limestone concaved downward the most at the compaction stage as compared to that of the other four kinds of rock, which means that there are more pores inside limestone than the others.

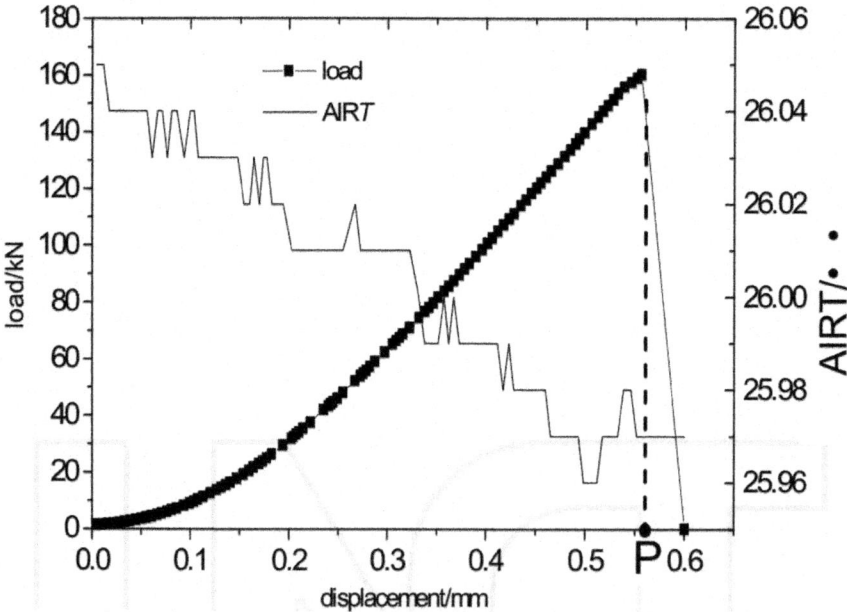

Figure 18. AIRT of uni-axially loaded limestone sample.

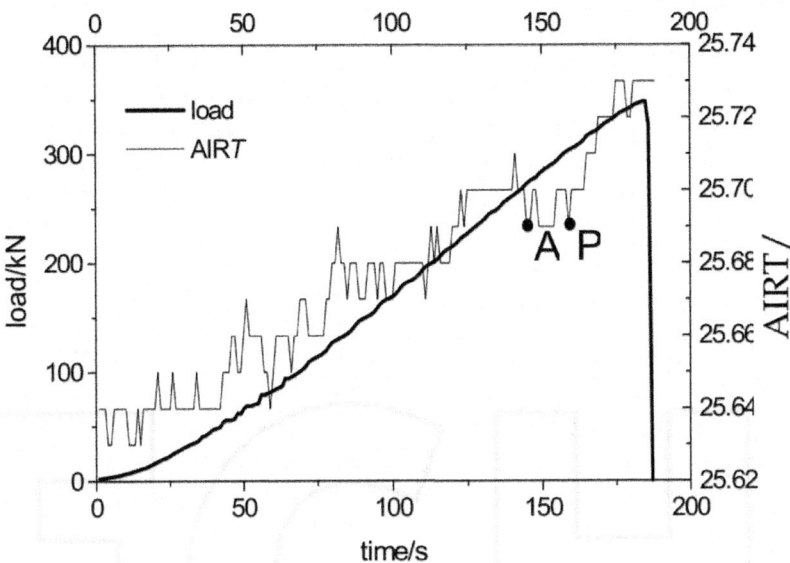

Figure 19. AIR*T* short-dropping precursor for uni-axially loaded gabbro sample.

The Classification of Precursors

Analysis to the evolution of AIR*T* curves discovered that a large amount of rock samples in condition of uni-axial loading, compressively-sheared loading and bi-axial loading had presented obvious precursors for rock fracturing. As referring to the general process of AIR*T* evolution, the AIR*T* anomaly precursors for rock fracturing and hazard could be classified as short-dropping, rapid-rising and dropping-to-rising respectively.

Short-dropping precursor

The AIR*T* curve rises with loading but has a short dropping at loading stage-IV; later, the AIR*T* curve will rise again. The bi-axially load on collinearly and non-collinearly disjointed faults had shown short-dropping precursors as in Figure 17, and the point P was suggested to be the precursor point of rock fracturing and rock hazard. Figure 19 shows another typical case of gabbro sample uni-axially loaded. Here, point A is the turning point of AIR*T* from rising to short dropping, and point P is another turning point from short dropping to rising again, which is suggested to be the precursor point of rock fracturing and rock hazard.

Rapid-rising precursor

The AIR*T* curve rises slowly with loading but turns to rise rapidly before rock fracturing, and the turning point is exactly the precursor point. Figure 15 and Figure 16a have this kind of precursor. Figure 20 shows another typical case of marble sample uni-axially loaded. Here, point P is the turning point of AIR*T* from rising slowly to rising fast, which is suggested to be the precursor point of rock fracturing and rock hazard.

Dropping-to-rising precursor

The AIR*T* curve drops slowly with loading but turns to rise just before rock fracturing, and the turning point is exactly the precursor point. Figure 16b has this kind of precursor. Figure 21 shows another typical case of marble sample uni-axially loaded. Here, point P is the turning point of AIR*T* from dropping slowly to rising fast, which is suggested to be the precursor point of rock fracturing and rock hazard.

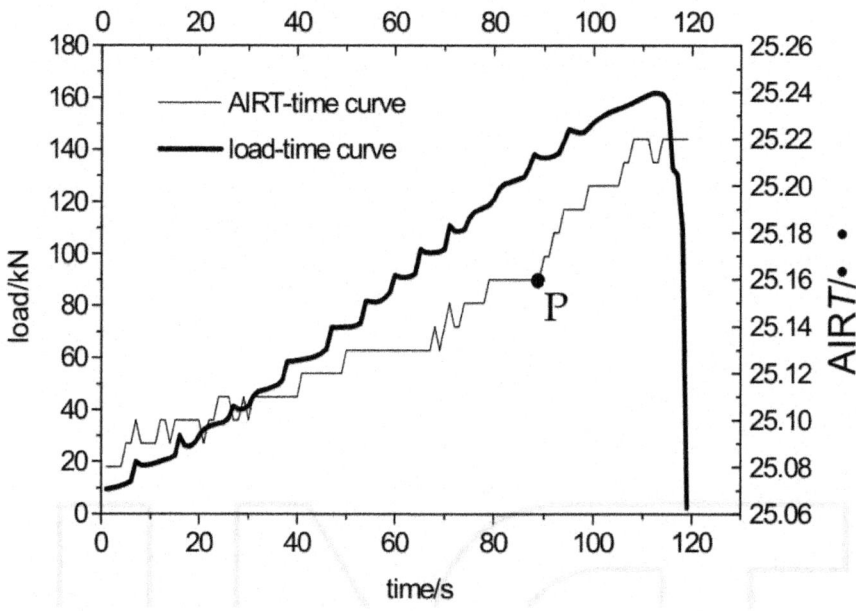

Figure 20. AIR*T* fast rising precursor for uni-axially loaded marble sample.

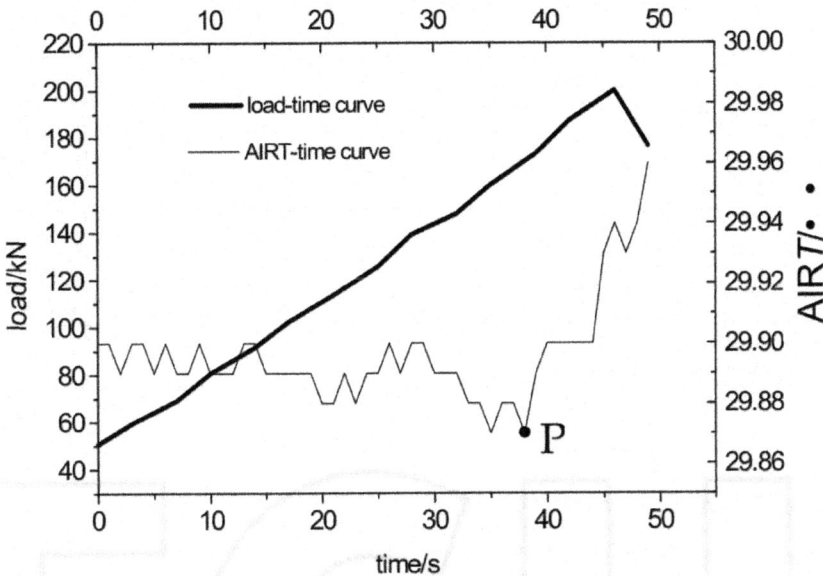

Figure 21. AIRT dropping-to-rising precursor for compressively sheared loaded marble sample.

The Temporal Features of Precursors

The occurrence moments of the precursors of AIRT of totally 52 tested rock samples are listed in Table 1(Wu et al, 2006b). Although the loading conditions and the rock samples are different, the precursor occurrence moment, were very similar as $0.77\sim0.94\sigma\,c\,(\sigma_p)$. Here, σ_c is the uni-axial compressive strength, and σ_p is the peak stress. The precursor occurrence moment of uni-axially loaded or compressive sheared rock sample is $0.79\sigma_c$ and $0.82\sigma_c$ respectively. It is worthy to mention that the precursor occurrence moment of bi-axially loaded collinearly disjointed faults and echelon faults are much different. That for collinearly disjointed faults was close to the peak stress, $0.87\sigma_p$, while that for echelon faults was far away from peak stress, $0.77\sigma_p$. It provides an important evidence for the complexity of study on tectonic EQ prediction on shock time, based on satellite infrared remote sensing and referring to the seismogenic mechanism.

Table 1. Statistic for precursor occurrence in different loading conditions

Loading condition		Sum of tested samples, St	Sum of samples with precursors, Sp	The ratio: (Sp/ St)×100%		Average of precursor occurrence moment
uni-axial loading		22	9	41%		$0.79_\sigma c$
compressively-sheared loading	70°	7	1	14%	58%	$0.94\sigma p$
	60°	8	6	74%		$0.82\sigma p$
	45°	11	8	73%		$0.77\sigma p$
Bi-axial loading for col-linearly disjointed faults		2	2	100%		$0.87_\sigma p$
Bi-axial loading for en echelon faults		2	2	100%		$0.77_\sigma p$

Large IRR at Fracturing Centre

As the recording rate of the infrared imager applied was 60f/s, the transient IRR temperature at the fracturing center could be snapped. In condition of uni-axial loading, the fracturing center of a brittle rock is usually at the "X"-shaped fracturing center. It is discovered that the transient IRR temperature at the fracturing center is much higher than that on rock surface, as in Figure 22, and it is positively related with rock strength and rock deformation. For some compressively sheared hard rock samples made from gabbro and gneiss, the transient IRR temperature at the fracturing center is higher than 155°C, which is the upper limitation of the 2nd temperature range (72~155°C) of the imager applied.

In condition of high angle, 60°and 70°, compressively sheared loading, the fracturing center is at the center of the fractured shearing zone. Since the ruptured upper block of rock sample was pushed apart from the steel platen immediately after the abrupt rupturing, usually be 1~2 s after the rupturing, the inside shearing zone got exposed to the imager immediately and the transient IRR temperature filed was snapped. It was discovered that the IRR temperature on the inside shearing zone is not only much higher than that of outside rock surface, but also inhomogeneous distributed, neither even nor centripetal, as in Figure 23. It means that much more mechanical energy had been converted into frictional thermal and IRR energy due to the intensive energy accumulation, the sufficient local deformation and the abrupt frictional sliding at the center of the shearing zone. In other words, the large IRR temperature at the inside shearing center had reflected the comprehensive effect of local concentrated energy conversion and frictional thermal.

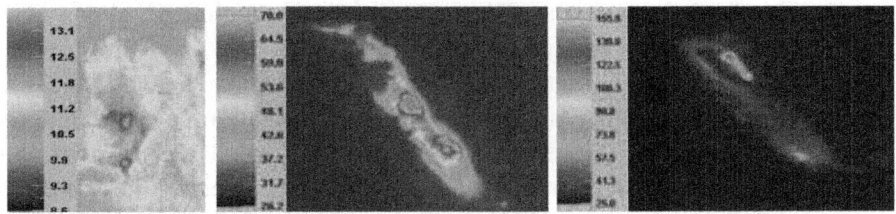

a) uni-axially loaded granite b) compressively sheared gabbro-1 c) compressively sheared gabbro-2

Figure 22. The transient IRR thermograme of fracturing rock samples.

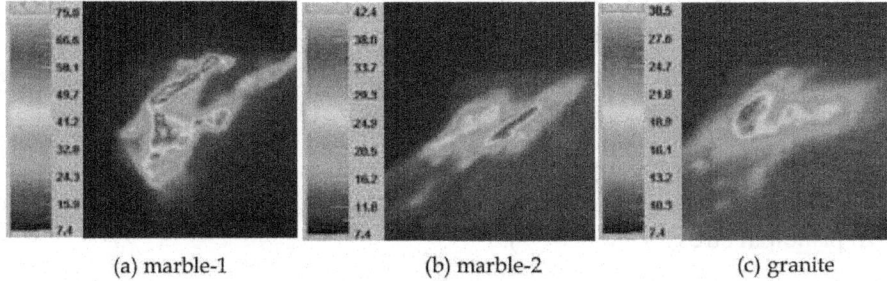

(a) marble-1   (b) marble-2   (c) granite

Figure 23. The inside IRR isothermal filed on the fracturing zone of compressively sheared rock samples.

Hence, it could be deduced that in condition of great tectonic stress, large deformation and/or abrupt frictional sliding, large temperature as high as hundreds or thousands of degree Celsius is possible inside crust rocks. The high temperature could cause partial melt of crust rock, which provides a scientific explanation for the existence of pseudotachylyte in some lager faults (Nicolas et al., 1977, Sibson et al, 1980) and for the failure-generated EQ lights (Martelli et al., 1989). Besides, we can deduce that the continuous shearing deformation or the abrupt fracturing of highly loaded rock/coal body in a coal mine is possible to cause local sheared heating of temperature hundreds of degree Celsius, which might be a potential ignition of local methane (the minimum ignition temperature is 595°C).

REMOTE SENSING ROCK MECHANICS MODEL (RSRM-MODEL)

Thermo-Mechanical Coupling Effect

Thermo-Mechanical Coupling in a Loaded Solid

The heat production inside a loaded solid is called as thermo-mechanical coupling effect. According to the material features and the different deformation stages of a loaded solid, the thermo-mechanical coupling is classified as thermo-elastic, thermo-plastic and thermo-viscous respectively for elastic deformation, plastic deformation and viscous deformation. Generally the rock is a hard brittle solid, its plastic and viscous deformation could be ignored, and the thermo-elastic effect and the frictional thermal are the two chief mechanisms of surface IRR from loaded rock. Kelvin coined the thermo-elastic theory in 1853 that the changed physical temperature of a loaded component is correlated to its changed stress as follows:

$$\Delta T / T = -K_0 \Delta \sigma \tag{1}$$

Here: T is the absolute temperature of a loaded component (K); ΔT is the changed temperature (K); K_0 is the thermo-elastic factor (MPa⁻¹); and Δ_σ is the changed sum of three principal stresses ($\sum \sigma_i, i = 1,2,3$, MPa).

As for an isotropic linear elastic solid loaded bi-axially with a free surface, the surface physical temperature variation is tightly correlated with the sum of two principal stresses ($\sum \sigma_i, i = 1,2$):

$$\Delta T = -\alpha / \rho C_p \cdot [T \cdot \Delta(\sigma_1 + \sigma_2)] \tag{2}$$

Here: T is the surface absolute temperature of a loaded solid (K); ΔT is the changed temperature (K); α is the factor of linear expansion (K⁻¹); ρ is the solid density (Kg m⁻³); Cp is thermal capacity of solid at normal atmosphere (J Kg⁻¹ K⁻¹); σ_1 and σ_2 are the two principal stresses (MPa). The thermo-elastic factor K is defined as K=−α/ρC_p.

For the mechanism of stress measurement with TSA and SPATE, the relationship between the stress increment and the IRR signal based on equation (2) is as follows (Mounatin and Webber, 1978):

$$\Delta(\sigma_1 + \sigma_2) = A_{th} \cdot \Delta S$$
$$\Delta S = \Delta(\sigma_1 + \sigma_2) \cdot A_{th}^{-1} \cdot \tag{3}$$

Here: A_{th} is a comprehensive factor called as corrective factor, which is a function of solid surface emissivity, solid surface physical temperature, solid thermo-elastic factor and three parameters related to the IRR detector, unit in MPa U⁻¹. ΔS is the increment of thermo-elastic voltage signal detected (U).

Changed IRR Temperature of Loaded Rock

If the slight change of rock surface emissivity, rock thermo-elastic factor and the physical parameters of IRR detector during rock loading could be ignored, and if the changed rock surface physical temperature cannot be ignored due to the thermal exchange and the frictional thermal, the relationship between A_{th} and changed rock surface physical temperature could be expressed as $A_{th}=\beta \cdot T^{-1}$. The detected IRR signal S is a direct representation of surface IRR temperature, i.e., $\Delta IRRT=\gamma \cdot \Delta S$. Hence, the following equation for changed IRR temperature could be deduced:

$$\Delta IRRT = \gamma \cdot \beta^{-1} \cdot T \cdot \Delta(\sigma_1 + \sigma_2)$$

(4)

Here: $\Delta IRRT$ is the changed IRR temperature (K); β is a constant correction factor related to rock surface's emisivity, rock thermo-elastic factor and three parameters of IRR detector, in unit MPa K U^{-1}; γ is a transfer factor between detected voltage signal and IRR temperature (K U^{-1}).

It means in equation (4) that the changed IRR temperature of rock surface is a direct reflection of the changed sum of the two principal stresses. If no frictional thermal produced and the thermal exchange is stable, the IRR temperature of rock surface is to rise with loading, and the spatial-temporal evolution of surface IRR image will be stable. If there is no frictional thermal produced but the thermal exchange is unstable, the surface IRR temperature will be unstable, and the spatial-temporal evolution of surface IRR image will also be unstable. If there is frictional thermal produced inside and conducted to rock surface, both the thermal exchange and the surface IRR temperature of rock surface will be unstable, and the spatial-temporal evolution of surface IRR image will get complicated, as in Figures 2~5, 10, 11.

Especially, in condition of compressive shearing and fictional sliding, there is a large amount of frictional thermal produced in the friction zone, which is to cause the rise of physical temperature in friction zone. The rock surface temperature will rise if the thermal conduction from the friction zone can reach to rock surface. Hence, the IRR image anomaly will be distinguished and sometimes be large as a combined effect of rock stress and frictional thermal, as in Figures 6~9.

Remote Sensing Rock Mechanics Model

Energy-related IRR from Loaded Rock

As a relatively independent closed system comprised of loader head, rock sample and environmental air, as in Figure 24, the rock deformation, rock fracturing and rock hazard are all of a complex process of energy input and consumption. If the possible chemical reactions inside a loaded rock can be ignored, the inputted energy of a loaded rock will include the mechanical work from loader and the heat input through positive thermal exchange from loader head and environmental air. The energy consumption by the loaded rock is much more complex including the energy accumulation in rock and the energy dissipation from rock.

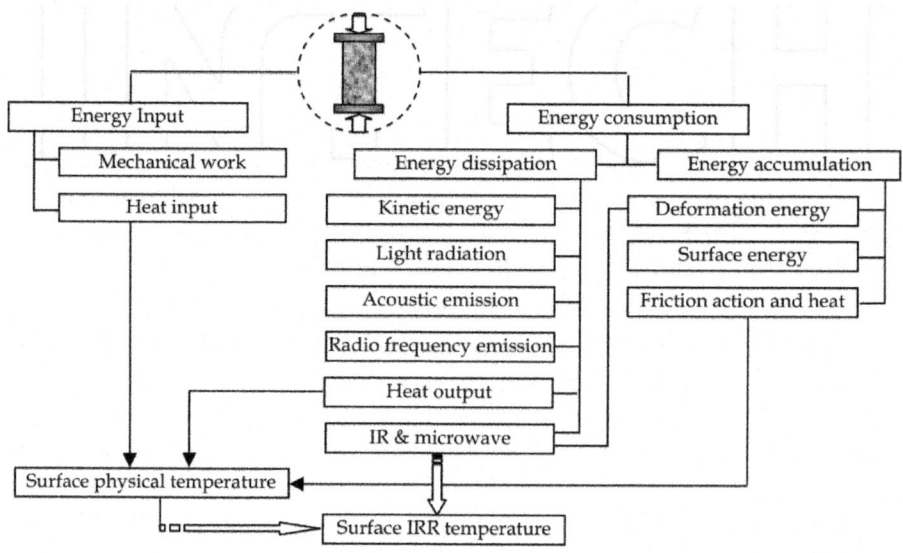

Figure 24. The IRR mechanism related to the energy accumulation and consumption of a loaded rock.

The energy accumulation in a loaded rock includes the positive elastic-plastic deformation energy of rock (the positive change of oscillation and rotation energy of mineral molecules), the surface energy of new produced fractures or fissures, and the friction actions between mineral molecules, grains, joints, fissures and fractures inside the rock as well as thus produced frictional thermal. The energy dissipation from loaded rock includes the negative thermal exchange with the loader head and/or environmental air (i.e., heat output), the kinetic energy of departed fragments of fractured rock, the light radiation, acoustic emission, radio frequency emission and IR & microwave radiation.

The thermal exchange and the friction action are to change the heat state of a loaded rock, and the rock surface physical temperature is a direct index reflecting the heat state of the loaded rock. Stephen-Boltzmann law states that the IRR strength (radiation flux density) of any material, at temperature above absolute zero degree, is biquadratic to its surface physical temperature. Crystal Physics states that the energy jump of molecules oscillation and/or rotation due to the change of molecules distance, resulting from deformation, is an important mechanism of electromagnetic radiation. Hence, rock surface IRR is a comprehensive effect of rock deformation and rock surface thermal state. Rock surface IRR temperature could be a detective index reflecting rock surface physical temperature and rock surface deformation field, which implicating the complex physical-mechanical process inside the loaded rock.

In spite of thermal exchange and plastic deformation, the thermo-elastic effect and the frictional thermal are two of the main mechanics of changed IRR from loaded brittle rock. In the stage of elastic deformation, the thermo-elastic effect is the main cause; while in the stage of plastic deformation or fracturing, the friction-thermal effect plays a great role. At the moment of rock fracturing or hazard, the fraction-heat effect gets more distinguished. The frictional thermal effect depends on two factors being frictional force (decided by normal stress and frictional coefficient) and frictional speed respectively. The larger the frictional force and the quicker the frictional speed, the more the frictional heat.

RSRM-Model Based on Independent System

The rock sample, load header and environment air could be taken as a closed independent system, as in Figure 25, and the energy of the loaded rock sample is in a balance state as follows:

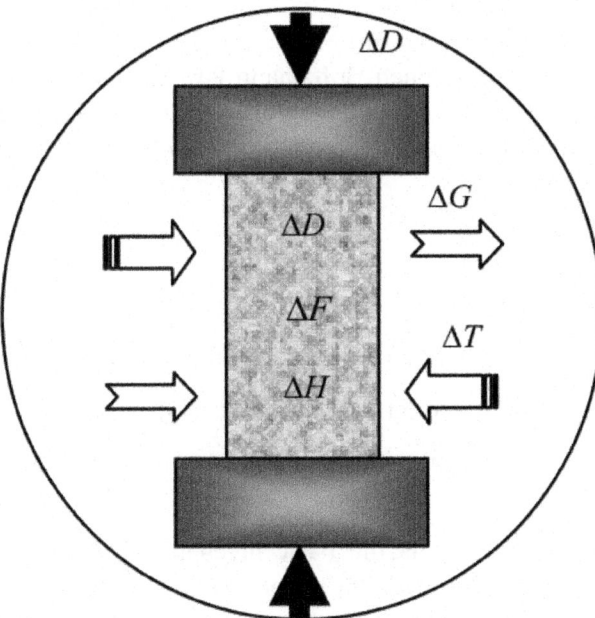

Figure 25. The energy balance of a loaded rock sample in a closed independent system.

$$\Delta M + \Delta T = \Delta D + \Delta F + \Delta H + \Delta G \qquad (5)$$

Here:

ΔM The inputted mechanical energy from loader, J; be positive;

ΔT The inputted thermal energy from loader and environment air, J; be positive;

ΔD The produced deformation energy of rock sample in elastic and plastic state, J; be positive;

ΔF The sum of consumed rock fracturing energy and formed fracture surface energy, J; be positive;

ΔH The heat energy increment of rock decided by its physical temperature, J; be positive if temperature rise or be negative if temperature drop;

ΔG The energy consumed by the desorbing and escaping of pore gas in rock samples, J; be positive.

From equation (5) we have:

$$\Delta H = (\Delta M + \Delta T) - (\Delta D + \Delta F + \Delta G) \qquad (6)$$

The change of heat energy will result in the change of physical temperature, surface radiation energy and IRR temperature of loaded rock samples. Referring to Stephen-Boltzmann law, the AIRT is a direct index of rock radiation energy, and the change of AIRT (ΔAIRT) must have certain a relationship with the physical temperature (T) of loaded rock sample as:

$$E_{IR} = f(\text{AIR}T) = \varepsilon\sigma T^4 \qquad (7)$$

Here:

E_{IR} The radiation energy of a loaded rock, J;

AIRT   The surface average IRR temperature of a loaded rock, K;

E   The radiation factor of rock sample, $0 < \varepsilon < 1$;

Σ The constant of Stephen-Boltzmann, $\sigma = 5.6679 \times 10^{-8}$, $J \times m^{-2} \times K^{-4}$;

Thermo-elastic effect is the basic mechanism of changed IRR from a loaded rock sample. Moreover, the desorbing and the escaping of pore gas, the expanding of initial fissures or joints, the friction between fissures, fractures, joints and grains, the thermal transfer between the rock, loader header and environment air, and the radiation from the environment all are to have thermal effect on loaded rock samples. The thermal state is comprehensively affected by the six factors which are rock stress, pore gas desorbing & escaping, rock fracturing, heat transferring, rock frictionating and environment radiation respectively. The equation could be expressed as follows:

$$\Delta AIRT = f[\Delta(\sigma_1 + \sigma_2), \Delta G, \Delta F_1, \Delta H, \Delta F_2, \Delta E]$$
$$= f_1(t) + f_2(t) + f_3(t) + f_4(t) + f_5(t) + f_6(t) \tag{8}$$

Here:

$\Delta AIRT$ The detected change of AIRT, K; be positive if rise, be negative if drop;

$f_1(t)$ IRR temperature change due to thermo-elastic effect ($\Delta(\sigma_1 + \sigma_2)$, K; be positive or negative;

$f_2(t)$ IRR temperature change due to pore gas adsorbing & escaping (ΔG), K; be negative;

$f_3(t)$ IRR temperature change due to the production of new fractures and the expansion of initial fissures, joints and new produced fractures (ΔF_1), K; be negative;

$f_4(t)$ IRR temperature change due to frictional thermal (ΔF_2), K; be positive;

$f_5(t)$ IRR temperature change due to heat transfer (ΔH), K; be positive or negative.

$f_6(t)$ IRR temperature change due to environment radiation (ΔE), K; be positive.

Thermo-elastic effect: $f_1(t)$

Referring to thermo-elastic theory and equation (4), $f_1(t)$ could be calculated as:

$$f_1(t) = \gamma \cdot \beta^{-1} \cdot T \cdot \Delta(\sigma_1 + \sigma_2) \tag{9}$$

In condition of uni-axially compressive loading, the load is to cause temperature rise in that σ_2 is constant zero and the positive σ_1 will linearly increase with loading. If $\Delta\sigma_1$ is positive, the $f_1(t)$ will be positive; If $\Delta\sigma_1$ is negative, the $f_1(t)$ will be negative. Hence, the $f_1(t)$ will be positive before the compressive stress peak, and will turn to be negative after the compressive stress peak.

In condition of uni-axially tensile loading, the load is to cause temperature drop in that σ_2 is constant zero and the negative σ_1 will linearly increase with loading. If $\Delta\sigma_1$ is positive, the $f_1(t)$ will be positive; If $\Delta\sigma_1$ is negative, the $f_1(t)$ will be negative. Hence, the $f_1(t)$ will be negative before the tensile stress peak, and will turn to be positive after the tensile stress peak.

In condition of compressively-sheared loading, the rock sample will be compressed by the normal component of the load. However, as to the central shearing plane, it will suffer not only compressive stress but also shearing stress. The σ_1 refers to the positive compressive stress normal to the shearing plane, while the σ_2 refers to the shearing stress which is actually negative tensile stress along the shearing plane. Hence, the ΔAIRT is decided by the sum of compressive stress and the tensile stress. If $\Delta(\sigma_1+\sigma_2) > 0$, the $f_1(t)$ will be positive; If $\Delta(\sigma_1+\sigma_2) < 0$, the $f_1(t)$ will be negative. As to Figure 1c, it's easy to know that $\Delta(\sigma_1+\sigma_2)$ will be positive if the shearing angle $\gamma 45^0$, $\Delta(\sigma_1+\sigma_2)$ will be zero if the shearing angle $\gamma=45^0$, $\Delta(\sigma_1+\sigma_2)$ will be negative if the shearing angle $\gamma 45^0$.

Pore gas desorbing & escaping effect: $f_2(t)$

Any rock has pores of different size more or less inside. Some rock, especially the sedimentary rock, may has certain gas, such as CH_4, CO_2, CO and O_2, enclosed in the pores or/and absorbed on the pore surface (Wang, 2003; Yang et al., 1999). Usually, most of the pores are enclosed and the gas molecules stay inside both in free gassy state and absorbed state. If the rock is loaded and suffers deformation, the volume of pores will decrease which results in the escape of gas from the pores. Once the load and deformation cause the enclosed pores getting fractured, the gassy molecules will escape firstly and the absorbed molecules will get desorbed to be gassy molecules and escape later. Both desorbing and escaping actions need to make use of heat energy from the rock, and thus will result in the AIRT drop of rock surface. Hence, the $f_2(t)$ is always negative, and the more the pores and gas enclosed, the more the negative effect of $f_2(t)$.

Fracture effect: $f_3(t)$

With loading and deforming, the rock is to get fractured. The new produced fractures together with the initial fissures and joints will extend both in width and length. The production of new fractures needs to consume energy, and the extension of fissures, joints and fractures also needs to consume energy. Hence, the $f_3(t)$ is always negative, and the more the fractures produced and fissures, joints and fractures extended, the more the negative effect of $f_3(t)$.

Frictional thermal effect: $f_4(t)$

With loading and deforming, the friction action is to occur between rock fissures, rock joints, rock grains, and new produced fractures. The friction action could be interpreted as: 1) at the beginning stage of loading, the friction may only be resulted from between rock fissures and between rock joints; 2) later, rock deformation increases with loading, new fractures are produced, and the friction between rock grains and between new produced fractures will join in; 3) finally, at the ending stage of loading, the rock deformation and fractures will be sufficiently developed, and the frictions between rock grains and between new produced fractures will be the chief contributors to the frictional thermal. In a word, the $f_4(t)$ is always positive no matter what is the principle friction factor. The more the friction, the more the positive effect of $f_4(t)$.

Heat transfer effect: $f_5(t)$

In the process of loading and inside the respectively independent loading system, the heat exchange is inevitable between the rock sample and the load header, the shearing platen or cushion-blocks, the surrounded atmosphere, etc. If the current temperature of rock sample is higher than the others, the heat of rock sample will be transferred out to whose temperature is lower. If the current temperature of rock sample is lower than the others, the heat of the others will be transferred into rock sample. Hence, the temperature of rock sample is a dynamic balance behavior between the heat transferred in and the heat transferred out. If the heat transferred in is more than that transferred out, the $f_5(t)$ will be positive; otherwise, it will be negative.

Environment-radiation effect: $f_6(t)$

The IRR detected by infrared imager includes not only the direct radiation from rock surface itself, but also the reflected radiation from environment. In laboratory, the chief environment radiations are the scattering sunshine, the moving human bodies and the illumination lamp. For the uncertain change of scattering sunshine, the movement of human bodies before the loaded rock sample, and the fluctuation of illumination light, the environment radiation effect on rock sample will be random. Hence, sometimes $f_6(t)$ may be positive, but sometimes $f_6(t)$ be negative. To eliminating the environment-radiation effect, the human bodies inside the laboratory were not permitted to move during testing process, the illumination lights were turned off, and the windows as well as its curtains were closed. Furthermore, some experiments were conducted in the evening so as to avoid the scattering sunshine completely.

Experiment Interpretation with RSRM-Model

Due to the comprehensive effects of $f_1(t) \sim f_6(t)$, the evolution of AIRT will be complex and will result in different possibility for abnormal AIRT precursors in different rock loading conditions, which including the loading scheme, rock type, and environment parameters. The following discussions are based on that $f_5(t)$ and $f_6(t)$ can be ignored.

Uni-axial Loading Experiments

For uni-axially loaded rock, $f_4(t)$ will take place only after that the rock has sufficiently deformed and the fractures have sufficiently developed.

At loading stage I and stage II, the rock surface AIRT are decided by $f_1(t)$ and $f_2(t)$. If the loaded rock is igneous rock or metamorphic rock, $f_2(t)$ is very rare since no gas absorbed in its pores usually, and the AIRT will rise with loading. If the loaded rock is sediment rock and with gas closed and absorbed in pores, $f_2(t)$ is inevitable, and the AIRT will rise if $f_1(t) > |f_2(t)|$, or be constant if $f_1(t)=|f_2(t)|$, or drop if $f_1(t) < |f_2(t)|$.

At loading stage III, fractures get sufficiently developed and pores get seriously damaged. The $f_3(t)$ begins to has more and more effect on evolution process of AIRT. AIRT will rise if $f_1(t) > |f_2(t)+f_3(t)|$, or be constant if $f_1(t)=|f_2(t)+f_3(t)|$, or drop if $f_1(t) < |f_2(t)+f_3(t)|$.

At loading stage IV, the friction action starts and $f_4(t)$ begins to have more and more effect on the evolution of AIRT. AIRT will rise if $[f_1(t)+f_4(t)] > |f_2(t)+f_3(t)|$, or be constant if $[f_1(t)+f_4(t)]=|f_2(t)+f_3(t)|$, or drop if $[f_1(t)+f_4(t)] < |f_2(t)+f_3(t)|$. Since $f_2(t)$ is very rare for igneous rock or metamorphic rock, the rise speed of AIRT of loaded igneous rock or metamorphic rock will get fast at stage IV, and the speed turning point is suggested to be the precursors point.

Compressive Shearing Experiments

For compressively sheared rock, not only $f_1(t)$ but also $f_4(t)$ is decided by shearing angle (γ). If $\gamma 45^0$, $f_1(t)$ will be positive for $\Delta(\sigma_1+\sigma_2) > 0$, and the friction action will be much strong in that σ_1, which is normal to the friction plane, is large. If $\gamma=45^0$, $f_1(t)$ will be zero since $\Delta(\sigma_1+\sigma_2)=0$. If $\gamma 45^0$, $f_1(t)$ will be negative for $\Delta(\sigma_1+\sigma_2) < 0$, and the friction action will be much week since σ_1 is slight.

Biaxial Loading Experiments

For bi-axially loaded rock, $f_1(t)$ will always be positive. As to bi-axially loaded en echelon faults, collinearly and non-collinearly disjointed faults, $f_2(t), f_3(t)$ and $f_4(t)$ will occur simultaneously, and the evolution of AIRT will be fluctuated. If $[f_1(t)+f_4(t)] > |f_2(t)f_3(t)|$, AIR$T$ will rise; if $[f_1(t)+f_4(t)] > |f_2(t)f_3(t)|$, AIR$T$ will keep in the same level, and if $[f_1(t)+f_4(t)] < |f_2(t)f_3(t)|$, AIR$T$ will drop. Usually, the fact is that $[f_1(t)+f_4(t)] < |f_2(t)f_3(t)|$ at the fracturing stage, and there is a short drop of AIRT, which is called as 'silence' before EQ (Ohtake et al., 1981). However, $f_4(t)$ will be an important factor at the later stage of loading for the concentrated formation of fractures and friction in the disjointed zone, and the final state of AIRT will be rise for $[f_1(t)+f_4(t)] > |f_2(t)f_3(t)|$.

EARTHQUAKE THERMAL INFRARED ANOMALIES

The prediction of EQ is very difficult, but it's not impossible. A number of signs warning of EQs, such as foreshock activities, peculiar animal behaviour, increased low frequency EM-noise, concentrations of radon in water and air, ionosphere and magnetosphere perturbation, radio frequency emissions, terrestrial gas emanations, EQ clouds, and satellite TIR anomalies, have been proposed and reported during the past centuries. Satellite TIR anomaly was firstly reported in 1989, and had been repeatedly verified in the world during the past 20 years. It is becoming a prospecting space observation technology for seismic activity monitoring and for EQ predicating.

General Features of EQ TIR Anomaly

Gorny (1988) firstly reported that there were large area of TIR anomalies in METEO satellite remote sensing images, spatial resolution being 5 Km and wave length being 10.5~12.5μm, before many moderate-strong tectonic EQs in the mid-Asia and the east-Mediterranean region. Tronin (1996) analyzed 10000 about TIR images of channel AVHRR-2 of NOAA in ten years for the mid-Asia, and reached that there existed average anomaly, 1~5°C, before the EQs at this region, and that there was obvious statistical relations between the EQ and TIR anomaly. Qiang (1990), Cui (1999), Liu (1999),Xu (2000), Zhang(2002), Ouzounov(2004), Arun (2005), Liu (2007), Wu (2008) also reported that there occurred TIR anomalies in satellite images (NOAA, FY, MODIS) days before more than 100 EQs in Asia (China, India, Iran, Japan, Kamchatka, Pakistan, Turkey) and Europe (Italy, Greece, Spain). Analysis to all the reported cases, it was uncovered that satellite TIR anomaly before EQ has the following features generally (Wu et al., 2009):

- Temporal features: Satellite TIR anomaly usually appears 1~26 days before shock, and reaches to its peak 1~2 days before shock, and will disappeared soon after shock.
- Spatial features: The spatial distribution and geometrical shape of TIR anomaly is tightly related with tectonic structures such as plate borders and active faults. With the EQ impending, the TIR anomaly will move to or extend to gradually the coming epicentre along the structures.
- Temperature features: the temperature of the TIR anomaly is usually 2 ～6°C higher than that of outside or surrounding the TIR anomaly.

- Magnitude features: there are positive relations somewhat between the TIR anomaly energy (the anomaly area times the anomaly temperature) and the magnitude of future shock.

Anomalies Interpretation with RSRM-Model

The RSRM experimental results are applied to analyze satellite remote sensed TIR anomalies before several strong EQs in Asia. Referring to the seismogenic mechanisms, the satellite TIR anomalies are in good accordance with the detected IRR anomalies from rock fracturing and seismogenic experiments with fault system being simulated with disjointed faults and intersected faults.

Dongsha Ms5.9 EQ 1992 in Taiwan, China

Dongsha Ms5.9 EQ occurred in Taiwan on Sept 14, 1992. The NOAA satellite images show that there was TIR anomaly appeared along the regional faults and downfaulted basins before shock as in Figure 26 (Wu et al., 2006c). There was an isolated and spoon-shaped high temperature area to the southwest of Taiwan Island 25 days before shock (Aug 19, 1992). The head of 'spoon' locates in the downfaulted basin-IV and the handle of 'spoon' distributes along fault-6. Satellite TIR image on Aug 22 shows that the high temperature on 'spoon head' diminished, but the TIR anomaly on 'spoon handle' became wide. Later, the anomaly moves gradually to the epicenter. TIR image on Sept 9 shows that a large area of high temperature had appeared around the coming epicenter, and the maximum temperature (in brown color) appeared south-close to the coming epicenter. It indicated that the TIR anomaly was consistent with the regional tectonic structures (faults and basins) in spatial position and geometry.

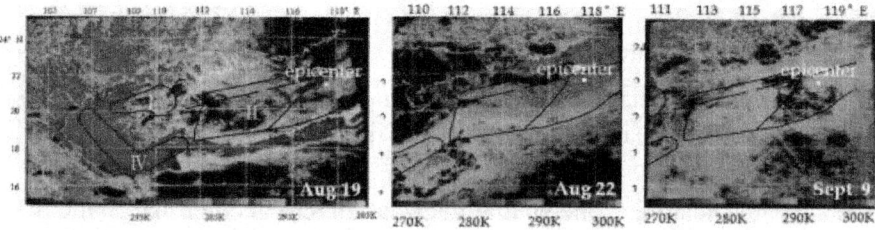

Figure 26. Satellite TIR anomaly images before Dongsha Ms5.9 EQ (Sept 14, 1992).

Zhangbei Ms6.2 EQ 1998 in China

Zhangbei Ms6.2 EQ occurred in China at 11:50 am, Jan 10, 1998. With the overlay of the active fault system investigated and deduced, in red lines, on night time NOAA satellite infrared images, as inFigure 27, it was discovered that (Wu et al., 2007b): 1) 18 days before shock, Dec 28, 1997, the TIR images on land and sea surface are basically normal, and the contours of TIR temperature field of Bohai Bay appeared to be along the coastline; 2) 5 days before shock, Jan 5, 1998, there occurred a positive TIR anomaly strip from Bohai Bay to Zhangbei passing through Beijing, its temperature was 3 C higher than that of outside, and the contours of TIR temperature field of Bohai Bay got offset the coastline; 3) 1 day before shock, Jan 8, 1998, the positive TIR anomaly strip had got wider and its temperature was 7 C higher than that of outside, and the contours of TIR temperature field of Bohai Bay got in accordance with the positive TIR strip; 4) 1 day after shock, Jan 11, 1998, the pattern of TIR images on the surface of land and sea turned to be normal again.

There are several disjointed active faults along the positive TIR strip, including a possible uncovered great active deep fault going from Bohai Bay to Zhangbei. Besides, to the southwest of Zhangbei, there are two small active faults pointing to Zhangbei. If extend the three faults respectively towards Zhangbei, it could be found that Zhangbei is exactly the intersection point. Hence, the tectonic background around the epicenter is basically comprised of three intersected active faults as two groups, i.e., the primary fault is the independent one from Bohai Bay to Zhangbei, and the secondary two act as another group. The two groups of faults split basically the regional crust into three geo-blocks (geo-block A, B and C_1+C_2) as the RSRM experiment model in Figure 28b (rock block A, B and C), and the secondary two faults act as the two sides of an acute wedge-shaped geo-block (geo-block C_2) as the RSRM experiment model in Figure 28d (rock block C). The geometric features of the faults and the spatio-temporal features of TIR anomaly were similar. Therefore, the mechanical mechanism of Zhangbei EQ should be classified to be the failure of an intersected active fault system, which with an acute wedge-shaped geo-block being its secondary active object.

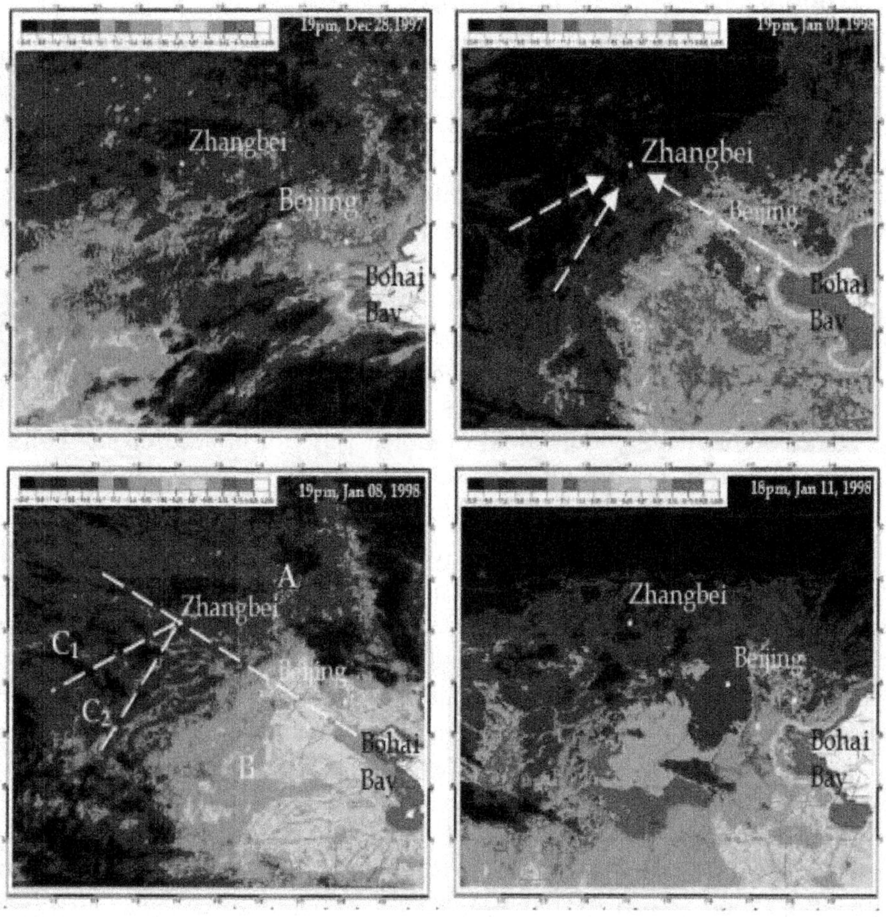

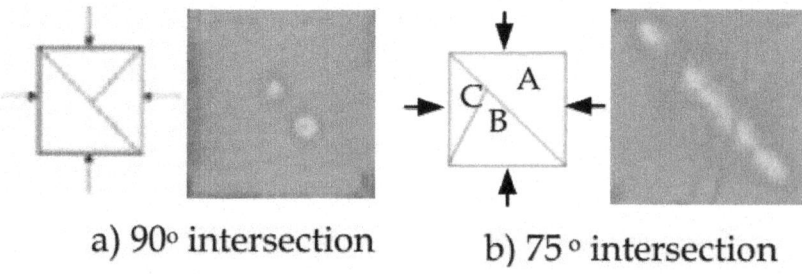

Figure 27. The TIR images overlaid with detailed active faults before and after Zhangbei Ms6.2 EQ (Jan 10, 1998).

a) 90° intersection   b) 75° intersection

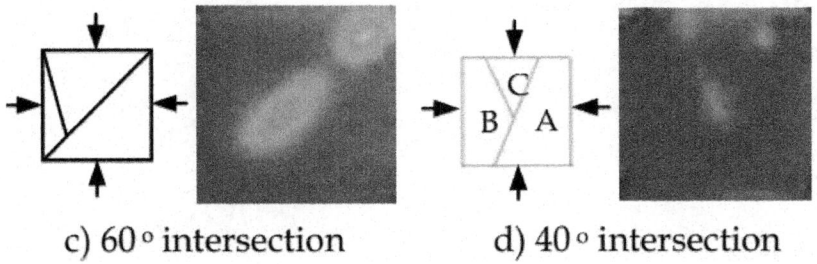

c) 60° intersection   d) 40° intersection

Figure 28. The IRR anomaly features of simulated tectonic activities due to bi-axially load (Wu et al., 2007b).

Izmit Ms 7.8 EQ 1999 In Turkey

Izmit Ms 7.8 EQ occurred in Turkey on Aug 17, 1999. In the epicenter zone, there were two en echelon tectonic faults, and the epicenter is 45km about to the west of the disjointed place of the two en echelon faults, as in Figure 29. With thermal image of Aug 1&2 being the reference, the differential thermal images from Aug 6 to Aug 26 were obtained, and it was revealed that there were NOAA satellite TIR anomaly at the disjointed zone of two en echelon faults since Aug 6, 11 days before shock (Tronin, 2000), as in Figure 29. The RSRM experiments on the simulation of tectonic EQ due to the fracturing of disjointed faults had revealed that there were IRR anomaly increment and concentrated deformation at the disjointed zone, as in Figure 30. Obviously, the spatial features of NOAA TIR anomaly before Izmit Ms 7.8 EQ were the same as that of IRR anomaly before the fracturing of disjointed faults.

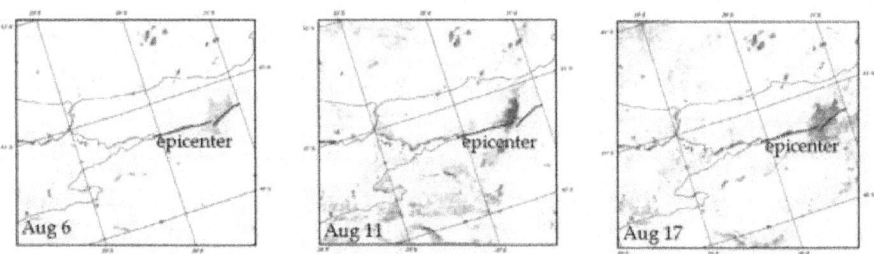

Figure 29. TIR anomalies 0~11 days before Izmit Ms 7.8 EQ (Aug 17, 1999).

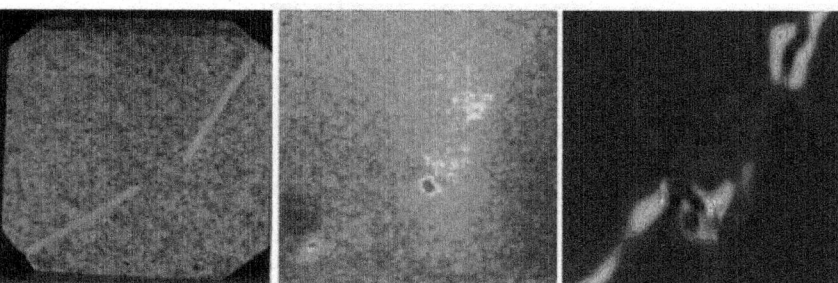

Figure 30. The IRR anomaly and deformation concentration of a bi-axially loaded granite sample with disjointed faults.

Hengchun Ms7.2 EQ 2006 in Taiwan, China

Hengchun Ms7.2 EQ occurred in Taiwan at 12:26pm, Dec 26, 2006. TIR images from stationary satellite FY-2 showed that there was TIR anomaly nearby the coming epicenter. The TIR anomaly appeared to the east of Philippines six days before shock, and moved gradually toward west to Philippines. Later, the anomaly changed its direction to the north, and the temperature increased 10°C about one day before shock. Figure 31 shows that the satellite TIR images around Hengchun on Dec 25, 2006 (Liu et al., 2007). A disjointed thermal strip (in orange color) appeared on the southwest of Taiwan at 1:00am, and developed gradually to be an X-shaped thermal anomaly zone at 10:00am. The epicenter located closely to the cross point of the X-shaped zones. The evolution process of satellite TIR images was very similar to that of X-shaped thermal IRR anomaly in RSRM experiments as inFigure 2.

Wenchuan Ms8.0 EQ 2008 in China

Wenchuan Ms8.0 EQ occurred in China at 14:28 am, May 12, 2008. The epicenter locates at the transferring zone of Tibetan Plateau to Sichuan Basin, which is only 92Km to the NW of Chengdu, the capital of Sichuan province. Analysis to FY-2C TIR images, as in Figure 32, it is discovered that there was an high temperature strip with length of 3 000 km appeared on April 23, 20 days before shock, which started from India Plate and developed to northeast along the east front of Tibetan and Loess Plateau (Wu et al., 2008). The cause might be the great accumulation of fictional sliding stress along the east foreland of Tibetan and Loess Plateau due to the subduction of India Plate to Euro-Asia Plate. The east foreland of Tibetan and Loess Plateau act as the fictional sliding plane between the west part of China (including Tibetan and Loess Plateau) and the east part of China (including the North, Middle and Southwest China).

The evolution of satellite TIR images was similar to that of detected strip-shaped IRR anomaly in the fictional sliding experiments as in Figure 8 and Figure 9, which shows the evolution of TIR anomaly along the sliding plane.

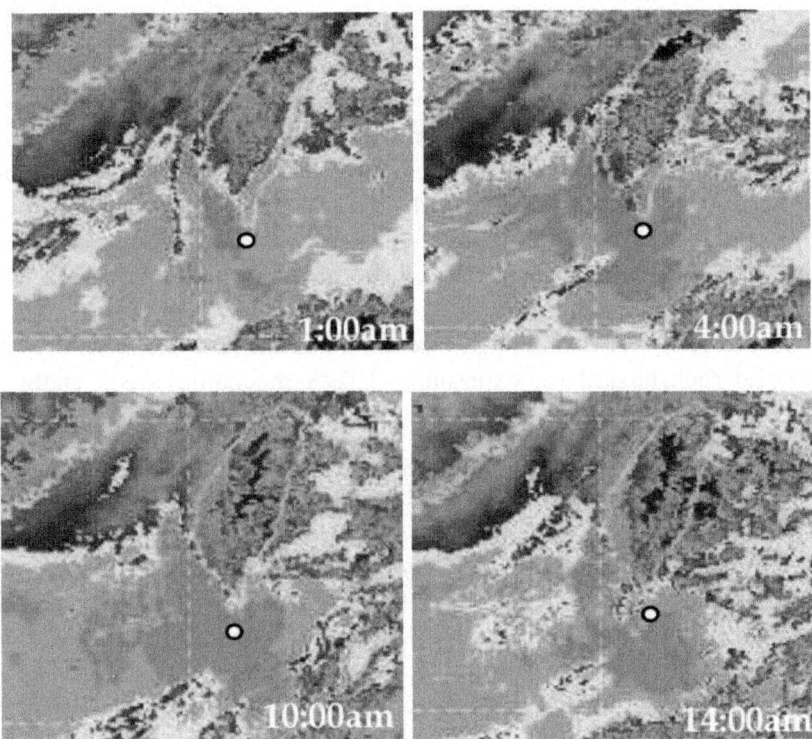

Figure 31. Satellite TIR images one day before Hengchun Ms7.2 EQ (Dec 25, 2006).

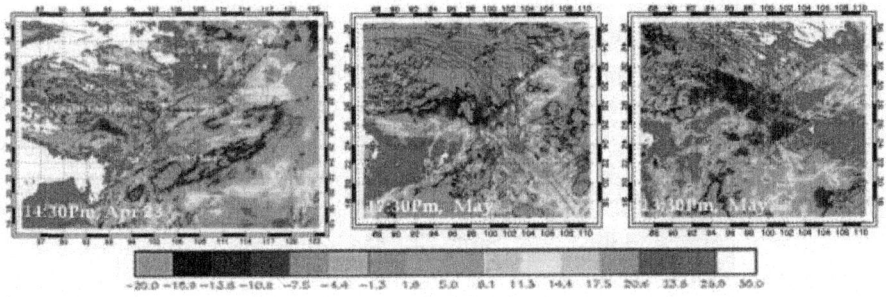

Figure 32. Satellite TIR images before Wenchuan Ms8.0 EQ (May 12, 2008).

FUTURE RESEARCHES

On RSRM Experiment

As a detectable remote sensing signal related with rock stress and physical temperature, IRR is a meaningful index for studying rock load, rock deformation, rock fracturing and rock hazard. The temporal evolution of surface IRR from loaded rock is the comprehensive effect of rock thermo-elastic acting, pore gas desorbing & escaping, fractures producing & extending, rock frictionating, heat transferring and environment radiation. The IRR image anomaly referring to the spatial-temporal evolution of IRR from loaded rock is an important precursor for rock fracturing, and will be meaningful for the predication of geo-hazards including tectonic EQ. For the practical application of RSRM, deeper research on IRR imaging detection quantitatively and specially on rock stress and rock hazard for experimental rock mechanics, rock engineering, tectonic activity and strong EQ is demanded.

The mechanism of experimental detected IRR anomaly can be theoretically interpreted by taking the load header, the rock sample and the environment to be a closed independent system in energy balance state. There are two of the main rock-physics mechanisms, respectively being thermo-mechanical coupling and frictional thermal due to tectonic stress, rock fracturing and fictional sliding, for the change IRR from loaded rock samples (Wu, et al., 2006c). Besides, positive hole (P-hole) activation due to piezoelectricity was suggested to be another mechanism of IRR from loaded quartz-bearing igneous rock (Freund, 2002), such as granite, basalt, diorite, and gabbro.

To search for scientific interpretation on the relations among satellite TIR anomaly, rock stress and experimentally detected IRR anomaly, the EM transferring process from underground rock body to satellite sensors, through lithosphere, earth surface coversphere (including soil layers, water bodies and vegetations), the atmosphere and lithosphere, should be systematically studied. Pulinets pointed out that the incubation of an EQ is to disturb the ionosphere (Pulinets, 1998), and it was suggested that lithosphere-atmosphere-ionosphere (LAI) coupling is the mechanism of satellite TIR anomaly before strong EQ (Molchanov et al., 2004). Nevertheless, the action of earth surface coversphere on the transferring and the magnifying of EM signals from underground loaded rock to atmosphere should not be ignored, even if its physical mechanism are not clear. For the scientific interpretation of satellite TIR anomaly before strong EQ, the lithosphere-coversphere-ionosphere (LCA) coupling is the key, while for the scientific interpretation of ionosphere anomaly, the lithosphere-coversphere-atmosphere-ionosphere (LCAI) coupling should be focused. However, present experiments on LAI, LCA and LCAI coupling are rather insufficient. Future experiments specially designed to uncover the mechanisms & laws and to construct the models & quantitative equations of LAI, LCA and LCAI couplings are expected.

On EQ Thermal Infrared Anomaly

Although there are uncertain influences from meteorological variation, satellite TIR anomaly has quite different identification features from that of unseismology-resulted TIR anomaly. Satellite TIR remote sensing is becoming a prospecting technique for monitoring tectonic activities and for predicting strong tectonic EQ, which provide a negativism to that EQ cannot be predicted. Anyway, the practical predication of EQ is not so easy. The regional tectonic background and the active fault system have extremely important affects on the incubation of EQ and the TIR anomaly. Especially, the intersected faults, compressively-sheared faults, and disjointed faults are to control the location and the routing of the spatial evolution of satellite TIR anomaly, and the brightest spot of TIR anomaly along the fault, or at the intersection point, or at disjointed zone of faults might foretell the possible epicenter (Wu et al., 2007b).

First of all, massive observation information including crust stress, land deformation, atmosphere components, underground water, surface and near-surface temperature, satellite remote sensing TIR anomaly and EM disturbance in ionosphere should be integrated together for data fusion and cross checking to analyze comprehensively the tectonic activity and rock fracturing process. A grid-based distributed database and analysis tools on TIR remote sensing images, with global and regional tectonic structures being its background, should be set up to assist the extraction of EQ TIR anomaly in different temporal and spatial scale. Besides, a quantitative model for tectonic activity analysis and for EQ magnitude predication based on TIR anomaly should be developed.

The GEOSS under construction is to provide an integral and integrated monitoring on earth environment, Geo-hazard and global disasters. A generalized remote sensing (GRS) based on the integration of space-based, aero-based, near-surface based, in-situ based and underground-based monitoring is forming in the world (Wu and Liu, 2007). The international broad and sincere cooperation, inside the framework of GEOSS without discipline exclusion and data privacy, between seismologists, remote sensing scientists, meteorologists, geophysical scientists, geochemist and spatial information scientists in good faith is expected. Besides, a powerful spatial information system, especially for EQ early warning and short-coming prediction based on GEOSS, should be developed. It should has powerful functions such as massive information classification, smart theme mapping, easy map-layer overlay, fast features extraction, effective data fusion, intelligent data mine, powerful knowledge discovery, and easy access and share.

A possible technical procedure based on GRS for satellite TIR anomaly monitoring, analyzing and early warning of tectonic EQ inside the framework of GEOSS might be that: 1) the seismological geology background being the foundation of satellite TIR anomaly analysis; 2) the long-term GPS continuous monitoring and D-InSAR measurement being the guidance of tectonic stress detection and active fault identification; 3) the underground water temperature, near surface air temperature, radon & green gas, structural cloud anomaly dairy monitoring being the forerunner for preliminary identification of coming EQ; 4) the anomaly analysis of satellite TIR, NCEP temperature and ionosphere disturbance being the dominant for early warning of temporal-spatial-magnitude parameters of EQ; and 5) the additional celestial stress on active faults being a special leading disturbance for possible tectonic EQ.

REFERENCES

1.   K. Arun, C. Swapnamita, 2005 Thermal remote sensing technique in the study of pre-earthquake thermal anomalies. J. Ind. Geophys. Union, 9 3 197 207

2.   B. Brady, G. Rowell, 1986 Laboratory investigation for the electrodynamics of rock fracture. Nature, 321 29 488 492

3.   C. Cui, J. Zhang, Q. Xiao, 1999 Monitoring the thermal IR anomaly of Zhangbei earthquake precursor by satellite remote sensing technique, Proc. 20th Asia Remote Sensing Congress, 1179 1184 , Hongkong

4.   M. Deng, J. Qian, J. Yin, et al. 2001 Research on the application of infrared remote sensing in the stability monitoring and unstability prediction of large concrete engineering. Chinese J Rock Mech. Engi., 20 2 147 50 .

5.   F. Freund, 2002 Charge generation and propagation in igneous rocks. J. Geodynamics, 33(4-5): 543-570.

6.   N. Geng, C. Cui, et. Deng, al, 1992 Remote sensing detection on rock fracturing experiment and the beginning of Remote Sensing Rock Mechanics. ACTA Seismologic SINICA, 14(supp) : 645 52 .

7.   V. Gorny, A. Salman, A. Tronin, et al. 1998 The earth outgoing IR radiation as an indicator of seismic activity, Proc. Acad. Sci. USSR, 30 1 67 69 .

8.   J. Hudson, J. Harrison, 1997 Engineering rock mechanics, Elsevier Science Inc., New York, 85 112

9.   D. Liu, K. Peng, W. Liu, et al. 1999 Thermal omens before earthquake. Acta Seismologica Sinica, 12 6 710 715

10. S. Liu, L. Wu, J. Li, et al. 2007 Features and mechanism of the satellite thermal infrared anomaly before Henchun earthquake in Taiwan Region. Science & Technology Review, 25 6 32 37

11. S. Liu, L. Wu, Y. Wu, et al. 2002 Quantitative study on the thermal infrared radiation of dark mineral rock in condition of uni-axial loading. Chinese J. Rock Mech. & Engi., 21 11 1585 89

12. M. Luong, 1990 Infrared thermovision of damage processes in concrete and rock. Engi. Fracture Mechanics, 35 (1~3): 127-35.

13. G. Martelli, P. Smith, A. Woodward, 1989 Light, radiofrequency emission and ionization effects associated with rock fracture. Geophy. J. Int., 98 2 397 401 .

14. O. Molchanov, E. Fedorov, A. Schekotov, et al. 2004 Lithosphere-atmosphere-ionosphere coupling as governing mechanism for preseismic short-term events in atmosphere and ionosphere. Natural Hazards and Earth System Sciences, 4 757 767

15. D. Mounatin, J. Webber, 1978 Stress pattern analysis by thermal emission (SPATE), Proc. Soc. Photo-Opt. Inst. Engi, 164:189.

16. A. Nicolas, J. Bouchez, J. Blaise, et al. 1977 Geological aspects of deformation in continental shear zones. Tectonophysics, 42 1 55 73

17. M. Ohtake, T. Matumoto, G. Latham, 1981 Evaluation of the forecast of the 1978 Oaxaca Southern Mexico earthquake based on a precursory seismic quiescence. In: Earthquake Prediction Maurice Ewing Series American Geophysics Union 4）, 53 62

18. D. Ouzounov, F. Freund, 2004 Mid-infrared emission prior to strong earthquakes analyzed by remote sensing data. Adv. Space Res. 33 268 273 .

19. S. Pulinets, 1998 Seismic activity as a source of the ionospheric variability. Adv. Space Res., 22 6 903 907

20. Z. Qiang, X. Xu, C. Dian, 1990 Thermal infrared anomaly-precursor of impeding earthquakes. Chinese Science Bulletin, 35 17 1324 1327

21. D. Renata, 1977 Electromagnetic phenomena associated with earthquakes. Geophsical survey, 3 2 157 174 .

22. R. Sibson, et al. 1980 Power dissipation and stress levels on faults in the upper crust. J. Geophysical research, 85(B11): 6239-6247.

23. W. Thomson, 1853 Trans. R. Soc. Edinburgh, 20 83 261 .

24. A. Tronin, 1996 Satellite thermal survey-a new tool for the studies of

seismoactive regions. J Remote Sensing, 17 8 1439 1455

25. A. Tronin, 2000 http://www.iki.rssi.ru/earth/ppt/tronin.ppt

26. X. Wang, 2003 A research on the inorganic carbon dioxide from rock by thermal simulation experiments. Advance in Earth Science, 18 8 515 20

27. L. Wu, S. Liu, 2007a Generalized remote sensing for solid Earth hazards under condition of GEOSS. Science & Technology Review, 25 6 5 11

28. L. Wu, J. Wang, 1998 Infrared radiation features of coal and rocks under loading. Int. J. Rock Mech. & Min. Sci, 35 7 969 976

29. L. Wu, C. Cui, N. Geng, et al. 2000 Remote Sensing Rock Mechanics (RSRM) and associated experimental studies. Int J Rock Mech Min Sci, 37 6 879 88 .

30. L. Wu, J. Li, S. Liu, 2009 Infrared anomaly analysis based on reference fields for earthquake remote sensing. Seismology Geology, 2009(in press)

31. L. Wu, J. Li, X. Xu, et al. 2007b Theoretical analysis to impending tectonic earthquake warning based on satellite infrared anomaly, Proc. 2007 IEEE Int. Geosciences &Remote Sensing Symposium, 3723 3727

32. L. Wu, S. Liu, W. Shi, et al. 2003 Experimental study on infrared anomaly of tectonic earthquake, Proc. SPIE, Remote Sensing For Environmental Monitoring, GIS Applications & Geology Ⅲ, 5239 376 387 .

33. L. Wu, S. Liu, Y. Wu, 2006c The experiment evidences for tectonic earthquake forecasting based on anomaly analysis on satellite infrared image, Proc. 2006 IEEE Int. Geosciences &Remote Sensing Symposium, 2158 2216

34. L. Wu, S. Liu, Y. Chen, et al. 2008 Satellite thermal infrared and cloud abnormities before Wenchuan earthquake. Science & Technology Review, 26 10 28 29

35. L. Wu, S. Liu, Y. Wu, et al. 2002 Changes in IR with rock deformation. Int. J. Rock Mech. & Min. Sci., 39 6 825 31

36. L. Wu, S. Liu, Y. Wu, et al. 2004a Remote Sensing Rock Mechanics (I) : laws of thermal infrared radiation from disjointed jointed faults fracturing and its meanings for tectonic earthquake omens. Chinese J Rock Mech. & Engi., 23 1 24 30

37. L. Wu, S. Liu, Y. Wu, et al. 2004b Remote Sensing Rock Mechanics (Ⅱ) : laws of thermal infrared radiation from bi-sheared faults friction sliding and its meanings for tectonic earthquake omens. Chinese J Rock Mech. Engi., 23 2 192 198

38. L. Wu, S. Liu, Y. Wu, et al. 2004c Remote Sensing Rock Mechanics (IV): laws of thermal infrared radiation from compressively-sheared fracturing rock and its meanings for earthquake omens. Chinese J Rock Mech. & Engi., 23 4 539 544

39. L. Wu, S. Liu, Y. Wu, et al. 2006a Precursors for rock fracturing and failure-part I: IRR image abnormalities. Int J Rock Mech Mining Sci, 43 3 473 482

40. L. Wu, S. Liu, Y. Wu, et al. 2006b Precursors for rock fracturing and failure-part II: IRR T-curve abnormalities. Int J Rock Mech Mining Sci, 43 3 483 493

41. L. Wu, S. Liu, X. Xu, et al. 2004d Remote Sensing Rock Mechanics (III) : laws of thermal infrared radiation and acoustic emission from friction sliding intersected faults and its meanings for tectonic earthquake omens. Chinese J Rock Mech. & Engi., 23 3 401 407

42. X. Xu, X. Xu, Y. Wang, 2000 Satellite infrared anomaly before Nantou Ms=7.6 earthquake in Taiwan, China. ACTA Seismologica SINICA, 22 6 656 659

43. I. Yamada, K. Masuda, H. Murakami, 1989 Electromagnetic and acoustic emission associated with rock fracture. Phys. Earth Planet. Inter., 57(1-2):157-168.

44. Y. Zhang, W. Shen, 2002 Satellite thermal infrared anomaly before the Xinjiang Qianhai boder Ms8.1 earthquake. Northwestern Seismological J., 34 1 1 4

Chapter 2

PERFORMANCE OF EVA-BASED MEMBRANES FOR SCL IN HARD ROCK

Karl Gunnar Holter

Department of Geology and Mineral Resources Engineering, Norwegian University of Science and Technology, Sem Sælands vei 1, 7491 Trondheim, Norway

ABSTRACT

The bonded property of multi-layered sprayed concrete tunnel linings (SCL) waterproofed with sprayed membranes means that the constituent materials will be exposed to the groundwater without any draining or mechanically separating measures. Moisture properties of the sprayed concrete and membrane materials are therefore important in order to establish the system properties of such linings. Ethyl-vinyl-acetate based sprayed membranes exhibit high water absorption potential under direct exposure to water, but are found to be significantly less hygroscopic and exhibit lower sorptivity (water absorption rate) than sprayed concrete. This material behavior explains the relatively dry in situ condition of the membrane that was observed. Measured in situ moisture content levels of the membrane material in tunnel linings have been found to vary within the range of 30–40 % of the maximum water absorption potential, and show a decreasing trend over the first 4 years after construction has been completed. A model for the mechanical loading, moisture condition and thermal exposure of the membrane and the resulting realistic parameters to be tested is presented. Laboratory testing methods for the membrane materials are evaluated considering possible loads, moisture and freezing exposure. Material testing of membrane materials was conducted with preconditioning to realistic moisture contents and under different temperature conditions including relevant freezing temperatures for tunnel linings. The main effects of the in situ moisture condition of the tested membrane materials are favorable tensile strengths in the range of 1.1–1.5 MPa and low risk of freeze–thaw damage. The crack bridging capacity of the tested membranes is found to be sensitive to temperature. With membrane thicknesses in the range of 3–4 mm, crack bridging capacity up to 4–6 mm opening of the crack width at 23 °C and approximately 1 mm opening at −3 °C was measured for the tested membranes. No significant reduction of the tensile bond strength could be demonstrated after 35 freeze–thaw cycles with −3 °C minimum temperature at the membrane location in the lining. Further work is required to verify the performance of the SCL system under exposure to high hydrostatic pressures and the effects of long term mechanical exposure.

INTRODUCTION

In hard rock environment in the Scandinavian countries permanent rock support linings are widely constructed with fiber-reinforced concrete and rock bolts (NGI 2013; NCA 2011; STA 2011). The final waterproofing and thermal insulation has normally been resolved by constructing a separate suspended shield structure (NPRA 2012; STA 2011). Modern requirements for service lifetime, serviceability and maintainability have raised concerns with the use of these shield lining systems. Cast-in-place concrete lining waterproofed with sheet membranes or pre-cast concrete segment linings for rail and road tunnels have therefore been proposed as the future technical solution in rail and road tunnels subjected to high traffic density (NPRA 2012; Holter et al. 2013).

SCL with sprayed ethyl-vinyl-acetate (EVA)-based membranes are being considered as a possible technical solution under certain conditions as an alternative to the well-established lining systems. The main benefit would be the reduced total lining thickness since the rock support lining based on sprayed concrete can be utilized as part of the final lining, and large concrete thicknesses can be avoided. Although such linings with spray applied membrane have been constructed for approximately a decade and have seen increased use in some countries, the main properties and function have yet to be fully understood.

SCL waterproofed with a sprayed membrane represents a continuously bonded multi-layered structure from the rock mass to the tunnel lining surface. The bonded property of the lining structure implies that the constituent materials of the lining will be exposed to the groundwater without any constructed draining or mechanically dividing measures. The construction process of spray-application produces continuous and bonded interfaces, which also can be assumed to be perfect hygric contacts between the different layers. The moisture properties of the membrane material and the concrete on either side of the membrane, as well as the exposure to any groundwater in the immediate rock mass will influence the moisture condition of the materials in the tunnel lining.

A research project in Norway has been carried out in order to assess the suitability of this lining system for modern rail and road tunnels. An important part of this research has been to conduct site and laboratory investigations in order to establish the function and properties of such linings. This work contains several main modules which have required detailed studies. The investigation of the in situ moisture condition and possible moisture transport mechanisms through sprayed concrete tunnel linings are published in Holter and Geving (2015) which forms the basis for moisture exposure during laboratory testing. The freeze–thaw resistance of the sprayed concrete in tunnel linings under realistic moisture contents and thermal exposure has also been investigated and will be published in a separate paper.

The scope of the present investigation is to study the properties of the membrane material, particularly the loading conditions for the membrane, evaluate testing methods and conduct testing of important parameters under varying climatic conditions. A conceptual model for the tunnel lining is presented in order to define the main items, its properties and the important processes for the waterproof SCL system. The in situ exposure conditions for the membrane will be substantiated from field investigations. The context which is considered in our study is a hard rock environment in which the primary rock support structure is considered stable and has no imposed ground induced loads or deformations on the bonded membrane and inner lining.

This study refers to EVA-based membranes with products from two different suppliers. The study contains the following main elements:

- Definition of material model based on the layout of the tunnel lining.
- Model for different loading scenarios of the membrane.
- Field investigations: moisture content, thermal exposure and in situ tensile bond strength.
- Evaluation of laboratory test methods for membranes in a SCL context.
- Laboratory investigations of hygroscopic properties of the concrete and membrane materials.
- Laboratory investigations of mechanical properties of the membrane material.
- Analyses of results.

The first findings of this study were presented at the World Tunnel Congress 2014 (Holter et al. 2014). Findings from additional field and laboratory investigations have been included. The recommendations of the ITAtech Design Guidance for Sprayed Membranes (ITA/AITES 2013) compiled by Dimmock (2014) will form the basis for the evaluation of test methods for membranes. Adjustments to these test methods will be discussed and made based on the loading model and the findings from the thermal and moisture analyses.

Possible degradation processes and long term durability under relevant mechanical loading and climatic exposure, as well as recommendations regarding testing details and acceptance criteria will be discussed based on the results.

CONCEPTUAL MODEL FOR TUNNEL

Current SCL designs and the application methodology for concrete and membrane materials form the basis for the conceptual model. The bonded and thus undrained interfaces of the multi-layered structure result in moisture transport processes governed by the hygroscopic properties of the constituent materials. For hard rock tunnels the tunnel lining structure consists of a primary lining based on fiber-reinforced sprayed concrete and rock bolts. In poor ground conditions in a hard rock environment reinforced sprayed concrete ribs are frequently used for permanent ground support (Grimstad et al. 2008; Mao et al. 2011). In order to produce a suitable substrate for the application of the membrane, a regulating layer of sprayed concrete is normally required. For these investigations the regulating layer has been applied using wet-mix fiber-reinforced sprayed concrete. Dry-mix sprayed concrete or mortar for use in the substrate for the membrane has been excluded from this investigation. The conceptual model for the waterproof SCL is shown in Table 1 and Figs. 1 and 2.

Table 1: Main items in the conceptual model for waterproof SCL

Main item in conceptual model	Situation, condition	Processes
Tunnel lining structure with sprayed concrete and membrane in walls and crown	Bonded undrained contacts from rock surface through all materials No thermally insulating materials in lining Concrete of different ages	Moisture transport through lining structure Differential shrinkage, of concrete on either side of membrane Thermal and shrinkage induced movement of cracks in concrete Exposure to water at cracks Exposure to movement at cracks caused by gravitation, shear displacement, stress release, swelling

Rock mass below GW table	Saturated jointed rock material Exposure to GW at the rock-concrete interface	Local saturation of concrete material at rock-concrete interface Water flow on joints into tunnel through invert
Tunnel space	Climate in tunnel Seasonal variations in temperature and relative humidity	Exposure of lining surface to tunnel climate Heat flux from rock mass to tunnel space Cyclic freezing and thawing of lining Change of properties of membrane and concrete

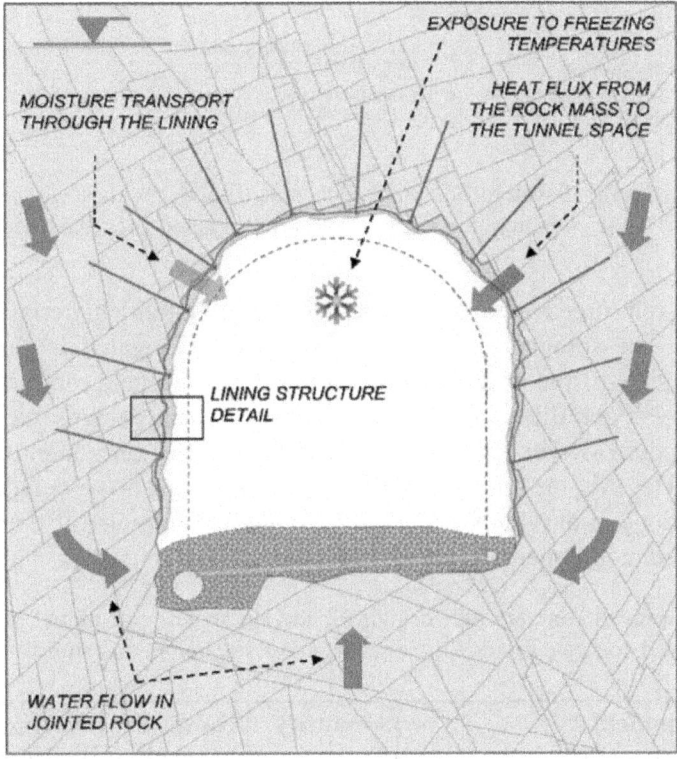

Figure. 1: Main elements in the conceptual model for a tunnel with permanent SCL based on fiber-reinforced sprayed concrete, sprayed waterproofing membrane and rock bolts constructed in hard rock. Detail is shown in Fig. 2.

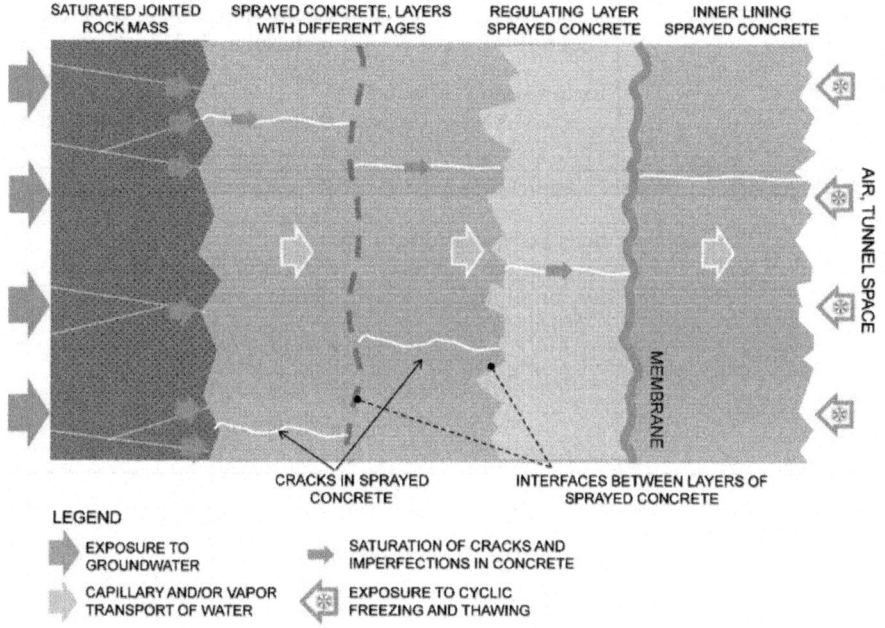

Figure. 2: Detail of waterproof SCL. Conceptual model with section of the lining structure with the constituent materials, moisture transport processes and exposure to freezing.

Fiber-reinforced sprayed concrete is the structural material in an SCL structure. The sprayed concrete mix designs investigated in this study all represent state-of-the-art developments in mix designs and robotic application technology. Four different sites with different mix designs were included in this study. However, the basic mix designs differ only slightly from one another. The mix design of the sprayed concrete used for detailed material investigations in our study is shown in Table 2. The range of the contents of the different components is also given.

Table 2: Sprayed concrete mix design for the Harangen road tunnel, and range for other sprayed concrete mixes for tunnel sites investigated in this study

Component	Quantity	Range for investigated concrete mixes from the other test sites
Cement CEM II A-V 42.5 17–18 % fly ash content	502 kg/m^3	488–513 kg/m^3

Micro silica fume	25 kg/m³	21–26 kg/m³
Water added with base mix	240 kg/m³	202–245 kg/m³
Water added with accelerator at spraying nozzle	17 kg/m³	17–20 kg/m³
Water/binder-ratio[a]	0.45	0.44–0.47
Aggregate 0–8 mm	1497 kg/m³	1497–1588 kg/m³, fractions used 0–4, 0–8, 0–10 mm
Superplasticizer	7 kg/m³	7–10 kg/m³, different suppliers
Fiber reinforcement, structural polypropylene (PP)	9 kg/m³ 1 % by volume	5–9 kg/m³ (PP) 0.5–1 % by volume 40 kg/m³ (steel) 0.6 % by volume
Binder paste content	0.43 m³/m³	0.41–0.44

[a]Considering equivalent binder content: weight of cement + two times weight of micro silica

All together five membrane products nominated M1 to M5 have been included. For the field investigations of the lining structure only M1 has been analyzed so far.

MECHANICAL LOADING, MOISTURE AND FREEZING

Mechanical Loading

Loads from the Rock Mass

A study of the possible loading of rock support linings based on sprayed concrete and rock bolts from the groundwater and rock mass has been undertaken (Holter 2014). The current practice with the use of rock mass classification according to the Q-system (NGI 2013) normally ensures rock stability with a high factor of safety. From this study it is concluded that in hard rock environment the stresses and loads which occur in tunnel linings are negligible in most cases. Even in severe weakness zones significant loading of the tunnel lining structure does not normally take place, other than local loads (Mao et al. 2011; Grimstad et al.2008; compilation by Holter 2014). Still, special design of the rock support is undertaken for severe weakness zones.

Ground Water Induced Loads

The waterproof SCL lining system represents an undrained structure. Hence, possible water pressures acting on the tunnel lining need to be considered. A study including monitoring of groundwater pressures around sprayed concrete tunnel linings with drained inverts has been conducted (Holter 2014). These results indicate water pressures lower than the hydrostatic pressure in the immediate vicinity of the tunnel lining.

Any occurrence of unfavorable ground water pressures in the immediate rock mass needs to be considered in the rock support design as well as evaluating the need for drainage measures where this is feasible. Ground water under a certain pressure can possibly saturate cracks and imperfections in the sprayed concrete in the primary lining. Under such circumstances a wet-crack situation with ground water pressure exposing the membrane locally can be hypothesized. The investigated sites in this study had a water pressure near the lining of maximum 2 bars. No deterioration of the lining structure was detected at any of the test sites. However, the wet crack problem at higher hydrostatic pressures cannot be assessed in detail from this study.

Loads from the Weight of the Tunnel Lining

The gravity induced stresses in the tunnel lining caused by the weight of the tunnel linings represent a constant static load. By considering a thickness of the inner layer sprayed concrete of 100 mm, and assuming that the concrete lining is "hanging" on the substrate a gravity induced tensile stress of 2 kPa in the center of the tunnel crown can be calculated.

Dynamic Loads from the Traffic Area of the Tunnel

Rail and road tunnels are exposed to fluctuations in air pressure caused by traffic. Highway and high speed rail tunnels in Norway have design requirements for expected maximum dynamic loads and number of loading events throughout the service lifetime. Current design requirements for rail and road tunnels (NPRA2006; NNRA 2012; STA 2014) are shown in Table 3.

Table 3: Dynamic loads in modern tunnels given as sudden change in air pressure per design traffic event

Tunnel type	Design speed (km/h)	Air pressure change per event (kPa)	Number of events in service lifetime	Time interval between each event, range
Highway, double carriage-way	140[a]	1.5	5×10^7	20–60 s
Rail, double track	250	10	1×10^7	2–5 min
Rail, single track	250	8	1×10^7	2–5 min

[a]20 km/h higher than legal speed limit

For a tensile loading consideration, the values for air pressure changes shown in Table 3 are considered changes in tensile stress. For a high speed double track rail tunnel a single design event for traffic induced air pressure change in the tunnel imposes tensile stresses with a factor five times higher than the calculated static gravity induced load from the tunnel lining. However, the dynamic air pressure induced loads are approximately a factor 100 times lower than measured in situ tensile bond strengths of the membrane-concrete interfaces. It is therefore considered very unlikely that this dynamic loading represents a dynamic fatigue scenario for a bonded SCL structure.

Deformations of the Membrane over Cracks in the Concrete and Shear Deformations along the Concrete-Membrane Interfaces

Deformations can occur in the sprayed concrete lining caused by the differential shrinkage of the concrete with different age on either side of the membrane, as well as thermally induced contraction of the concrete material due to fluctuations in the temperature. Such deformations are illustrated in Fig. 3. Each layer of concrete will exhibit a set of shrinkage cracks which will normally not persist across layers with different age. The membrane represents a deformable and ductile material, which is designed to bridge the cracks in the concrete. The two concrete layers, one on either side of the membrane may be applied with a time gap of several weeks or up to several months. From a load consideration perspective, the full shrinkage potential from the covering layer of concrete is assumed. Shrinkage properties of sprayed concrete has been subject to a recent Swedish study (BeFo 2014; Bryne et al. 2014a). Free (unrestrained) shrinkage of fiber-reinforced sprayed concrete after approximately 120 days was found to be in the range of 0.045–0.055 %, or 0.45–0.55 mm per m on laboratory sprayed slab specimens subjected to norm climate conditions (storage at RH

50 % and 20 °C, following 7 days of initial curing under water). In a restrained context such as bonding to rock as well as the unilateral exposure to moisture on the rock side and drying on the air side, precise assessments of shrinkage are difficult to make. Effects of surface drying may cause high shrinkage locally at the concrete surface. The shrinkage will result in the cracking of the concrete material. The use of fiber reinforcement and the restraint caused by the bond to the membrane will have some crack width reduction effect. Measurements of crack widths in the concrete lining has been conducted (Sect. 4.5 in this paper) in order to substantiate typical crack widths.

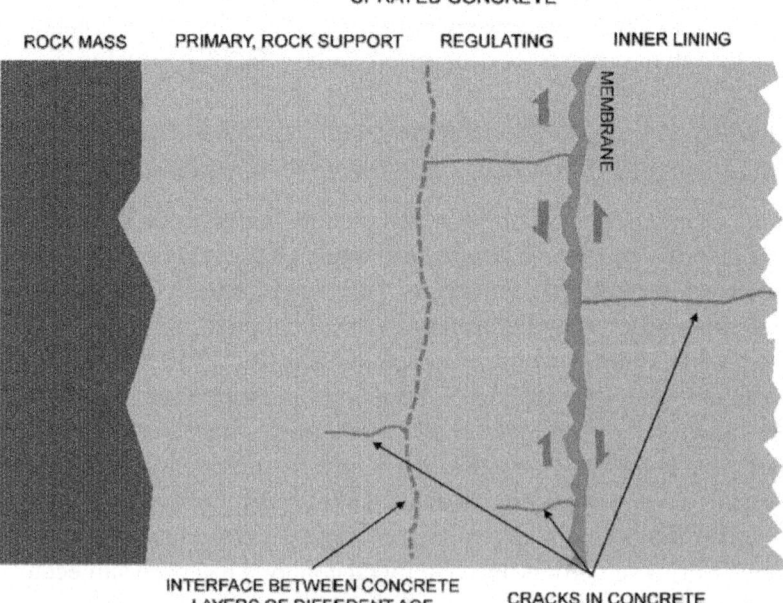

Figure. 3: Shear deformation and elongation at cracks of the membrane in the lining structure. *Top* photo showing persistent shrinkage crack in the secondary lining being bridged by the membrane. *Bottom* model for cracks and shear deformation.

Exposure to Moisture

The continuously bonded property of the waterproof SCL system implies that there is an exposure to the groundwater at the interface between the rock mass and the sprayed concrete. Both the constituent materials concrete and membrane exhibit capillary and hygroscopic properties. Thus, the in situ moisture content of the lining materials and its effect on the mechanical properties need to be accounted for. Moisture properties of the lining materials are shown in Sect. 6.1 in this paper. The measured in situ moisture content in the investigated tunnel linings is shown in Sects. 4.2 and 4.3.

Consideration of the Monolithic Character of the Lining

The mechanical performance of a continuously bonded SCL depends on the performance of the weakest element in the lining structure. A tunnel lining based on two layers of sprayed concrete separated by a bonded membrane should ideally be considered as one structure for the entire lining thickness. For this reason, the tensile bonding strength of the membrane-concrete interfaces should not be significantly lower than the tensile bonding strength between the rock surface and the sprayed concrete. Tensile bonding strengths for sprayed concrete interfaces to the rock substrate vary highly depending on rock type and the type of surface, as well as the application and material details of the sprayed concrete. Measured values for tensile bonding strength for the interface between sprayed concrete and rock vary between 0.2 and 1.8 MPa (NCA 2011; BeFo 2014; Bryne et al. 2014b). The gravity induced tensile stresses would be approximately a factor of 100 times lower than the lowest recorded tensile bond strength of concrete against rock. For this reason it is reasonable to propose an acceptance criterion for tensile bonding strength for the membrane which is in the magnitude of relevant tensile bonding strength between rock and sprayed concrete. Norwegian and Swedish standards propose 0.5 MPa as a minimum required tensile bond strength between rock and sprayed concrete. The ITAtech Design Guidance (ITA/AITES 2013) for sprayed membranes proposes an acceptance criterion of 0.5 MPa for tensile bonding strength.

Exposure to Freezing

The basic SCL design in our study has no insulating layers to avoid freezing exposure. The aim of this study is to determine the possible damage or reduction in performance caused by realistic freezing exposure. Given a membrane thickness of 3–4 mm, the thermal conductivity of the concrete material in secondary lining will be decisive for the thermal exposure. Each tunnel will represent an individual case with respect to freezing exposure based on the rock mass temperature, the winter climate, the ventilation of the tunnel and the location in the tunnel considered.

FIELD INVESTIGATIONS

Overview, Goal

The main goal of the field investigations was to substantiate as much as possible the loading conditions for the membrane (moisture, thermal and crack situation), as well as carrying out in situ measurements of the tensile bonding strength of the interfaces between the membrane and the concrete. We have included the investigations carried out on the large scale laboratory lining structure as part of the field investigations in this paper since this investigation context has proven to cover comparable conditions to in situ tunnel. The field investigations were carried out in the period 2012–2014. Table 4 shows an overview of the conducted field investigations with locations and main purposes. The 4 locations and type of test sites are described in further detail in Holter and Geving (2015).

Table 4: Overview of conducted field investigations for sprayed waterproofing membranes

Investigation	Location, test site	Main purpose of investigation
Moisture exposure and in situ moisture content, development over time	Gevingås Harangen Ulvin Karmsund	Basis for moisture conditioning during laboratory testing. Basis for assessment of degradation mechanisms
Mapping of cracks in sprayed concrete linings	Gevingås	Obtain realistic crack data for sprayed concrete
In situ tensile bonding strength	Ulvin Gevingås Laboratory lining structure	Tensile bonding strength under real exposure
Freezing exposure parameters	Ulvin Laboratory lining structure	Thermal profile through lining during severe freezing exposure

Moisture Content of the Tunnel Lining Structure

A detailed study of the moisture content and moisture transport mechanisms in waterproof SCL sections has been reported by Holter and Geving (2015). The findings of this study serve as an important basis for the analysis of the performance of the membrane described in this paper. The field investigations of the moisture content were carried out in three different tunnel sites at yearly intervals with ages up to 4 years. Several consistent observations were made during these investigations. The main features are:

- High degree of capillary saturation (DCS) of the concrete material, close to 100 %, at the rock-concrete interface.

- A gradient with degreasing DCS towards the lining surface.

- Depending on the lining thickness, the DCS of sprayed concrete on either side of the membrane is found to vary between 80 and 95 %. For primary (rock support) lining thickness of approximately 150 mm the DCS of the concrete at the membrane was found to be around 95 %.

The found moisture condition of the investigated linings can be explained by moisture transport processes from common building physics principles. Further details are given in Holter and Geving (2015).

Moisture Content in the Membrane Material

In addition to the investigations of the concrete material in the tunnel linings, also the membrane material was analyzed. Immediately after splitting of the cores, samples of the membrane material were removed from the concrete and tested for moisture content. Membrane samples from four different tunnel lining locations have been taken from 5 up to 38 months after construction. The measured values for moisture content, given as weight of water in % of dry weight of the membrane material are shown in Fig. 4. Dry weight of the membrane refers to weight after drying of 3–4 mm thick specimens at 105 °C for a minimum of 2 days.

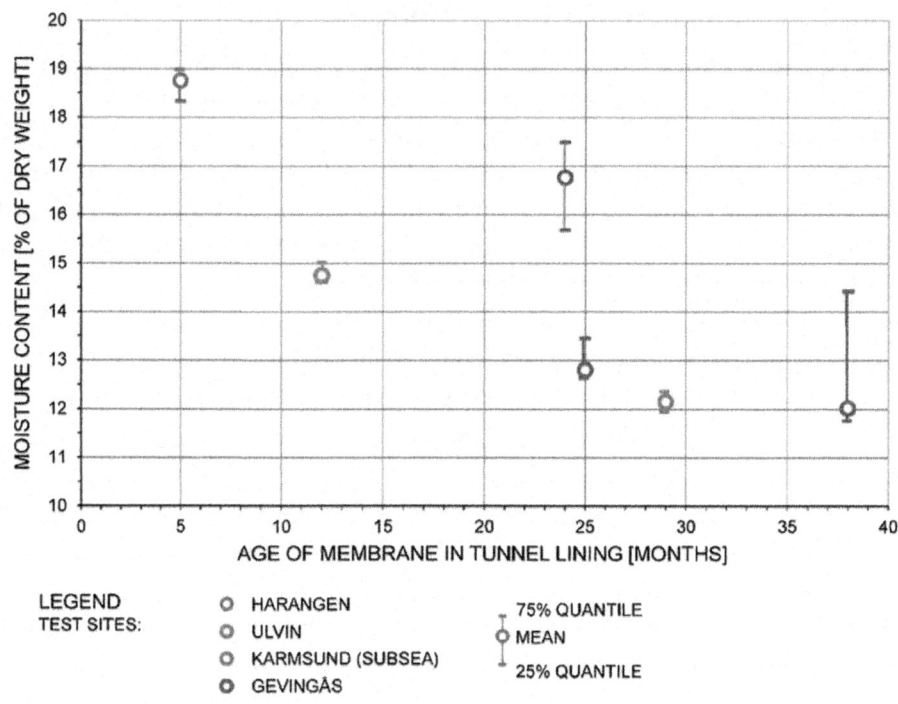

Figure. 4: Measured development of in situ moisture content in sprayed membrane material in tunnel linings.

A trend with decreasing moisture content in the membrane material with increasing time after construction can be observed, in spite of high degrees of capillary saturation of the concrete on either side of the membrane. These data refer to four different hard rock tunnel projects indicated with the different colors. The tunnel linings in all four cases were constructed with drained inverts and waterproof undrained SCL in the walls and crown. The Karmsund case is a subsea road tunnel located at approximately 70 m below the groundwater table. Hence a complete saturation under hydrostatic pressure of the rock mass, and higher saturation of the imperfections of the concrete is likely to have taken place.

Thermal Exposure to Tunnel Linings

The rock and concrete materials exhibit thermal conductivities which govern the temperature profile form the rock mass to the lining surface under a given thermal exposure in the tunnel space. In this study, monitoring of temperatures under freezing exposure at full scale conditions, measurements of thermal conductivities of rock and concrete materials and thermal calculations were carried out. Thermal conductivities for sprayed concrete and intact rock were measured in a separate study (NTNU 2013). Some of the findings are shown in Table 5.

Table 5: Measured thermal conductivities

Material	Density (kg/m^3)	Thermal conductivity (W/m K)	COV
Rock, dark gneiss (Ulvin site)	2616	2.95	0.3–0.5 %
Rock, granodiorite (Trondhjemite, freezing laboratory)	2657	2.77	0.2 %
Sprayed concrete, Ulvin site, steel fiber[a], dry[b]	2138	1.64	0.5–1 %
Sprayed concrete, Ulvin site, steel fiber, saturated[c]	2214	1.85	0.2–0.5 %
Sprayed concrete, Gevingås site, PP-fiber[d], dry	2211	1.65	0.5–1 %
Sprayed concrete, Gevingås site, PP-fiber, saturated	2281	1.85	2–3 %

[a]Steel fiber, dosage 35 kg/m^3, 0.5 % by volume

[b]DCS 70 %

[c]DCS 100 %

[d]Structural polypropylene fiber, dosage 7 kg/m^3, 0.8 % by volume

Thermal monitoring with freezing exposure was carried out in the full scale lining section at the Ulvin test site and the lining structure in the freezing laboratory. The main goal of this monitoring was to measure temperature profiles under realistic conditions. The temperature at the location of the membrane can then be assessed. The two monitoring cases are explained in Fig. 5.

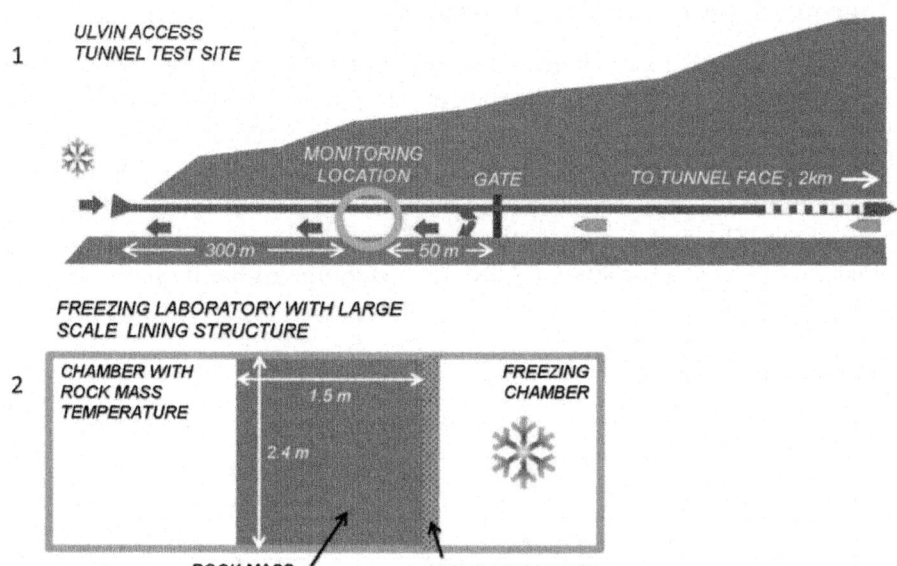

Figure. 5: Longitudinal vertical sections with configuration of the Ulvin test site for in situ measurements (case 1) and the laboratory lining structure (case 2) for controlled thermal exposure.

The two test sites for thermal monitoring have the following main characteristics:

- Case 1: Test site Ulvin, an access tunnel under construction with a test field of 90 linear m with SCL, with lining thickness 300 mm and membrane location at 150 mm from lining surface. The ventilation at the monitoring location was arranged with a gate in the tunnel so that air with a constant temperature of approximately 2 °C from the tunnel face (located more than 2 km in rock mass with constant temperature) could be alternated with cold air from outside. In this way an exposure to cold air at approximately 50 m distance from the portal could be applied experimentally in full scale. The field test at the Ulvin site could only be carried out in a short period of time for one severe freezing cycle over 36 h and hence, provided no information of long term freezing exposure.

- Case 2: Laboratory test facility with large scale lining structure constructed on a rock mass of homogenous granodiorite blocks with lining thickness 240 mm with membrane location at 110 mm. Controlled freezing exposure was applied to the lining surface, simulating different freezing scenarios. Exposure modes included cyclic loads for accelerated freeze–thaw testing of the lining materials, and as isothermic exposure in order to simulate the effect of long term cooling of the lining.

Findings from an investigation conducted in the Glödberget rail tunnel in north Sweden indicate that low air temperatures can penetrate far into the tunnel (STA 2012). The first 200–300 m from the portal can be exposed to air temperatures in the range of −15 to −20 °C in severe cases. The design of the tunnel lining for thermal insulation in portions with such severe exposure need to be evaluated in each single case based on local climate conditions and ventilation of the tunnel under operation during winter season.

The field test at the Ulvin site was arranged to produce a cooling of the tunnel lining by running the ventilation at approximately 1 m/s air flow with cold air from outside. Temperatures in the tunnel air at the test location in the range of −7 to −9 °C were achieved. After 36 h the test had to be terminated due to the tunnel construction cycle. A profile of the tunnel lining with the measured temperatures after 36 h is shown in Fig. 6. The calculated temperatures at steady state conditions with −7 and −9 °C in the tunnel space are indicated. The measurements indicate a background temperature of the rock mass at the location of the tunnel of approximately 7 °C.

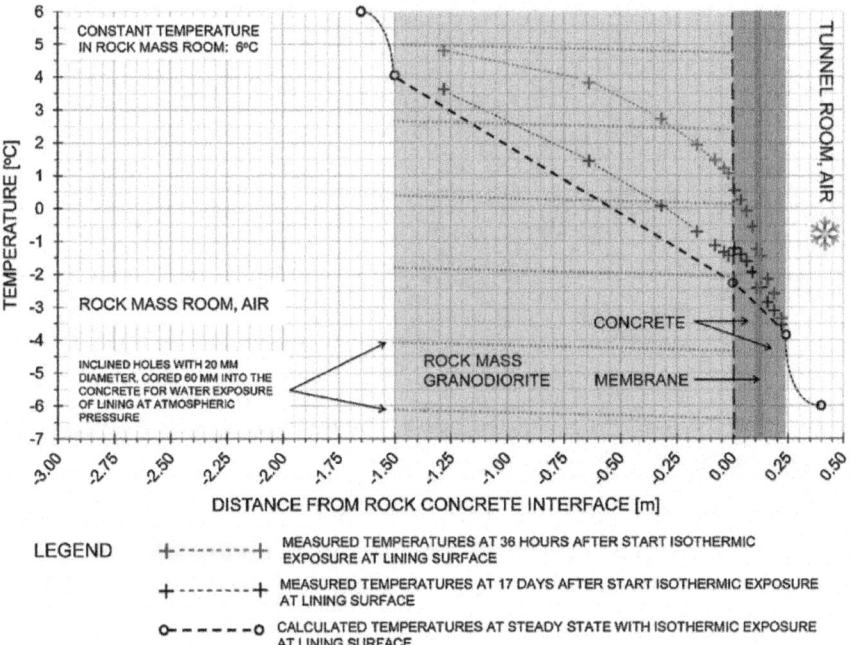

Figure. 6: Large scale laboratory simulation: measured and calculated temperature profiles through rock mass and lining structure.

A large scale simulation of isothermic freezing exposure with −6 °C in the tunnel space was carried out on the lining structure in the freezing laboratory (case 2, Fig. 5). This exposure was held constantly for 30 days with continuous thermal monitoring. The results are shown in Fig. 7. The measured temperature profiles after 36 h and 17 days together with a calculated temperature profile at steady state are indicated. The measured values refer to three different sets of sensors in the lining—rock mass structure, and hence exhibit a slight scatter due to precision of location.

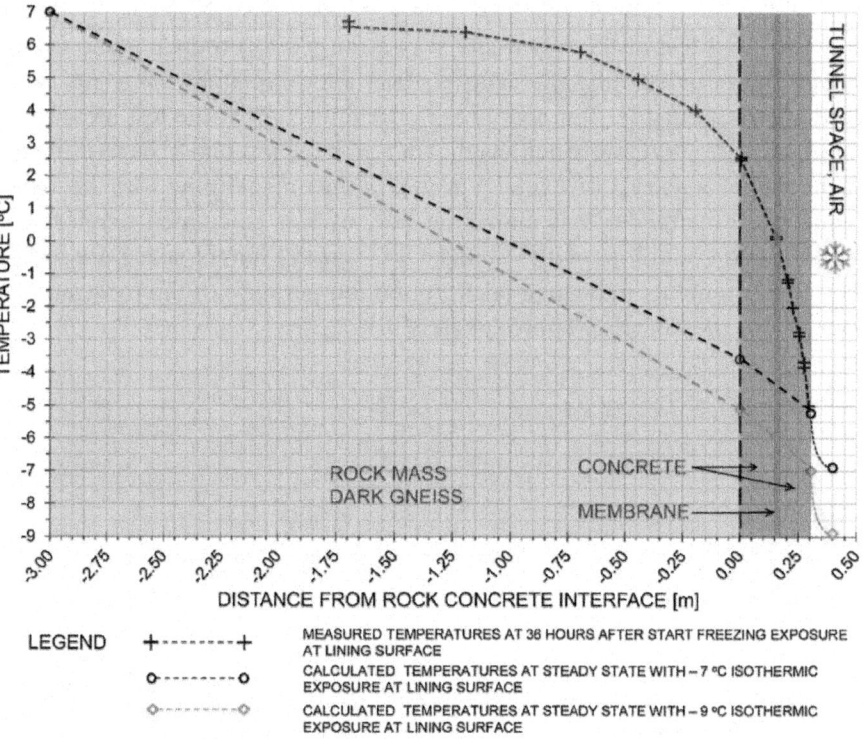

Figure. 7: In situ tunnel, Ulvin test site: measured and calculated temperature profiles through rock mass and lining structure.

Based on the conducted freezing exposure tests, and calculation of temperatures at steady state conditions, the minimum temperature exposure at the membrane at given lining thicknesses can be assessed, Table 6.

Table 6: Temperatures at the location of the membrane in an SCL structure based on measurements and thermal calculations at steady state

Air temperature in tunnel space (°C)	Thickness of covering layer of sprayed concrete over membrane (mm)	Temperature at membrane (°C)
−6	110	−3.5
−7	150	−4.5
−9	150	−6

Mapping of Cracks in the Sprayed Concrete

A mapping of cracks was carried out in the Gevingås rail tunnel on the 2nd August 2013, after an extended period of warm weather with maximum outdoor temperatures in the range of 25–30 °C. The temperature of the tunnel lining at 10 mm depth measured during the mapping of the cracks was 12 °C. Approximately 210 cracks were mapped and marked using a concrete crack width gauge (Fig. 8) in a systematic manner in order to re-record the same cracks later. Hence, the crack mapping was repeated at the exact same location in February 2014 when the temperature was 6 °C at 10 mm depth.

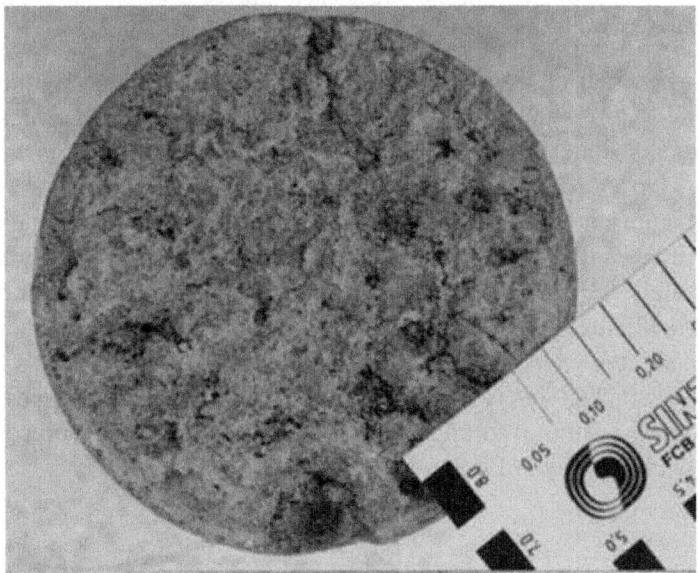

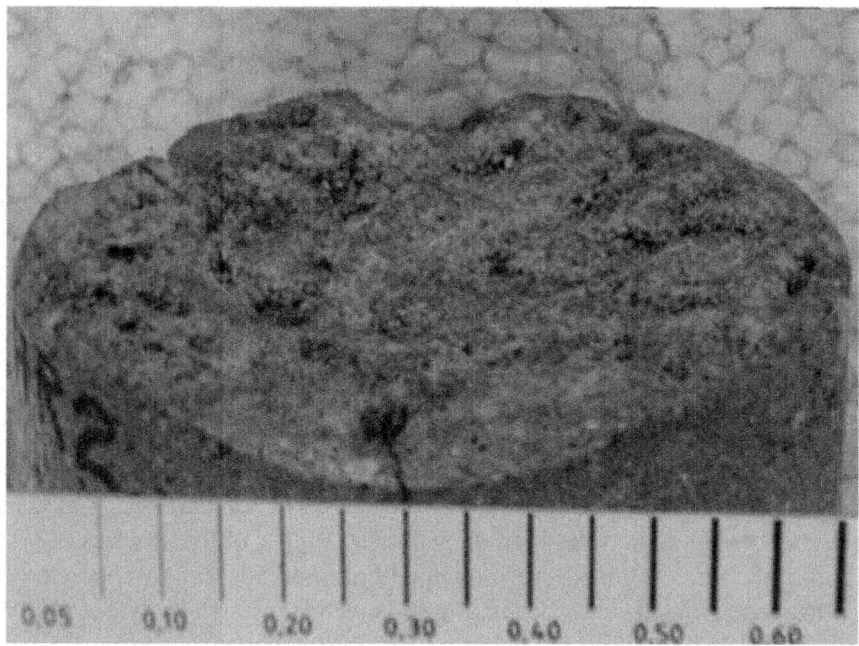

Figure. 8: Example of recordings of cracks in a sprayed concrete surface using a concrete crack measurement gauge.

The measured crack widths are shown in Fig. 9. Crack widths ranging from 0.05 to 0.2 mm account for 78 % of the recorded cracks for the measurements done in August 2013. The crack measurements in February 2014 show an increase in crack width, and a larger scatter of the recordings. The most represented crack width for the measurements conducted in February is approximately 0.3–0.35 mm. Thus, an average increase in crack width of approximately 0.2 mm with a temperature decrease of 6 °C is observed. A typical crack pattern was obtained by observing the sprayed concrete lining surface in an area which exhibited leaks and showed mineral stains from leaks through wet cracks. This is shown in Fig. 10. Visible crack distances vary from approximately 0.2 m up to approximately 1.5 m. The most represented crack distance is in the range of 0.7–1 m.

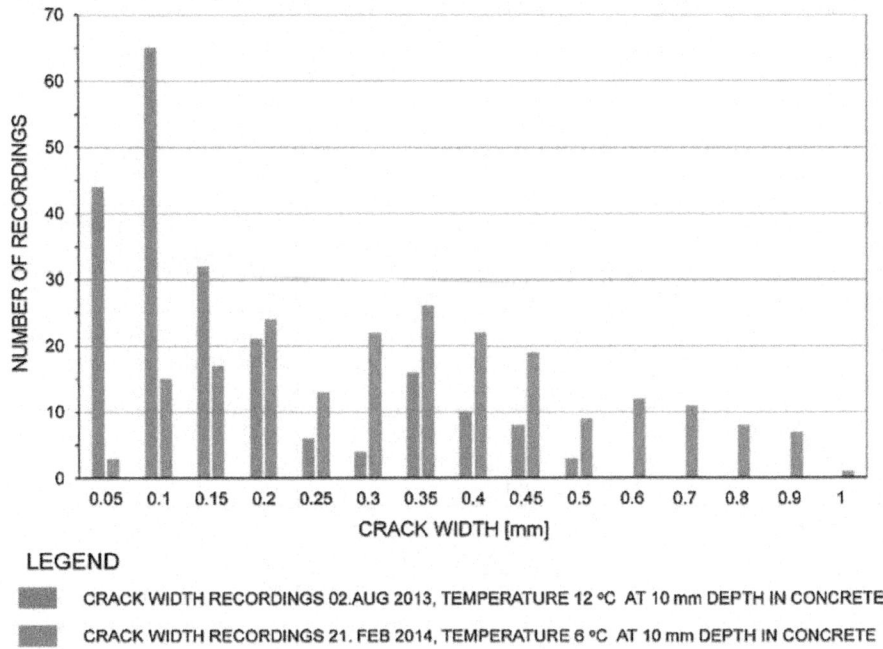

LEGEND

■ CRACK WIDTH RECORDINGS 02.AUG 2013, TEMPERATURE 12 °C AT 10 mm DEPTH IN CONCRETE

■ CRACK WIDTH RECORDINGS 21. FEB 2014, TEMPERATURE 6 °C AT 10 mm DEPTH IN CONCRETE

Figure. 9: Measured crack widths in the sprayed concrete lining surface at the same location in August 2013 and February 2014.

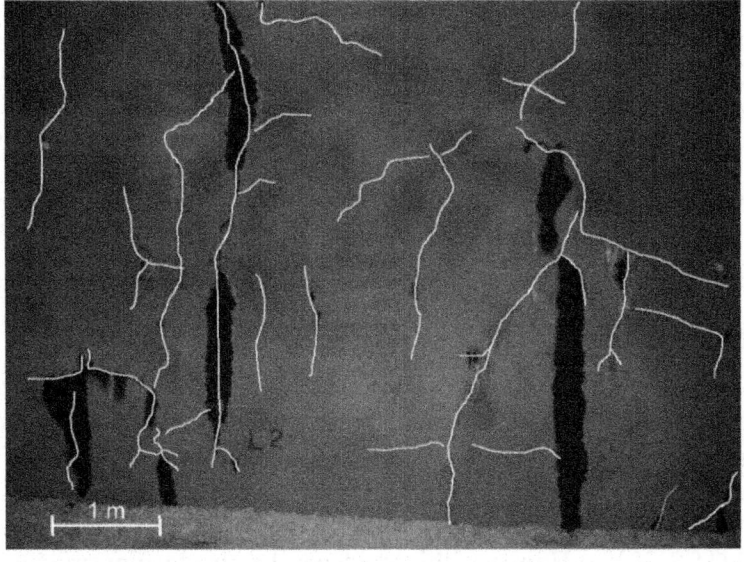

Figure. 10: Surface observations of cracks in sprayed concrete.

Summary of Field Investigations, Verification of Loading Model

The loads which expose the membrane considered in this study are summarized in Table 7.

Table 7: Compilation of loads on the membrane considered in this study

Load type	Relevant value/size	Implication for laboratory testing
Rock mechanical loads	None. Only local loads	None
Groundwater pressure induced loads	Very unlikely for the investigated cases. Not considered	None for cases with low or no hydrostatic pressure
Dynamic loads from traffic area	10 kPa amplitude of air pressure (pressure + suction loads)	None
Tensile loads	Gravity induced load from inner lining: 2 kPa	Not realistic requirement
Maximum crack width in concrete and thermal opening/closing	Typical crack width range: 0.1–0.3 mm, maximum 0.8 mm	Testing of elasticity under relevant temperatures and moisture contents required
	Thermally induced opening and closure : 0.6 mm	Crack bridging performance at 1 mm crack width proposed
Shear deformation along interfaces	0.5–0.6 mm/m	1 mm shear deformation within linear shear elasticity behavior
Moisture exposure	15–18 % moisture content range in the membrane material	Pre-conditioning of membrane to relevant moisture content
Thermal exposure	Possible temperature range +15 to −6 °C at membrane location in tunnel lining	Testing at realistic temperatures

EVALUATION OF LABORATORY TEST METHODS

The main purpose of the laboratory test methods is to conduct material testing of the membrane under realistic loading and climatic exposure. There are several standardized test procedures for building materials which may be used for membrane materials. The most updated compilation of suggested tests is given in ITAtech Design Guidance for Spray Applied Waterproofing Membranes (ITA/AITES 2013). However this guidance has no loading models, neither any guidelines for mechanical, thermal nor moisture exposure testing of the membrane material. In this section the loading model (Sect. 3) and findings from the field investigations (Sect. 4) will be used to substantiate details in the laboratory test methods and relevant acceptance criteria.

Moisture Properties of Lining Materials

Due to the hygric continuity of the lining structure and the direct exposure to groundwater, the moisture properties of the lining materials need to be included in order to substantiate the realistic moisture condition for testing. Recent reported testing of membranes for waterproof SCL (Su et al. 2013; Su and Bloodworth2014; Nakashima et al. 2015) have not included the moisture condition and moisture properties of the constituent materials in the lining. Testing of moisture properties of membrane and concrete materials have yet to be included in guidance for design and testing of spray applied membranes. Standard test methods for sorptivity and moisture content at equilibrium commonly used for concrete are adopted in this study. Thus, comparison to findings from other studies of concrete is possible.

Elasticity and Crack Bridging Properties of the Membrane

Preventing water flow through the lining by the bridging of cracks is the main waterproofing function of the membrane. Testing of the membrane's elasticity can be done by a pure elastic test or by a functional test of the resistance to rupture over a discontinuity in the substrate. Rupture of the membrane over a crack with increasing width is found to be the main failure mechanism in the loading model (Sect. 3). Hence, the crack bridging test as proposed in the ITAtech guidance (ITA/AITES 2013) guidance directly addresses a relevant failure mode. A pure elasticity test does not account for the bonding of the membrane to the substrate. It is difficult to quantify a requirement in terms of pure elasticity which translates to the required crack bridging capacity. However, the elasticity test is simple and can give an indication of the elasticity of the membrane material in order to reject unsuitable materials without conducting costly testing.

Elasticity Testing According to DIN 53504

This is a simple test conducted on specimens with standard dimensions which are stretched to failure while measuring tensile deformation and tensile force. Standard dog bone shaped specimens, shown in Fig. 11, have normally been used for this purpose. The sensitivity of EVA-based membranes to moisture content means that details regarding storage and conditioning as well test procedure for such membranes needs to include details regarding humidity and temperature. Sprayed specimens are preferred to molded specimens in order to test realistic membrane material. However, sprayed specimens are more difficult to produce with even thicknesses for the purpose of reproducing consistent standard dimensions for laboratory testing.

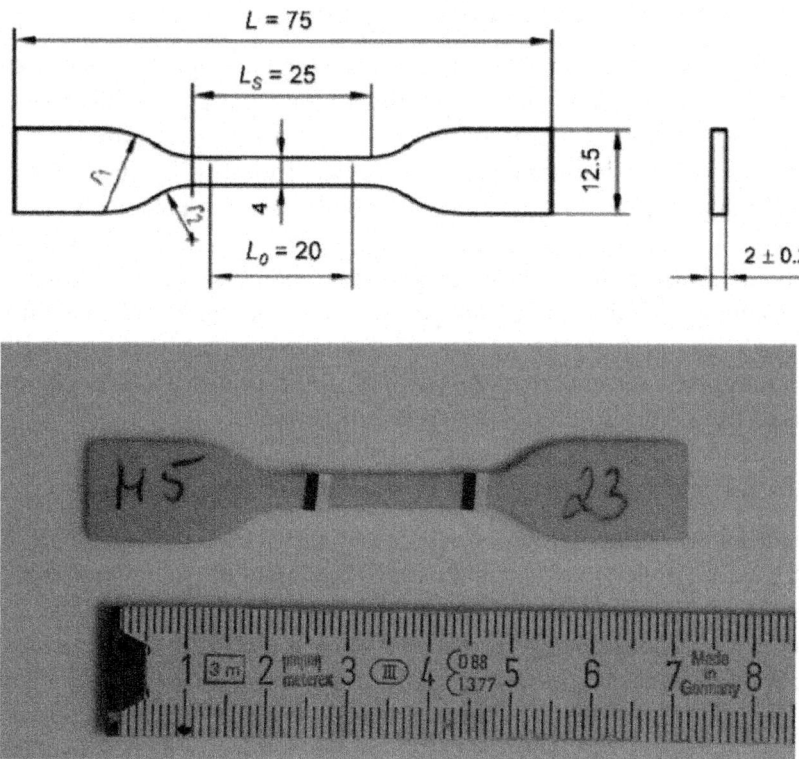

Figure. 11: Specimen for elasticity test according to DIN 53504. *Top* dimensions of the S2-type dog bone shaped specimen used for this purpose, with figures in mm. *Bottom* photo of specimen with markings of the length L_0 area for precise elongation measurement with video-extensometer.

Crack Bridging Performance

The proposed test method for crack bridging performance according to ITA/AITES (ITA/AITES 2013) is a static crack bridging test and is designed for the testing of coating materials on exterior surfaces of masonry and concrete (DIN EN 1062-7:2004). With this test method one can basically test only one crack width, although it would be possible to include a few increments in the crack width before the maximum crack width, given by the geometry of the test jig, is reached. The main features of this test method are shown in Fig. 12.

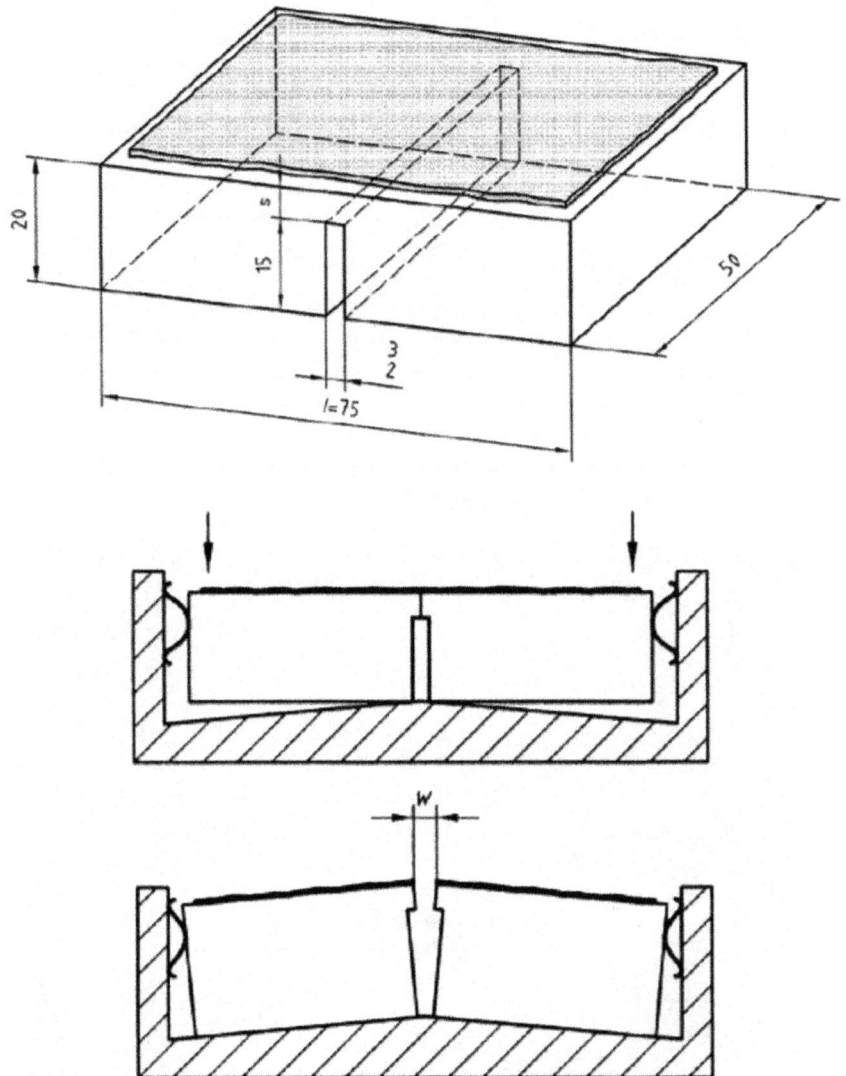

Figure. 12: Conceptual illustrations showing the static crack bridging test procedure according to DIN EN 1062-7 Annex C1, proposed by ITA/AITES 2013.

An adjusted crack bridging test has been considered in order to apply a more controllable opening of the crack and hence enable a precise determination of the crack width at rupture. This adjusted test method has many similarities to the dynamic tensile test described in DIN EN 1062 Annex C4. The adopted procedure is shown in Figs. 13 and 14.

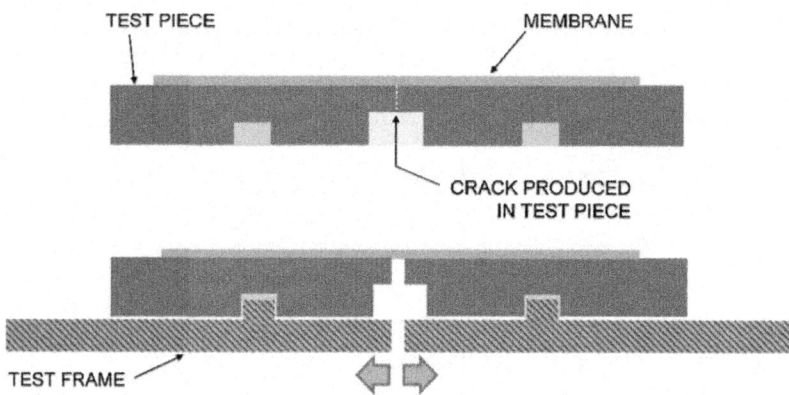

Figure. 13: Conceptual illustrations showing the loading mode during the adopted version of the dynamic crack bridging test.

Figure. 14: Crack bridging testing with adopted procedure in progress in climate chamber. (Courtesy by Wacker Chemie AG).

Tensile Bond Strength

Testing of tensile bond strength, also referred to as pull-off strength or adhesion, of sprayed membranes has been conducted by pulling the membrane off the substrate (Ozturk and Tannant 2010). This procedure uses a disc shaped plate mounted on an elevator bolt which is glued to an over-cored section of the membrane and subsequently pulled in a controlled manner. This method can prove useful for a temporary test of the membrane's tensile bond strength before the inner lining concrete is applied. Testing of tensile bonding strength of the membrane in the lining structure as proposed by ITA/AITES (2013) is a standard pull-off test for adhesion including the entire lining structure, according to EN ISO 4624 section 9 (2003). The principle of the test is shown in Fig. 15. However, the wet core drilling, the risk of applying unfavorable bending and tensile loads during the core extraction, inconsistent moisture conditioning and details in the test setup might influence the results significantly.

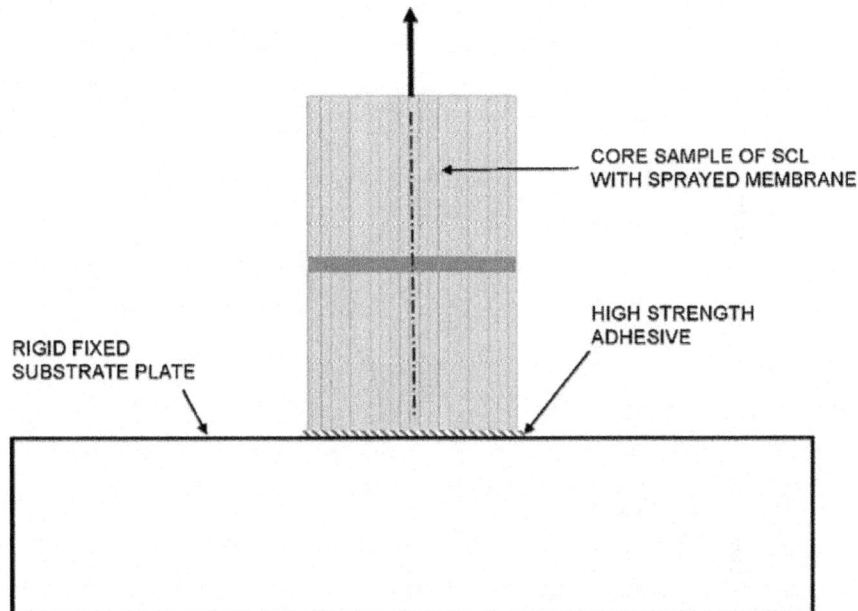

Figure. 15: Testing of tensile bonding strength according to EN-ISO 4624.

A procedure to measure tensile bond strength without extracting core samples was adopted. The main purpose of this procedure was to test the tensile bond strength under as realistic conditions as possible. The procedure was laid out as an in situ test in which the test specimen consisted of an over-cored part of the lining structure. The layout of the pull-off details were arranged in order to achieve a perfect axial alignment to the core specimen. The adopted test is shown in Fig. 16. The testing device used in this investigation could only record maximum tensile strength.

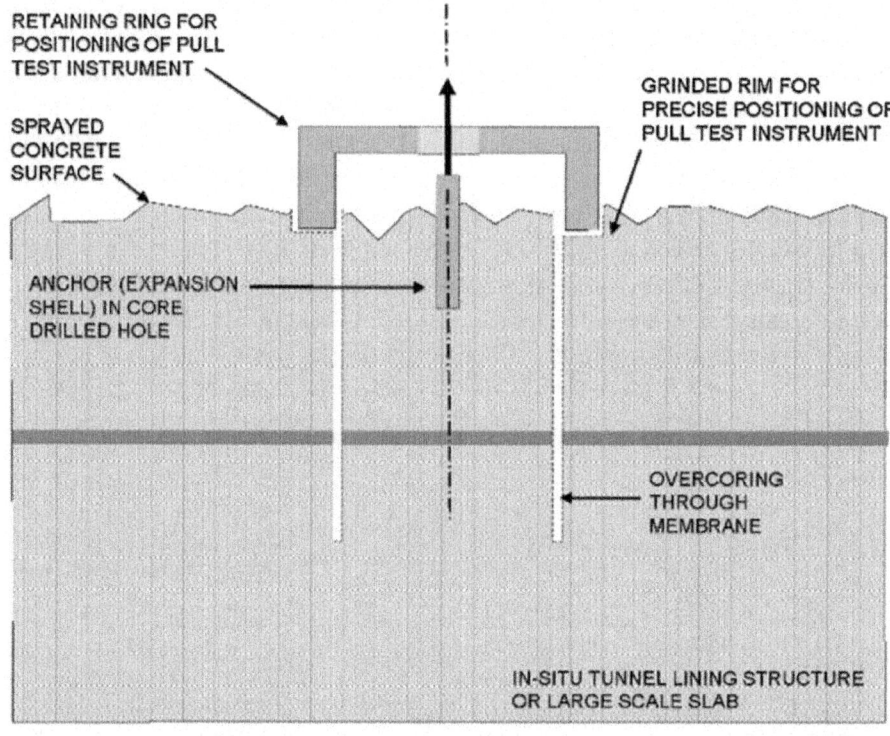

RETAINING RING FOR
POSITIONING OF PULL
TEST INSTRUMENT

SPRAYED
CONCRETE
SURFACE

GRINDED RIM FOR
PRECISE POSITIONING OF
PULL TEST INSTRUMENT

ANCHOR (EXPANSION
SHELL) IN CORE
DRILLED HOLE

OVERCORING
THROUGH
MEMBRANE

IN-SITU TUNNEL LINING STRUCTURE
OR LARGE SCALE SLAB

Figure. 16: Adopted procedure for in situ measurements of tensile bonding strength. *Top* conceptual diagram showing the layout of the test. *Middle* Preparation with over-coring with grinding of the test specimen. *Bottom* in situ specimen after testing.

Shear Performance of the Membrane-Concrete Interfaces

Direct shear testing is not proposed by ITA/AITES (2013). Direct shear testing is included in this study in order to establish the membrane's ability to perform under shear deformation which can occur between the substrate and inner lining concrete layers. Previous direct shear testing of such membrane concrete interfaces has been reported by BASF (TU Graz 2008), Su et al. (2013) and Su and Bloodworth (2014). These investigations refer to EVA based sprayed membranes in which the specimens were tested in dry state without any pre-conditioning to relevant moisture content. A study of the composite action of EVA-based membranes for SCL was carried out by Nakashima et al. (2015). However, this study also considers a lining structure and the membrane material in dry condition.

For our study a large scale shear box with constant normal load was available. Controlling the normal load in order to apply constant normal deformation or constant normal stiffness was not possible in our study. The test procedure and moisture conditioning of the specimens is explained in Sect. 6.6 together with the obtained results.

CONDUCTED LABORATORY INVESTIGATIONS

The laboratory investigation program was based on the conceptual model, the results from the field investigations and the evaluation of testing methods. The main goal of the laboratory investigations was to verify the conceptual model, establish detailed performance properties of the lining, as well as providing a basis for the acceptance of a membrane product under certain conditions.

Moisture Properties of Sprayed Concrete and Membrane Materials

Specimens

The specimens for the sprayed concrete testing of the moisture properties were all obtained from the tunnel lining from the same site and the same location, the Harangen test site (details in Table 1). Hence, the specimens represent the same concrete in terms of age, curing history, mix design and spray application. The membrane specimens were all produced from sprayed sheets, and tested after complete curing for approximately 18 months. The investigation of moisture properties in our study covers water content after immersion at atmospheric pressure, sorptivity (water absorption rate) and moisture bearing capacity in the hygroscopic range. These are illustrated in Fig. 17.

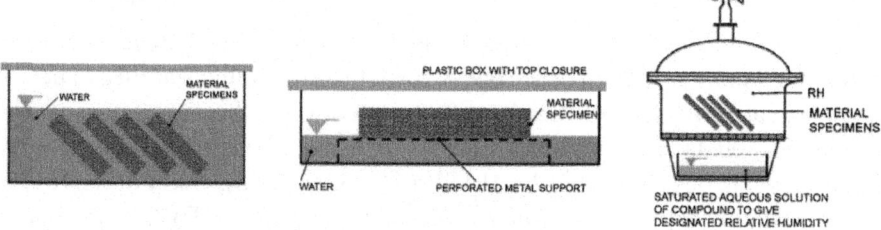

Figure. 17: Sketches of the three water absorption modes which have been tested in the laboratory. *Left* water content at complete immersion. *Middle* sorptivity (water absorption rate) at unidirectional water exposure. *Right* moisture bearing capacity at equilibrium in the hygroscopic range.

Water Content at Immersion at Atmospheric Pressure

Membrane specimens obtained by core drilling specimens from sprayed panels as well as sprayed membrane specimens from spray sheet panels were tested. Results are shown in Fig. 18 and Table 8. For the membranes M1 and M5 the maximum water uptake potential is found to be approximately 42 and 30 % of the dry weight of the material. The water content at immersion for concrete at atmospheric pressure is assumed to be equal to the complete saturation of the suction porosity of the concrete.

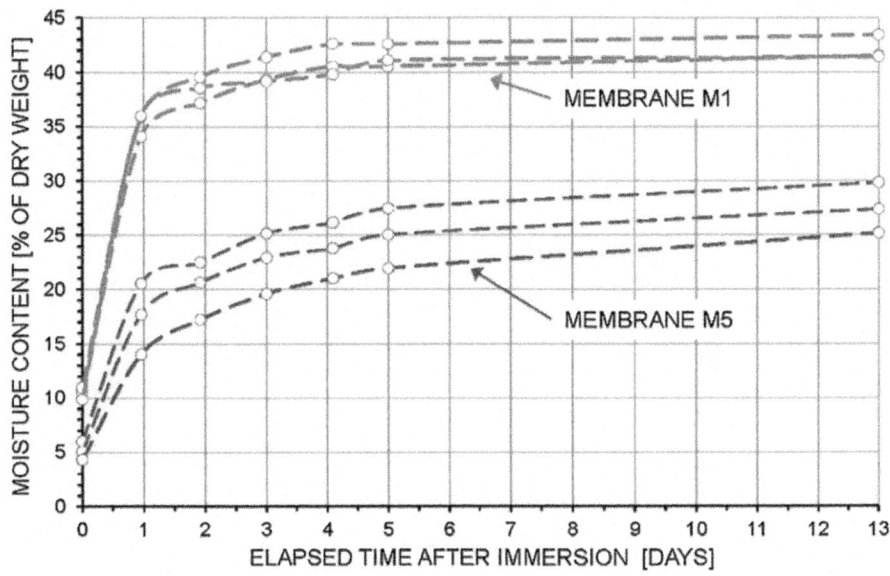

Figure. 18: Moisture content of membranes measured after complete immersion at atmospheric pressure for sprayed specimens of membranes M1 and M5.

Table 8: Moisture contents at immersion for two series of specimens for membranes M1 and M5

Membrane product	Moisture content at immersion (weight % of dry weight)	
	Specimens obtained from lining structure slabs (Fig. 19)	Specimens from sprayed membrane sheets
M1	41.5	42.4
M5	27.4	30.3

Water Absorption Rate (Sorptivity)

Sorptivity expresses the water absorption rate under unilateral and unidirectional water exposure, and was investigated by Holter and Geving (2015). A compilation of the results for concrete and membrane M1 is shown in Fig. 19. The measured water absorption rate of the two materials exhibit a significant contrast within the moisture content range which is found in tunnel linings.

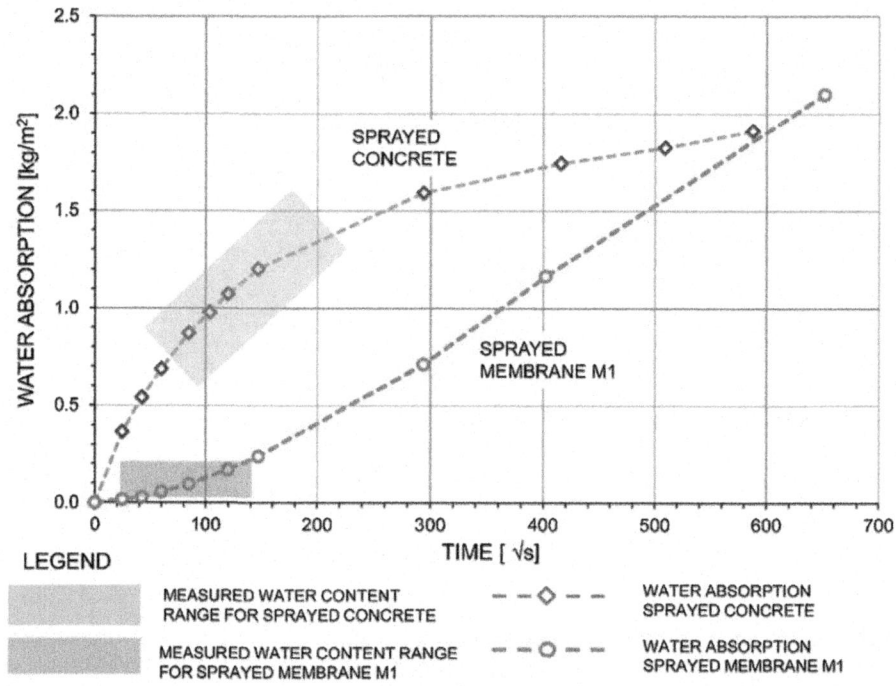

Figure. 19: Water absorption rate (sorptivity) of sprayed concrete and sprayed membrane (M1) represented as water absorption versus square root of time (compiled from Holter and Geving 2015).

Moisture Content at Equilibrium in the Hygroscopic Range

The investigation of the moisture content at equilibrium for concrete was presented by Holter and Geving (2015). Testing of the membranes M1 and M5 with moisture content at equilibrium obtained by isothermic desorption has been added in this study. The desorption isotherms show moisture content represented as degree of saturation at immersion for the materials versus relative humidity. A compilation of the results is shown in Fig. 20. The difference in behavior when taking the material from immersion to RH 100 % is noteworthy. Concrete loses approximately 10 % of its moisture content, whereas the membranes lose approximately 50 % of their moisture content. The desorption isotherms (Fig. 20) show that sprayed concrete is a much more hygroscopic material than the membrane material.

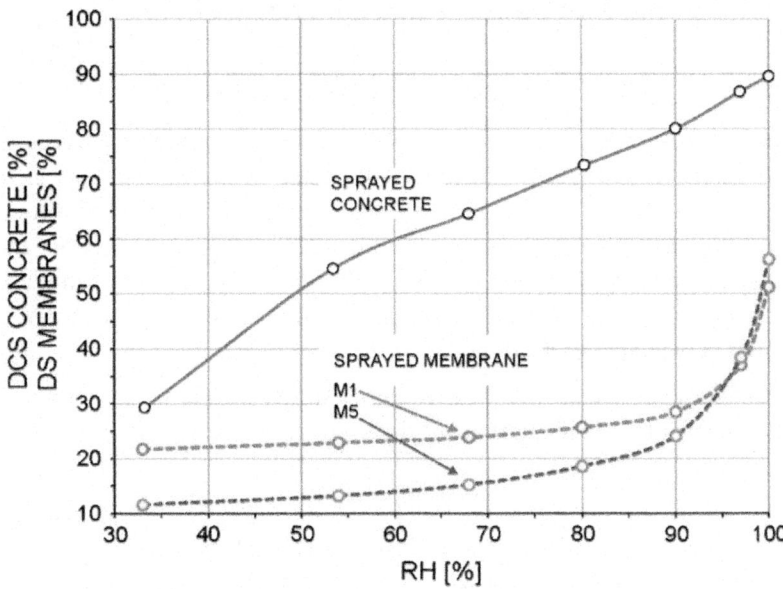

Figure. 20: Desorption isotherms obtained at 25 °C for sprayed concrete and sprayed membrane, showing the moisture content represented as degrees of saturation at equilibrium versus different values for RH.

Thermal Expansion of Sprayed Concrete

Linear thermal expansion was measured on cut prism samples of sprayed concrete with dimensions 70 × 70 by 280 mm in different temperature intervals from 3 to 34 °C. The measured values for the temperature interval 3–19 °C are found to be the most relevant and are shown in Table 9. The mix design of the sprayed concrete is shown in Table 2, Sect. 2.

Table 9: Measured linear thermal expansion coefficient for steel fiber-reinforced sprayed concrete

Parameter	Mean (m/m K)	COV (%)	Temperature interval (°C)	Number of specimens
Thermal expansion coefficient	1.27×10^{-5}	1.6	3–19	3

Elasticity of the Membrane Material, DIN 53504

Specimens, Conditioning and Testing Temperature

Specimens from both sprayed and molded sheets of membrane were prepared. Five membrane products were tested in three different test series, in three different laboratories. Altogether five membrane products were tested. Conditions during testing covered humidity conditioning at RH 50 and 95 % and at specific temperatures 23, 0, −3, −8 and −12 °C. A test of the polymeric content elasticity which could be related to the elasticity performance of the membranes was also carried out.

Our findings using this test method show that it is difficult to obtain consistent results when comparing different test series. The main limitations were:

- Difficulty in preparation (spray application) of specimens with even thicknesses required for reproducible laboratory tests.
- The thickness and evenness of the membrane specimens influences the result significantly.
- Different storage conditions after application and the precise conditioning details influence the absolute measurement values.
- Different interpretation of the test standard regarding testing details.

Findings

For the first elongation test, the polymeric content of the membrane products was tested using a thermo-gravimetric analysis with Argon as test gas in the test vessel. The membrane material was heated to 800 °C. Hence, it was possible to record the weight loss due to pure evaporation of the components, considered to be the pure organic polymeric content. The results are shown in Fig. 21. The two colors indicate the two different suppliers of the membrane products.

The results of the initial test are illustrated in Table 10. Cyclic freezing and thawing in air has the effect of a slight increase on the elongation performance. When thawing under water between each freezing cycle membrane M1 shows approximately 50 % reduction in elongation performance, whereas the other membranes are unaffected by the freeze–thaw exposure.

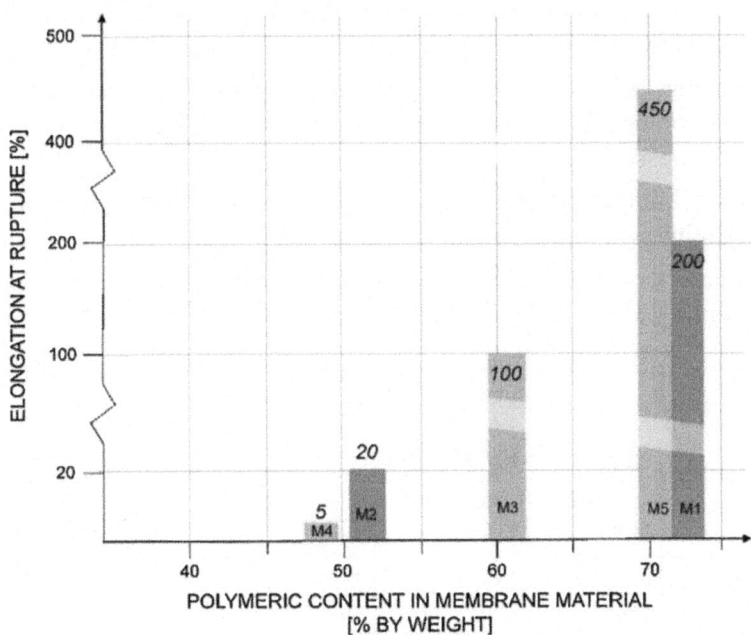

Figure. 21: Elongation performance of sprayed membrane samples versus polymeric content in the base (powder) membrane products M1–M5 measured by thermo-gravimetric analysis (TGA) with argon as test gas. The elongation was measured in one single test series (laboratory 1) with specimens which had undergone identical treatment from application to testing.

Table 10: Initial test series of elongation carried out in Laboratory 1

Membrane (all sprayed)	Measured elongation at failure, mean (%)		
	No freezing, storage at 23 °C RH 95 %	6 cycles[a], freezing to −20 °C in air, thawing at 23 °C RH 95 %	6 cycles[a], freezing to −20 °C in air, thawing at 23 °C immersed in water
M1	194	212	94
M2	20	22	18
M3	131	135	131
M4	4.3	4.9	6.2
M5	438	457	408

[a]Freezing 24 h, thawing 24 h

The main findings from the conducted elongation testing can be summarized as follows:

- Within the same test series, a consistent trend of significantly decreasing elasticity with decreasing temperature has been observed.
- A relatively large scatter is caused by varying thicknesses of the membrane within the same specimen as well as specimens with different thicknesses.
- Conditioning at RH95 % gives higher measured elasticities compared to specimens conditioned at RH50 %.
- Elasticity mainly increases with increasing polymeric content (shown in Table 11).

Table 11: Measured values for elongation for membrane according to DIN 53504 for two test series

Test location	Membrane sprayed/ molded	Pre-condi- tioning	Measured elongation (strain) at failure, mean (%)				
			23 °C	0 °C	−3 °C	−8 °C	−12 °C
Laboratory 2	M1 sprayed	RH 95 %	45		38	10	5
	M5 sprayed		685		89	20	8
Laboratory 3	M1 sprayed	RH 50 %	20	9	6	0.6	
	M5 sprayed	RH 50 %	242	56	14	1.4	
	M1 molded	RH 50 %		71	31		
	M1 molded	RH 95 %	242	76	55		
	M5 molded	RH 50 %		21	9		
	M5 molded	RH 95 %	446	40	13		

Crack Bridging

Specimens

Specimens for crack bridging testing were produced by applying the membrane material on pre-fabricated test pieces of porous artificial sandstone (Fig. 22). Both spray-applied and molded membrane specimens were prepared. Three series of specimens, shown in Table 12, were prepared in order to cover a range of temperatures and moisture contents.

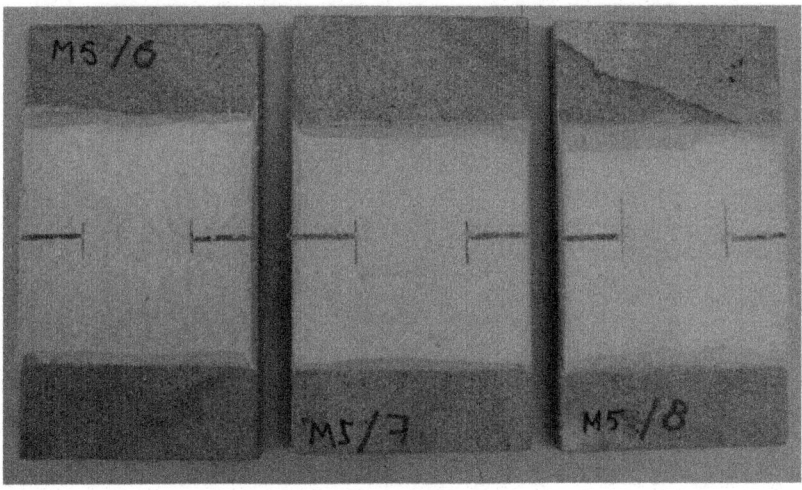

Figure. 22: Specimens for crack bridging test with membrane applied on the surface of test pieces. The dimensions of the test pieces are 100 mm × 200 mm.

Table 12: Matrix for conducted testing of crack-bridging

Test series number	Membrane, sprayed/ molded	Condi- tioned at RH (%)	Temperature at testing, number of speci- mens tested			
			23 °C	0 °C	−3 °C	−8 °C
1	M1 sprayed	95	3	2	3	3
	M5 sprayed	95	3	3	3	3
2	M1 molded	50		3	3	
	M5 molded	50		3	2	
3	M1 molded	95[a]		3	3	
	M5 molded	95[a]		3	3	

[a]Cured in RH 95 % for 28 days immediately after molding prior to testing

Procedure

The test method stated in DIN EN 1062 annex C4 dynamic tensile test, with the modifications developed by Wacker Chemie AG, described in Sect. 5.2.2 in this paper was followed. The crack width was increased in increments of 0.2 mm every 5 min. Immediately before a crack width increase was applied to the specimen, the membrane surface was visually inspected and deemed intact or ruptured. Any sign of initial rupture was interpreted as membrane failure (Fig. 23).

Figure. 23: Crack bridging testing in progress. *Top* specimen in testing apparatus for precise measurement of crack aperture at rupture. *Bottom* definition of rupture with visible initial damage of the membrane.

Results

The results of test series 1 and 2 are shown in Fig. 24. For the specimens with spray applied membrane (series 1) values are shown as mean with 25 and 75 % percentiles. For the molded specimens (series 2) only average values are shown, since there were too few satisfactory readings to obtain statistical data. The specimens had slightly differing membrane thicknesses. Therefore the value for rupture was given as the ratio between crack width at rupture and the membrane thickness which varied from 1.9 to 5.5 mm. The variation of the membrane thicknesses was largest for the specimens with spray applied membrane. The results (Fig. 24) show that the crack bridging capacity decreases with decreasing temperature with the given conditioning of the specimens. Both membrane products M1 and M5 were found to bridge cracks with an aperture in the range of 0.4–0.8 times the thickness of the membrane at −8 °C. At 23 °C the two membranes M1 and M5 were to bridge cracks with an aperture in the range of 1.3–1.6 times the thickness of the membrane. Series 3, tested only at 0 and −3 °C with molded specimens cured and pre-conditioned at RH 95 % for 28 days after application, did not exhibit any rupture at 11 mm which is the maximum crack width which the machine could produce. This indicates that the immediate curing after application at RH95 % leaves sufficient water in the membrane to act as a "softener" with resulting high elasticity.

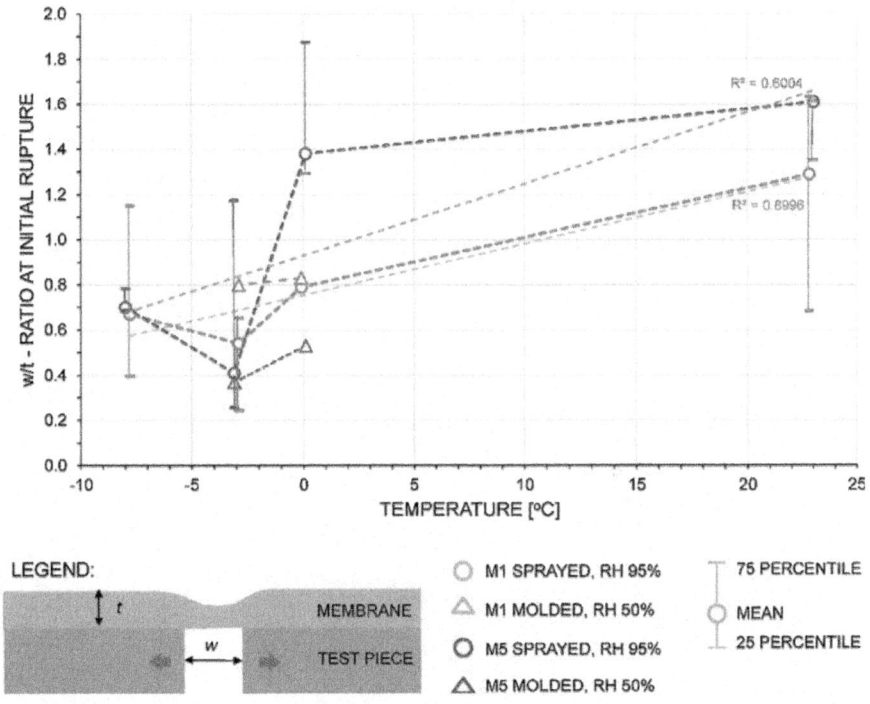

Figure. 24: Results from crack-bridging testing for membranes M1 and M5 conducted at different temperatures and pre-conditioned at RH 50 and 95 %.

Direct Shear Tests, Shear Bond Strength

Specimens

The specimens were produced from square panels using poured concrete with typical sprayed concrete mix design and spray applied membrane. In this way regular interfaces were achieved. The panels were stored under water for 5 months before specimens with 74 mm diameter were core drilled. The core specimens underwent another 30 days of storage under water. Throughout the storage under water a 40 mm wide strip strong tape was applied around the core completely covering the membrane and protecting it from direct water exposure. In this way the membrane only received exposure to water through the concrete pores.

Procedure

After water storage the specimens were prepared for shear testing by mounting them in a steel frame assembly. The process of preparing a series of three specimens in the test assembly and conducting the shear tests could be undertaken in 1 day. The assembly of the specimens is shown in Fig. 25 and the steps in the procedure are shown in Fig. 26.

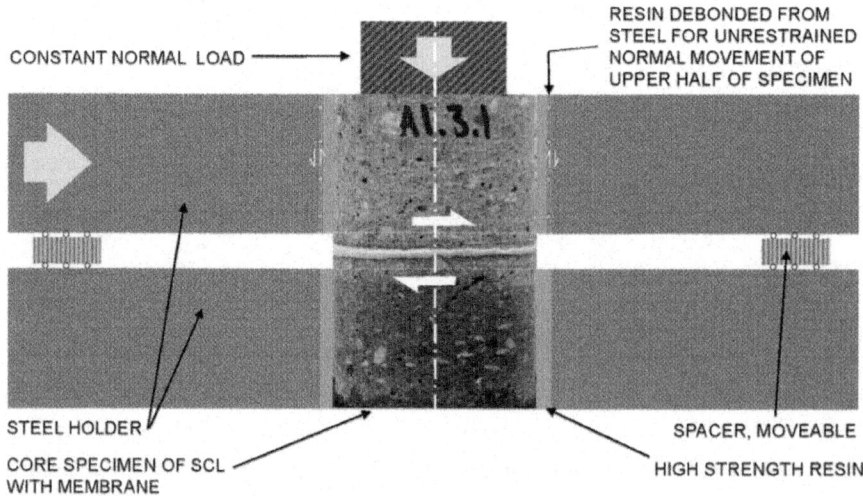

Figure. 25: Conceptual diagram showing a section of the assembly with core specimen mounted in the steel frame used for direct shear testing.

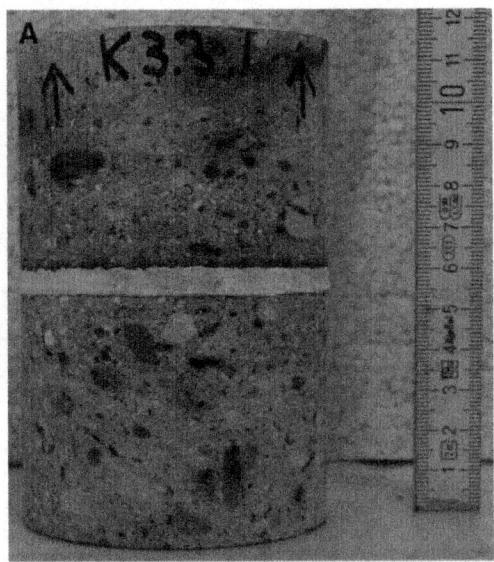

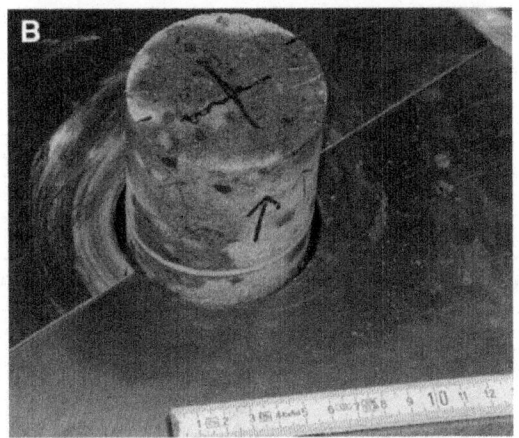

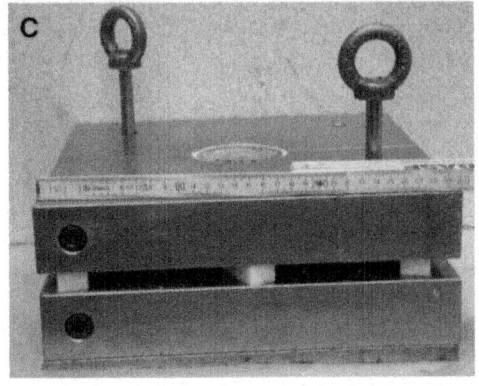

Figure. 26: Direct shear testing of SCL structure. **a** Typical core specimen with 74 mm diameter. **b** Specimens placed in steel holder ready for fixing with high strength resin. **c** Specimen in complete assembly ready for testing. **d** Placement of test assembly into shear box.

Details regarding the testing procedure are shown in Table 13.

Table 13: Conditions for shear testing

Parameter	Condition
Shear loading	Constant displacement, 0.5 mm/min
Normal loading	Constant normal load, 0.45 MPa
Measured parameters during test	Shear displacement, normal displacement, shear load
Age of specimen at testing	180 days

Results

The results are presented as shear-stress versus shear displacement diagrams, providing the following information:

- Peak shear stress for the specimen.

- Shear displacement at peak stress.

- Maximum shear displacement within linear elastic behavior.

- Shear stiffness during linear elastic behavior.

The results for membrane M1 are shown in Figs. 27 and 28 and for membrane M5 in Figs. 29 and 30. A compilation of the recorded data is shown in Table 14.

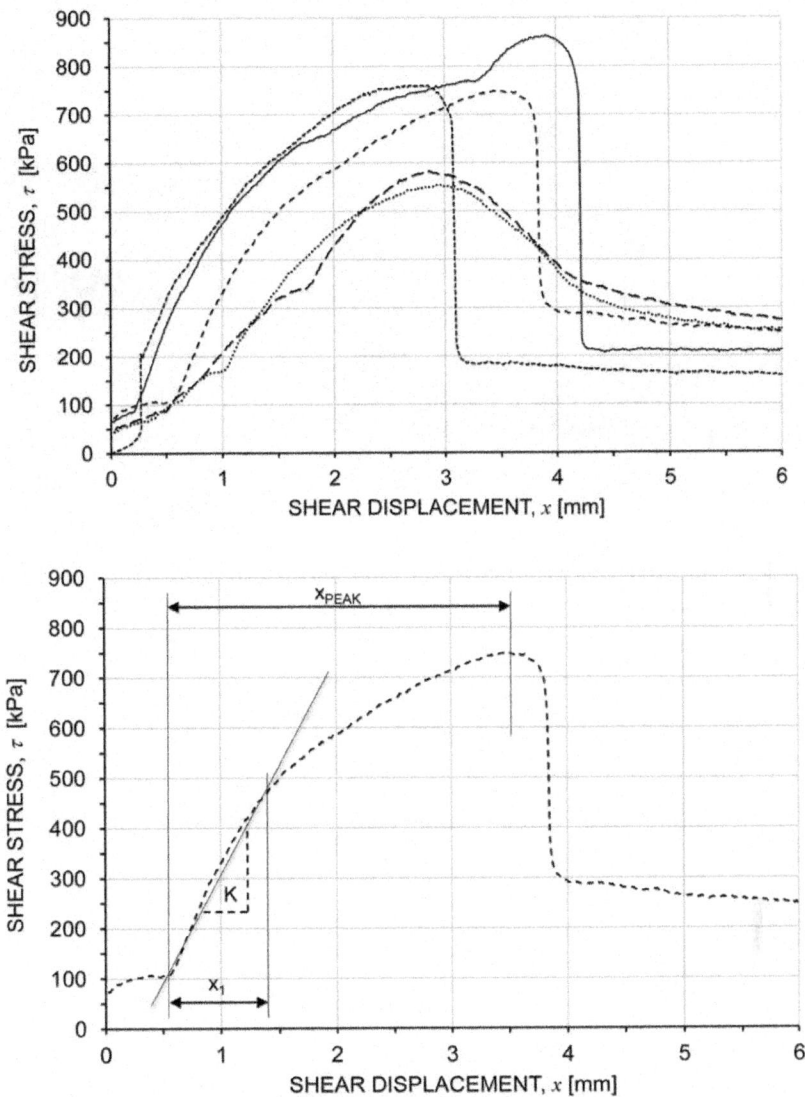

Figure. 27: Results from shear testing for membrane M1 represented as shear stress versus shear displacement. *Top* diagram showing results for all five specimens. *Bottom* results for one specimen with recorded parameters, shown in Table 16.

Figure. 28: Specimen of membrane M1 after shear testing exhibiting a debonding (adhesive) failure between membrane and the substrate concrete.

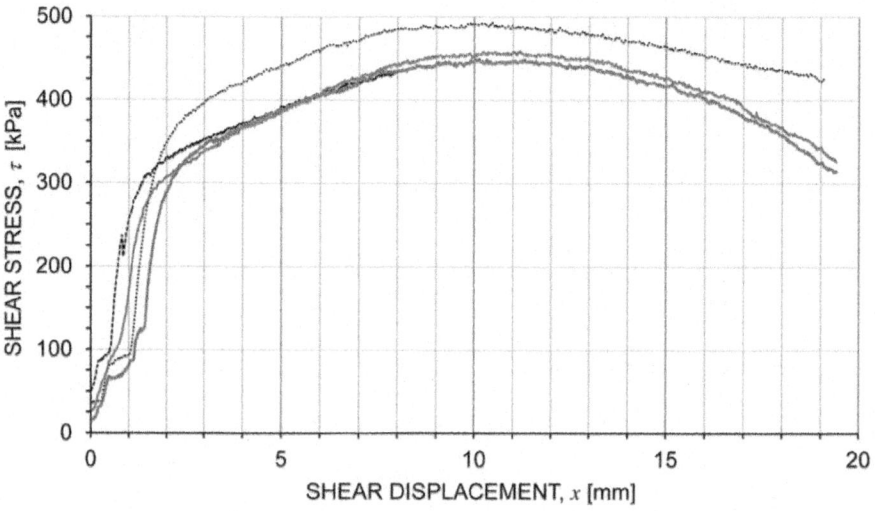

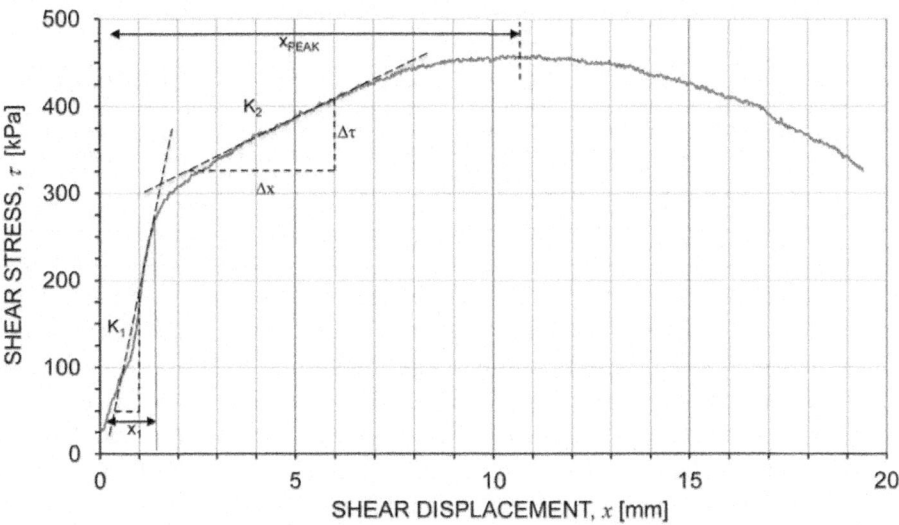

Figure. 29: Results from shear testing for membrane M5 represented as shear stress versus shear displacement. *Top* diagram showing results for all five specimens. *Bottom* results for one specimen with recorded parameters, shown in Table 15.

Figure. 30: Specimen of membrane M5 after shear testing exhibiting shear (cohesive) failure in the membrane material.

Table 14: Compilation of results from direct shear testing of membranes M1 and M5

Membrane (number of speci- mens)	Shear displace- mentx: mean, mm (COV, %)	Shear displace- mentx $_{PEAK}$: mean, mm (COV, %)	Shear stiffness K1: mean, MPa/m (COV, %)	Shear stiffness K2: mean, MPa/m (COV, %)	Peak shear stress: mean, kPa (COV, %)	Membrane thickness range, mm	Measured moisture content in membrane, % (COV)
M1 (5)	1.0 (33)	2.5 (25)	350 (20)	n.a.	745 (16)	3.5–4	15.7 (10)
M5 (4)	1.1 (18)	9.0 (7)	297 (20)	19 (9)	450 (5)	4–6	14.5 (12)

Both membranes exhibit almost the same behavior within the initial deformation, showing linear shear elasticity up to approximately 1 mm shear deformation. Membrane M1 exhibits a slightly higher shear stress at this point, corresponding to the higher shear stiffness K1 compared to M5. M1 exhibits a clear bonding (adhesive) failure (Fig. 28) with peak shear stresses in the range of 0.55–0.85 MPa, after approximately 3–4 mm shear deformation. After the initial zone of shear elasticity, the two membranes have very different behavior. Membrane M1 exhibits increasing strain softening behavior and membrane M5 exhibits a bi-linear behavior with increased displacement. In the latter phase a lower shear stiffness can be observed. After approximately 7–8 mm horizontal displacement membrane M5 exhibits almost perfect plasticity and reaches a peak shear stress of approximately 0.45–0.5 MPa. No clear failure could be observed during the shear testing with the M5 specimens. The tests for M5 were terminated after 19 mm of shear deformation. When removing the specimens from the test assembly, failure in the membrane could be observed since the upper and lower part of the specimen could easily be separated (Fig. 30).

Tensile Strength of Membrane-Concrete Interface

Specimens

Testing of tensile strength of the membrane-concrete interface was carried out in 4 different series (explained in Table 15), including laboratory tensile testing of 74 mm diameter core specimens drilled from panels, pull testing on panels with lining structure and in situ pull testing from full scale tunnel linings. The moisture content of the concrete and membrane materials of the specimens was measured whenever possible. The testing of the linings at the Gevingås and Ulvin tunnel sites was conducted parallel to the moisture condition sampling and testing. For the large scale laboratory lining structure (Fig. 7, Sect. 4.4) the moisture content of the sprayed concrete and membrane which was achieved after 6 months of moisture conditioning was found to be very close to the moisture content measured in tunnels (Holter and Geving2015). Hence, the laboratory lining structure could be used for controlled freeze–thaw testing with realistic moisture content.

Table 15: Matrix for the different test series for tensile strength

Test series	Type of lining structure for specimens	Testing method	Membrane products tested	Conditioning and exposure of lining structure or specimen	Age at testing
1	Sprayed panels 600 mm × 600 mm	Drilled core specimens, tested in pull test machine	M1 M2 M3 M4	Dry Saturated by immersion for 14 days Frozen/thawed	12–14 months
2	Sprayed panels 600 mm × 600mm	Laboratory In situ pull tests conducted on panels	M1 M5	Dry Saturated by immersion for 60 days Frozen/thawed 6 times, −20/+20 °C Testing in frozen and thawed condition after freeze–thaw exposure	18–19 months
3	Full scale lining in tunnel (Gevingås and Ulvin test sites)	In situ pull tests in tunnel lining	M1	In situ moisture exposure in rock mass	Ulvin: 29 months Gevingås: 37 months
4	Laboratory lining structure on rock mass with water exposure and freezing	In situ pull tests in lining structure	M1	Moist without freezing exposure Moist frozen-thawed to −3 at membrane, tested after 20 and 35 cycles In frozen condition after 35 cycles	19–25 months

Procedure

Both testing procedures described in Sect. 5.3 in this paper were followed. Test series 1 was conducted with the original method stated in EN-ISO 4624 on core specimens which were moisture conditioned by immersion for a minimum of 14 days and subsequently tested in a laboratory tensile pull machine. Test series 2 was conducted with the in situ tensile test method on slabs cut from slabs of lining structure. Prior to testing, the slabs received different pre-treatment types with moisture exposure at immersion and cyclic freezing and thawing. Series 3 and 4 comprise the tensile testing which was done on full scale lining structures, either in situ in tunnels or on the large scale laboratory lining structure. The adopted in situ tensile test method described in Sect. 5.3 was used for this purpose.

Results

The results for the first test series, which included the membranes M1, M2, M3 and M4 are shown in Fig. 31. With this testing method membrane M1 exhibits a range of strength reduction between 1.1–1.5 MPa tensile bonding strength (comparing dry specimens to saturated) and frozen/thawed specimens showed even more reduced tensile strengths. The scatter in measured tensile strength using this test method is relatively high. The membranes M2, M3 and M4 exhibit relatively low tensile strengths close to or below the recommended requirement of 0.5 MPa tensile strength.

The membrane M5 was introduced as a substitute for M1 and M2. M3 was discontinued for further testing. The results for series 2 which only includes membranes M1 and M5 are shown in Fig. 32. This testing context shows that saturation through 60 days immersion of the entire slab gives a significant reduction of the tensile strength for M1 and a slight reduction of tensile strength for M5 compared to dry specimens. Eight freeze–thaw cycles to −20 °C result in a reduction of tensile strength from approximately 0.7–0.5 MPa for M5. For M1 no readings were possible after the freeze–thaw cycles due to jamming of the drilling equipment at the membrane. For both membranes the tensile strength measured in frozen condition was significantly higher than for the measured strengths in dry or saturated condition.

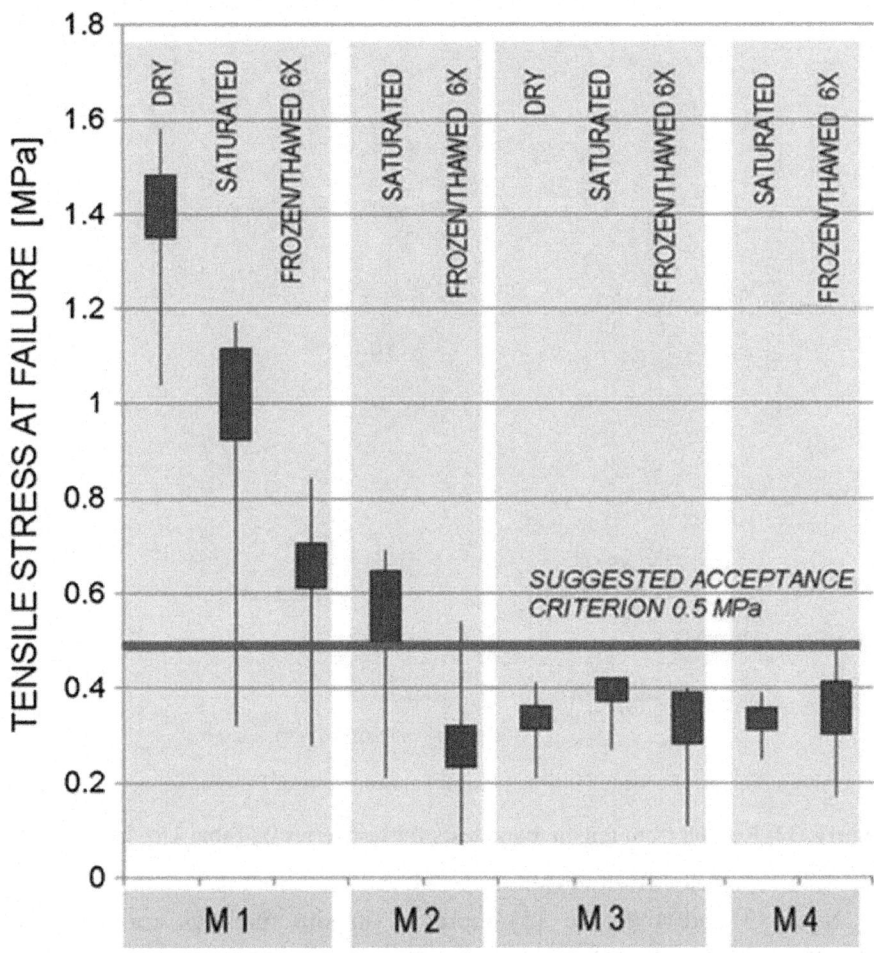

Figure. 31: Results from tensile bond tests for test series 1 (Table 15), drill core specimens with diameter 74 mm. Membrane numbers M1–M4 are explained in Sect. 6.3.2.

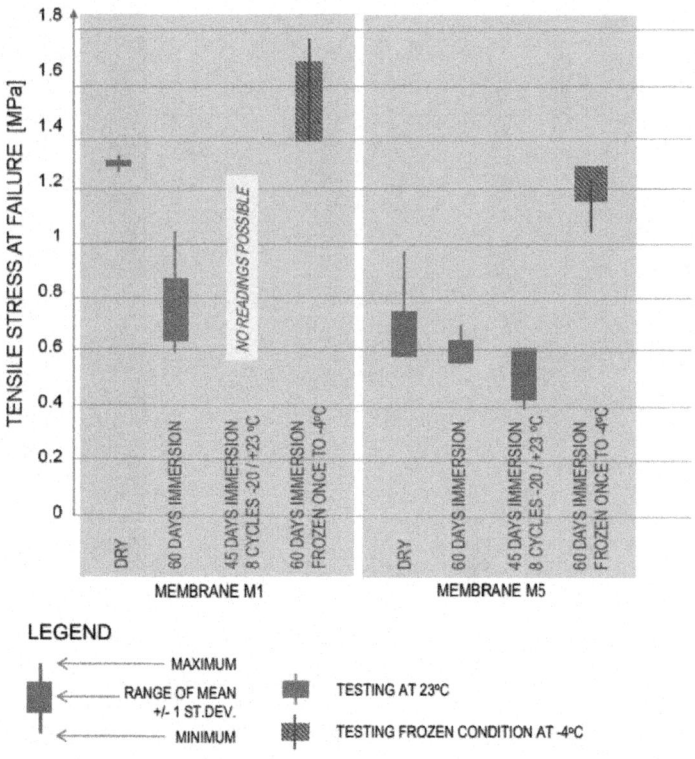

Figure. 32: Results from tensile bond tests for test series 2 (Table 15), in situ pull test method.

Series 3 and 4 (Table 15) represent in situ readings conducted with horizontal drilling on tunnel walls in different lining sections ranging from complete tunnel to large scale lining structure in a laboratory (Fig. 7). Exposure to moisture took place through the substrate sprayed concrete. Results from the tunnel test sites exhibit high tensile strengths in the range 1.1–1.6 MPa, with 1.3 MPa as the mean value (Fig. 33, left part). Two readings could be made at a wet crack (defect) in the inner lining, at which 0.8 MPa tensile bond strength was measured. Since realistic moisture contents were achieved in the lining structure at the SINTEF freezing laboratory, the effect of cyclic freezing and thawing on tensile bond strength could be measured (Fig. 33, right part). An initial tensile bond strength of 1.4 MPa at realistic moisture content was measured. After 20 and 35 freeze–thaw cycles with −3 °C minimum temperature at the membrane during each cycle, a slight reduction to respectively 1.15 and 1.1 MPa tensile strength could be measured. A tensile strength in the range of 1.1 to 1.3 MPa was measured in frozen condition at

−3 °C at the membrane. An additional freeze–thaw exposure with 20 cycles with a minimum temperature of −7 °C at the membrane was conducted after the first 35 cycles to −3 °C. Tensile strengths ranging from 0.4 to 0.7 MPa were measured after this exposure. Difficulty in conducting the pull tests was experienced during the last series at −7 °C due to damage in the outer part of the concrete lining.

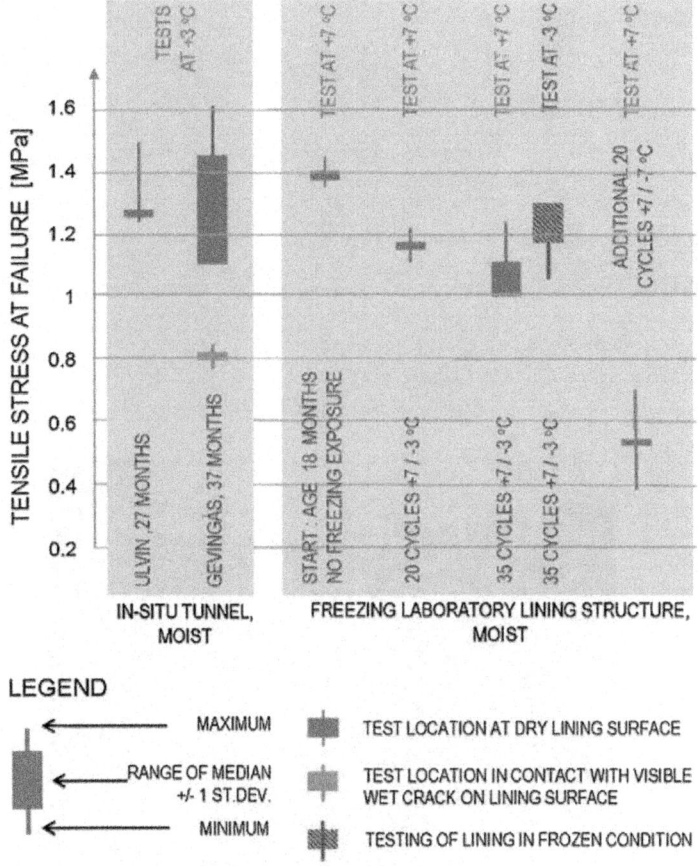

Figure. 33: Results from tensile bond tests for membrane M1, test series 3 and 4 (Table 15), in situ pull test method.

Microscope Analyses of the Interfaces between Concrete and Membrane

Scanning electron microscope (SEM) analysis of the interfaces of the membrane-concrete structure was conducted on specimens obtained from slabs which had been constructed with realistic application methods of both materials. The main purpose of this analysis was to study any visual characteristics or significant differences between the two interfaces, illustrated in Fig. 34 which could be of importance for the interpretation of the tensile bond and shear strength test results.

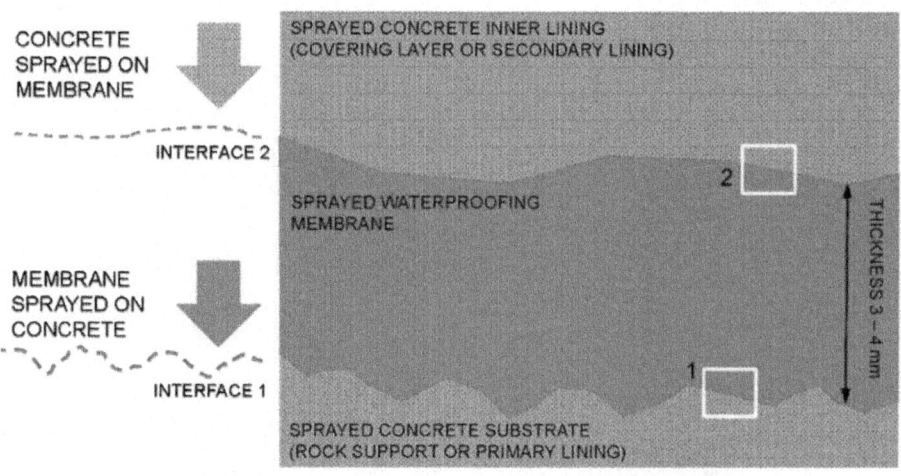

Figure. 34: Principle sketch and photo of the two interfaces between membrane and sprayed concrete. The interfaces 1 and 2 are shown respectively in Figs. 35 and 36 (courtesy by Wacker Chemie AG).

The interface on which the membrane has been applied on the substrate concrete (interface 1), shown in Fig. 35 exhibits a sharp contrast between membrane material and concrete material. Membrane material can be seen filling the irregularities of the sprayed concrete surface. The two materials exhibit distinct phases with no visible transition zone.

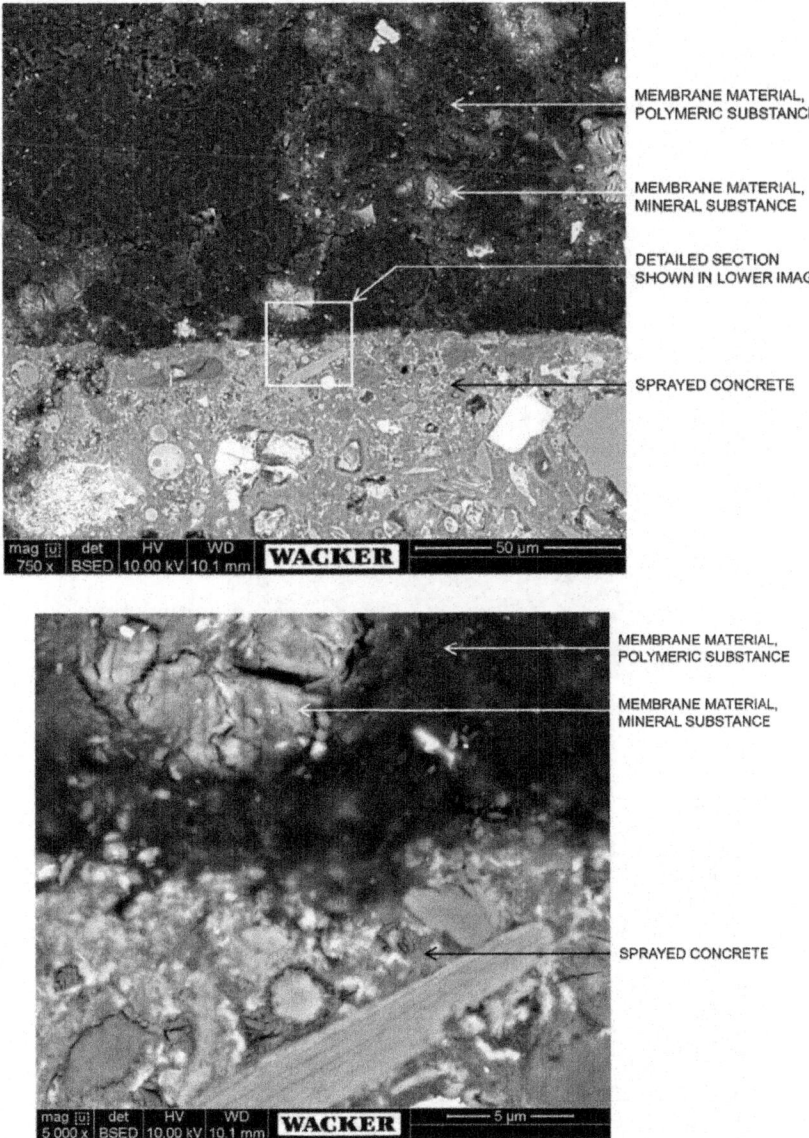

Figure. 35: Images obtained by SEM-microscopy of interface 1, with membrane spray applied onto a primary lining (rock support) sprayed concrete substrate. *Top* 750× enlargement. *Bottom* 5000× enlargement (courtesy by Wacker Chemie AG).

The interface on which the secondary lining concrete has been applied onto the membrane (interface 2) shown in Fig. 36 exhibits a different morphology than interface 1. A transition zone of approximately 15–25 µm thickness with visible effects of the impact of the sprayed concrete on the membrane can be clearly seen. This transition zone consists of a mineral phase with visible needle shaped crystals which separates the membrane material from the sprayed concrete material. With spectral analysis the mineral substance at the interface was found to be mainly composed of calcium carbonate $CaCO_3$.

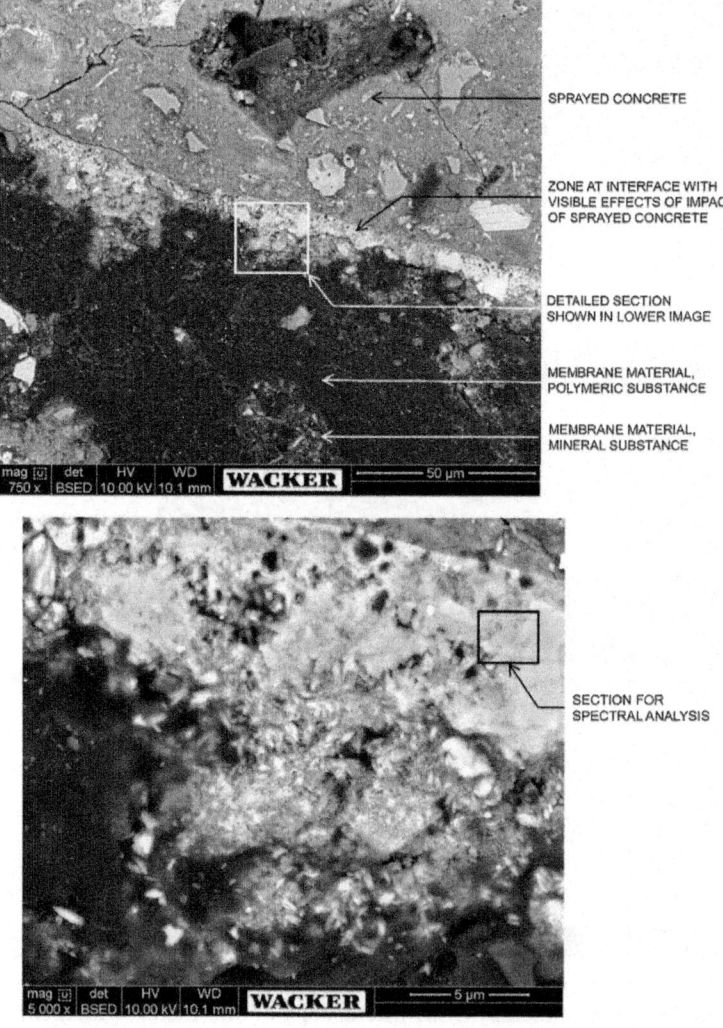

Figure. 36: Images obtained by SEM-scanning of interface 2, with sprayed concrete applied onto the sprayed membrane surface. *Top* 750× enlargement. *Bottom* 5000× enlargement (courtesy by Wacker Chemie AG).

DISCUSSION OF RESULTS

General

The testing of deformability and mechanical strength in this study contain accelerated or short term tests with main aim of simulating a loading scenario which takes place in the tunnel lining. The loading scenarios caused by thermally induced deformations considered in the loading model take place over several months. Possible effects of small and very slow deformations such as healing of the material, creep or reduction of strength have not been taken into account for short term tests.

Effect of Moisture Conditions of Specimens

Establishing realistic moisture condition of specimens for laboratory testing of this category of membranes is important and difficult. A test result should always be reported and evaluated with respect to its moisture condition. The results from in situ tensile bond strengths (with realistic moisture exposure) compared to specimens which are moisture conditioned by immersion, show that immersion very likely represents a too severe exposure to water, and gives lower strength values than realistic values. A consequence of this is that a complete testing program for membranes should include the construction of a full scale lining section in order to verify properties under realistic conditions in addition to findings from laboratory tests.

Thermal Exposure

The measured and calculated temperature profiles in the lining structure in this study cover freezing exposure with temperatures in the tunnel air in the range of −6 to −9 °C. Such temperatures in the tunnel air will occur under severe winter climate with outside air temperatures below −10 °C over sustained periods of time (STA 2012). The rock mass temperature at shallow locations in Scandinavia is normally 6–8 °C. This was measured in two Swedish studies (STA 2012) as well as the measured rock temperatures at the Ulvin site (Fig. 6, Sect. 4.4), as well as in the Gevingås rail tunnel (Holter and Geving 2015). The precise air temperature conditions at a given location in a tunnel needs to be assessed in each single case based on the local meteorological conditions and the ventilation characteristics of the tunnel.

Effect of Geometry of Specimens

The interfaces between membrane and sprayed concrete in tunnel linings ex-hibit surfaces with a certain degree of roughness. The measured mechanical properties shear and tensile strength will be influenced on the geometry of the interfaces which in turn will cause increased scatter. Our laboratory testing of specimens with planar surfaces represents ideal and unfavorable geometrical conditions, and takes no account for effects of surface roughness. For the specimens prepared from slabs with lining structure, the concrete surfaces were prepared by floating in order to produce the same geometry for all specimens. Our test results obtained from specimens with planar surfaces cannot be directly translated to the in situ properties. The peak stresses obtained in the laboratory for tensile and shear testing are likely to be lower than what would be the case under realistic interface conditions regarding moisture and geometry. On the other hand, specimens with the realistic roughness of the sprayed concrete surfaces would have introduced scatter making the interpretation of the results difficult, as well as limiting the reproducibility of the tests. For the tensile strength, our in situ measurements are the most representative. Such in situ measurements should be included in a test program in order to obtain values from realistic surfaces in addition to simplified or idealized surfaces during laboratory testing.

In situ Mechanical Loading of Lining and Membrane

The construction sequence of a waterproof SCL structure normally implies that the membrane and inner lining be applied after tunnel breakthrough, or several months after excavation and construction of the primary lining. Therefore the primary lining needs to be designed to be stable and designed for any rock mechanical loads before the membrane is applied. In our study we have therefore only included loads which can be imposed to the membrane by the possible effects of the membrane itself or the inner lining sprayed concrete.

Elongation and Crack Bridging

Elongation performance of a sprayed membrane according to DIN53504 (2009) can only give an indicative figure for the required elasticity for a tunnel lining purpose. This elongation performance exhibits significantly higher sensitivity to lower temperatures compared to the crack bridging. Conclusions based only on elongation results for temperatures 0 and −3 °C would likely deem the membrane unsuitable for such thermal exposure. The crack bridging results show significant performance at freezing temperatures, although a decreasing performance at temperatures 0 °C and below is observed.

We have applied strain loads on the membrane within a range to be expected by the effects of shrinkage and thermal expansion. Thermal fluctuations from approximately −3 to 15 °C can be expected at the membrane location within the lining structure. Thermally induced crack opening with an average crack distance of 1 m and a thermal change of 18 °C can be calculated to be in the order of 0.2 mm based on the thermal expansion coefficient. Our in situ crack measurements suggest a 0.2 mm crack opening for a drop in temperature of 6 °C. With a total thermal change over the year of approximately 18 °C in the lining structure at the membrane, 0.6 mm crack opening can be assumed. The possible shear deformations along the membrane interfaces caused by differential shrinkage or thermal expansion are in the same order of 0.5–0.6 mm. Our suggestion to use 1 mm as a crack bridging requirement and 1 mm as a critical shear deformation magnitude is therefore likely to be on the conservative side. Our crack bridging testing takes no account for any hydrostatic exposure at the cracks. We have only included the effects of high moisture content achieved by conditioning at RH 95 %.

Shear Performance

Shear testing of the membrane can contain several sources of error such as the loading rate, the normal loading mode and a possible oblique membrane plane relative to the shear direction. The applied loading rate during the test in the laboratory was 0.5 mm per minute whereas an in situ shear straining of the membrane most likely would take several months. Effects of creep and self healing therefore most likely would occur. Such effects are not accounted for in our short term test. The specimens for our tests had floated sprayed concrete substrate surfaces and were moisture conditioned by immersion. This likely represents an overexposure to moisture compared to in situ conditions. Hence, our laboratory findings for peak shear stress and shear stiffness are likely to be lower than values in realistic moisture exposure conditions.

Tensile Bond Strength

The site testing show consistent high values for tensile bond strength (Sect. 6.6.3, Fig. 33). At testing these lining sections had a history of several thermal expansion cycles as well as the exposure to the differential shrinkage between the two concrete layers. The test results from the Ulvin site also include one complete freeze–thaw cycle to approximately −3 °C at the membrane location. The measured high values for tensile strength indicate no in situ degradation of the lining after 4 years.

The laboratory testing on core specimens possibly contains three main sources of error: the geometry of the interfaces, the moisture condition of the specimen and the alignment of the pull direction parallel to the core axis. In addition the effect of a short term test with a duration of a few minutes might fail to account for all long term effects. Float finished concrete surfaces will normally result in a locally higher water/binder content and consequently higher porosity and possibly higher permeability. A slightly higher water exposure at the interface between membrane and concrete with a specimen with float finished surfaces compared to non-floated surfaces is therefore possible. The alignment of the pull equipment based on visual assessment will sometimes be difficult. Specimens without a perfect alignment in the testing machine might receive partial bending loads, and hence exhibit lower peak stress during the test. The in situ pull test method described in Sect. 5.3, Fig. 15, eliminates the three afore mentioned sources of error. However, with the available equipment a controlled loading speed could not be precisely applied. The effect of wet core drilling for either of the methods is unavoidable. Water exposure will soften the membrane at the core surface. When drilling in a downwards vertical direction on a slab of lining structure, the drilling water will fill the core groove and expose the membrane to water immediately before testing. When drilling horizontally in a lining structure the exposure to water will be less.

Performance under Freeze–Thaw Exposure

Our investigations pertaining to freeze–thaw durability comprise tensile bonding strength, elongation and crack bridging. The findings from the tensile testing after freezing exposure to −3 °C indicate that no significant damage occurs at this temperature. The likely explanation for this is the unsaturated condition of the concrete and membrane materials. This allows the volumetric expansion during the freezing of water to buffer into air filled voids without creating damage. For temperatures lower than −3 °C at the membrane location, thermally insulating measures need to be considered.

Durability and Service Lifetime

Prediction of service lifetime under freezing exposure is an important question. Our testing of tensile bond strength contains accelerated freeze–thaw tests in order to simulate a slightly more severe exposure with a high number of cycles which can be related to a period of service time. The number of freeze–thaw cycles that occur per year will vary from year to year in addition to the characteristics of the location. For tensile bond we have conducted 35 cycles to −3 °C at the membrane with 48 h per cycle which resulted in only minor reduction of tensile bond strength. Effects of healing between each freeze–thaw cycle are not accounted for in such an accelerated test layout. This indicates that real exposure would be less severe than our testing, and that our findings with high tensile strength after freezing exposure is likely to be realistic, or even conservative. Only when exposing the lining structure to 20 freezing cycles to −7 °C at the membrane location following the 35 cycles at −3 °C, a significant reduction in tensile bond strength could be observed. A precise service lifetime prediction is not possible based on our results. However, when this lining system is used in tunnels with moderate freezing exposure, with a lowest temperature of -3 °C at the location of the membrane, a service life time of 100 years or more is likely.

Recommended and Planned Further Work

The time dependent effects of in situ moisture exposure will be investigated further with continued sampling and testing of moisture content as well as in situ tensile bonding strength at the test sites.

Our study includes cases with low hydrostatic pressures. The effects of higher hydrostatic pressures (more than approximately 2 bars at the interface between rock and concrete) cannot be substantiated based on our results. Further material testing and large scale model investigations verified by field testing in order to substantiate the detailed behavior at the water filled cracks which expose the membrane are required. Such testing should account for all relevant material properties.

The detailed shear load characteristics need to be investigated in further depth. The main issues are: effects of long term, slow loading, creep and the normal loading mode, as well as the normal stiffness (with respect to the membrane surface) of the secondary lining structure.

CONCLUSIONS

A study of the properties of sprayed membranes for SCL in hard rock has been carried out with the following main scope:

- Assessment of loading, moisture and freezing exposure conditions based on field and large scale laboratory investigations.

- Evaluation of laboratory investigation methods.

- Conducting of laboratory investigations.

- Assessment of membrane properties including performance under freeze–thaw exposure.

The main findings from this study are the following:

- The main mechanical loading mechanisms on the membrane have been found to be represented by movement over cracks in the sprayed concrete and shear straining caused by differential shrinkage and thermal expansion.

- The moisture exposure to the membrane through the interfaces with the sprayed concrete leads to an in situ moisture content corresponding to 30–40 % of the membrane's maximum water uptake potential.

This has been found to be governed by the moisture properties of the membrane and concrete materials and the bonded contacts between these two materials.

- Membrane products with low polymeric content (below 70 %) exhibit low elasticity and are most likely unsuitable for tunnel waterproofing purposes in a bonded SCL context.

- Testing methods need to include details regarding moisture preconditioning and moisture exposure in order to test realistic materials and substantiate statements on in situ performance.

- A test program should include field testing in order to assess the relevance of the findings from laboratory testing.

- Testing of tensile bond strength on core or slab specimens in the laboratory which are conditioned by immersion, tend to give slightly lower measured values compared to site or large scale model testing.

- Testing of crack bridging shows decreasing performance at decreasing temperature. With 3 mm membrane thickness bridging of 1 mm crack opening at −3 °C at the location of the membrane in the lining has been found possible.

- Testing of shear properties indicate linear shear elasticity up to approximately 1 mm shear deformation.

- Testing of tensile strength show high in situ tensile bond strengths in the range of 1.1–1.5 MPa after 4 years.

- Exposure to cyclic freezing-thawing shows no significant reduction of the tensile bond strength at −3 °C at the membrane location.

- Further work is required to substantiate the performance of an SCL lining structure exposed to high hydrostatic pressures as well as effects of long term mechanical exposure.

ABBREVIATIONS, DEFINITIONS AND TERMS

SCL

Sprayed concrete lining. Permanent tunnel lining system based on fiber-reinforced sprayed concrete as the structural material with different possible waterproofing measures which are integrated into the sprayed concrete structure. Such linings may also include rock bolts for rock reinforcement

EVA-based sprayed waterproofing membrane

Ethyl-vinyl-acetate copolymer material used in the category of sprayed waterproofing membranes referred to in this paper

DCS

Degree of capillary saturation (%). Degree of saturation of concrete with respect to total suction porosity, equal to the ratio of water content of a given concrete specimen to its water content at saturation at immersion at atmospheric pressure at mass equilibrium

RH

Relative air humidity (%)

COV

Coefficient of variance, ratio of standard deviation to mean value

ACKNOWLEDGMENTS

The Norwegian Public Roads Administration, The Norwegian National Rail Administration, BASF Construction Chemicals Europe AG, Orica International Ltd. and The Norwegian Tunnelling Society NFF are acknowledged for assistance during the laboratory and field investigations and for financial support of this research project. The author wishes to thank Mrs. Köster and Mr. Bonin of Wacker Chemie AG, Burghausen, Germany for assistance regarding testing of membrane materials and the SEM analysis of the membrane-concrete structure. Mr. Anders Beitnes of Beitnes Consulting, Trondheim, Norway, Mr. Knut Garshol of K. Garshol Rock Engineering Ltd., Uddevalla, Sweden, Professors Bjørn Nilsen and Stig Geving of NTNU and Dr. Peter Schubert of iC-consulenten AG, Salzburg, Austria are acknowledged for critical review of the manuscript.

REFERENCES

1.   BeFo Stiftelsen Bergteknisk Forskning, Swedish Rock Engineering Research Foundation (2014) Investigation and development of material properties for shotcrete for hard rock tunnels. BeFo Report 133, Befo

2.   Bryne LE, Ansell A, Holmgren J (2014a) Investigation of restrained shrinkage cracking in partially fixed shotcrete linings. Tunn Undergr Space Technol 41:136–143

3. Bryne LE, Ansell A, Holmgren J (2014b) Laboratory testing of early age bond strength of shotcrete on hard rock. Tunn Undergr Space Technol 42:113–119

4. Dimmock R (2014) ITAtech harmonized best practice guidance for spray applied waterproof membranes. In: Proceedings of the 7th international symposium on sprayed concrete—modern use of wet mix sprayed concrete for underground support, Sandefjord

5. DIN 53504 (2009) Testing of rubber—determination of tensile strength at break, tensile stress at yield, elongation at break and stress values in a tensile test

6. DIN-EN 1062-7 (2004) Coating materials and coating systems for exterior masonry and concrete, part 7: determination of crack bridging properties

7. EN-ISO 4624 (2003) Paints, varnishes and plastics. Pull-off test for adhesion (ISO 4624: 2002)

8. Grimstad E, Tunbridge L, Bhasin RK, Aarset A (2008) Measurements of forces in reinforced ribs of sprayed concrete. In: Proceedings of the 5th international conference on wet-mix sprayed concrete for rock support, Tapir

9. Holter KG (2014) Loads on sprayed waterproof tunnel linings in jointed hard rock: a study based on Norwegian cases. Rock Mech Rock Eng 47:1003–1020

10. Holter KG, Geving S (2015) Moisture transport through sprayed concrete tunnel linings. Rock Mech Rock Eng. doi:10.1007/s00603-015-0730-1

11. Holter KG, Nermoen B, Buvik H, Nilsen B (2013) Future trends for tunnel lining design for modern rail and road tunnels in hard rock and cold climate. In: Proceedings of the world tunnel congress, Geneva

12. Holter KG, Nilsen B, Langås C, Tandberg MK (2014) Testing of sprayed waterproofing membranes for single-shell sprayed concrete tunnel linings in hard rock. In: Proceedings of the world tunnel congress 2014, Iguassu Falls

13. ITA/AITES International Tunnelling Association (2013) ITAtech Report No 2. Design guidance for spray applied waterproofing membranes. International Tunnelling Association

14. Mao D, Nilsen B, Lu M (2011) Analysis of loading effects on reinforced shotcrete ribs caused by weakness zone containing swelling clay. Tunn Undergr Space Technol 26:472–480

15. Nakashima M, Hammer AL, Thewes M, Elshafie M, Soga K (2015)

Mechanical behaviour of a sprayed concrete lining isolated by a sprayed waterproofing membrane. Tunn Undergr Space Technol 47:143–152

16. NCA Norwegian Concrete Association (2011) Publication No 7. Sprayed concrete for rock support. Norwegian Concrete Association, Oslo

17. NGI Norwegian Geotechnical Institute (2013). Using the Q-system. Rock mass classification and support design. Norwegian Geotechnical Institute, Oslo

18. NNRA Norwegian National Rail Administration (2012) Design guide for rail tunnels. Jernbaneverket; Underbygning, prosjektering og bygging/ tunneler, fra teknisk regelverk 6(1):2012

19. NPRA Norwegian Public Roads Adminstration (2006) Handbook No 163. Water and frost insulation of tunnels (Norwegian). Norwegian Public Roads Administration, Oslo

20. NPRA Norwegian Public Roads Administration (2012) Report No 127. Major research and development projects. Modern road tunnels 2008–2011. Norwegian Public Roads Administration, Oslo

21. NTNU (2013) Tunnel. Isotropic thermal conductivity and specific heat capacity of various rock and concrete samples determined with the transcient plane source technique. Report 13. October 2013. NTNU Department of Energy and Process Engineering, Trondheim

22. Ozturk H, Tannant DD (2010) Thin spray-on liner adhesive strength test method and effect of liner thickness on adhesion. Int J Rock Mech Min Sci 47:808–815

23. STA Swedish Transport Administration, Trafikverket (2011) Technical requirements for tunnels Publication no 2011:087. Trafikverket, Borlänge (Swedish)

24. STA Swedish Transport Administration, Trafikverket (2012) Temperature flow in rail tunnels—the Glödberg tunnel. Status report 2010. Publication no 2012:095. Trafikverket, Borlänge (Swedish)

25. STA Swedish Transport Administration, Trafikverket (2014) Design of underground structures in rock. Publication no 2014.144. Trafikverket, Borlänge (Swedish)

26. Su J, Bloodworth A (2014) Experimental and numerical investigation of composite action in composite shell linings. In: Proceedings of seventh international symposium on sprayed concrete—modern use of wet mix sprayed concrete for underground support. Sandefjord, Norway

27. Su J, Bloodworth A, Haig B (2013) Experimental investigations into the interface properties of composite concrete lined structures. In:

Anagnostou G, Ehrbar H (eds) Proceedings of world tunnel congress, Geneva. Taylor & Francis, London

28. TU Graz (Graz University of Technology, Austria) (2008) Laboratory Report Direct Shear Test Results Masterseal 345. Report GZ:95188, Institute of Rock Mechanics and Tunnelling, Graz

Chapter 3

AN EXPERIMENTAL INVESTIGATION OF SHALE MECHANICAL PROPERTIES THROUGH DRAINED AND UNDRAINED TEST MECHANISMS

Md. Aminul Islam and Paal Skalle

Department of Petroleum Engineering and Applied Geophysics, Norwegian University of Science and Technology, Trondheim, Norway

ABSTRACT

Shale mechanical properties are evaluated from laboratory tests after a complex workflow that covers tasks from sampling to testing. Due to the heterogeneous nature of shale, it is common to obtain inconsistent test results when evaluating the mechanical properties. In practice, this variation creates errors in numerical modeling when test results differ significantly, even when samples are from a similar core specimen. This is because the fundamental models are based on the supplied test data and a gap is, therefore, always observed during calibration. Thus, the overall goal of this study was to provide additional insight regarding the organization of the non-linear model input parameters in borehole simulations and to assist other researchers involved in the rock physics-related research fields. To achieve this goal, the following parallel activities were carried out: (1) perform triaxial testing with different sample orientations, i.e., 0°, 45°, 60°, and 90°, including the Brazilian test and CT scans, to obtain a reasonably accurate description of the anisotropic properties of shale; (2) apply an accurate interpretative method to evaluate the elastic moduli of shale; (3) evaluate and quantify the mechanical properties of shale by accounting for the beddings plane, variable confinement pressures, drained and undrained test mechanisms, and cyclic versus monotonic test effects. The experimental results indicate that shale has a significant level of heterogeneity. Postfailure analysis confirmed that the failure plane coincides nicely with the weak bedding plane. The drained Poisson's ratios were, on average, 40 % or lower than the undrained rates. The drained Young's modulus was approximately 48 % that of the undrained value. These mechanical properties were significantly impacted by the bedding plane orientation. Based on the Brazilian test, the predicted tensile strength perpendicular to the bedding plane was 12 % lower than the value obtained using the standard isotropic correlation test. The cyclic tests provided approximately 6 % higher rock strength than those predicted by the monotonic tests.

INTRODUCTION

Most borehole stability problems occur during drilling in overburden shale or shale-like materials, such as mudstones. Borehole instabilities are typically associated with high pore pressures (PPs) in shale located immediately above the hydrocarbon reservoir. Worldwide, there is an increasing tendency to go toward deeper high-pressure/high-temperature (HPHT) reservoirs. Consequently, the drilling window margin (collapse–fracture) is reduced, and more emphasis is placed on borehole stability predictions to avoid borehole collapse or fracture. Moreover, field observations imply that borehole collapse is strongly dependent on the wellbore orientation, especially in areas where the formation is strongly bedded and where the stress field is non-isotropic. Under such circumstances, accurate information on the rock strength and rock failure behavior in overburden shale is crucial to improve drilling safety. Knowledge of the mechanical properties of shale is vital to implement any 3D shale anisotropy borehole instability model (Crook et al. 2002; Søreide et al. 2009). As an example, to implement the modified Cam Clay model for the shale-related borehole instability assessment, mechanical properties and matrix anisotropic parameters, which consist of 21 attributes in total, are required (Søreide et al. 2009). Most of the parameters are related to the mechanical characterization of shale. However, a wide range in their magnitude was observed specifically for the Young's modulus, Poisson's ratio, friction angle, and the shear and bulk moduli. An accurate interpretation, together with sound engineering judgment, appear necessary to evaluate and quantify the mechanical properties of shale. This paper has addressed these concerns by interpreting the results obtained from extensive laboratory testing of shale.

It has previously been assumed that shale is not a reservoir rock and is, thus, not interesting in terms of hydrocarbon production (Hofmann and Johnson 2006). In general, the reservoir sections are the primary target, so shale samples from the overburden sections are limited. Some previous publications on the mechanical properties of tested shale were reported by Closmann and Bradley (1979), Steiger and Leung (1992), and Hoang and

Abousleiman (2010). However, the scarcity of deep shale testing samples has lead to a lack of published shale elasticity data. Meanwhile, the interest in shale has increased over the past few years, which was understandably triggered by recognition by the oil industry that the primary migration of oil is incomplete in some shale. Potential challenges to the understanding of shale are shale geochemistry, anisotropy, petrophysical properties, and the response of the seal to pressure as a barrier. The growing interest in shale-gas and caprock integrity for CO_2 storage has resulted in a rising demand for fundamental rock property data. Performing work on stimulation wells to create a complex fracture network in a shale-gas field is a difficult task, particularly the drilling of horizontal wells. However, this is the only viable method at this stage to develop a drainage strategy in the shale-gas reservoirs. Fluid flow modeling to develop an analytical solution in this particular event is one current issue where accurate shale property knowledge is essential. Moreover, infill drilling is a practice used by the industry to penetrate interbedded shale-sand formations, requiring the most accurate rock mechanical models to correctly assess the mud weight operational window. Accurate study of all issues requires a database of shale characteristics and its mechanical properties.

The prime consideration when evaluating and modeling borehole stability problems in overburden shale or in an interbedded reservoir is the lack of relevant test data to describe shale properties accurately. In reality, cutting a real shale specimen from an interbedded sand-shale layer is a challenging, costly, and time-consuming task. Experimental data for real shale specimens at the maintained downhole conditions are rare, while data on weak shales (i.e., Pierre-1) are fairly abundant. It is, therefore, a well-accepted approach to execute experimental investigations on outcrop shale and use the data to provide the necessary material data sets for the fundamental borehole stability model. In practice, fundamental models are calibrated against field cases and readjusted. Many investigators (Johnston 1987; Schmitt et al. 1994; Horsrud et al.1994; Horsrud 1998; Spaar et al. 1995; Holt et al. 1996; Claesson and Bohloli 2002; Crook et al. 2002; Søreide et al. 2009; Fjær et al. 2008) have, instead, used outcrop shales to calibrate borehole stability models. The present work deals with data sets for Pierre-1 shale at room temperature and at the high confining pressure present at downhole conditions. It was our intention to compile published shale outcrops test data. This was impossible due to the inconsistency in the reported information and evaluation techniques. It was also unclear as to which types of outcrop batch were used.

Shale is a rock with laminated structure, and the bedding plane orientation determines the mechanical properties of each shale sample. It is a well-accepted statement and a well-defined concept in soil and rock mechanics textbooks that the shale bedding plane leads to borehole instability. Jaeger and Cook

(1979) reported the laboratory test results of mechanical properties of layered rocks during failure. A comprehensive study, together with field evidence on the plane of weakness and anisotropy, were introduced and reported in the oil industry by Bradley (1979), Maury and Santarelli (1988), and Aadnøy and Chenevert (1987). Aadnøy and Chenevert (1987) investigated the effects of wellbore inclination, anisotropic stresses, and anisotropic rock strength by modeling highly inclined boreholes. Under certain conditions, the rock would fail along the planes of weakness. Because of the geomechanical properties of the weak plane (alignment of phyllosilicates due to overburden digenesis), slip surfaces may exhibit a significantly higher potential to fail compared with stronger rock units, such as limestone and sandstone. However, this statement is valid only for particular geological formations and may not be applicable in other geological arenas. For example, the same author performed a separate study (Aadnøy et al. 2009) on a tightly folded structure in British Columbia, and it was observed there that the planes of weakness in bedded rocks may lead to borehole collapse. However, in a three-dimensional space, there were combinations along wellbore inclinations and azimuths, instead of along the weak planes, where the sample failed. The bedding plane is, thus, a key issue in this study. In general, the critical parameters that cause instability are the planes of weakness, anisotropy, the relative normal stress values on the borehole, the well trajectory, and the relative angle between the borehole and the bedding plane (Bradley 1979; Aadnøy and Chenevert 1987; Aadnøy et al.2009; Crook et al. 2002; Fjær et al. 2008; Søreide et al. 2009; Islam et al. 2009).

As discussed earlier, many researchers have studied the properties of clay and shale. Selected test results of mechanical shale properties under drained conditions are published and reported by Closmann and Bradley (1979), Horsrud et al. (1998), Crook et al. (2002), Søreide et al. (2009), and Hoang and Abousleiman (2010). These previous investigations can also be useful for this study to some degree. However, determination of the mechanical properties of shale under drained and undrained test conditions with variable confinement pressure is currently insufficient. To overcome this limitation, this study performed an extensive test program to explore shale heterogeneity and its directional properties. The following parallel activities were performed:

- Characterization and evaluation of shale heterogeneity through a drained and undrained stress path. The mechanical properties and postfailure material behavior were determined under different sample orientations, confinement pressures, and loading steps.
- Determination of the estimated and quantified mechanical properties of shale.

- Proposition of proper data sets for a borehole stability model.

Both compressibility and triaxial tests were utilized, with emphasis on evaluating the elasticity, yielding, and failure response as a function of the confining pressure and bedding plane orientation. The petrophysical testing included porosity, permeability, and bulk- and solid-density measurements. A total of three triaxial tests with different sample orientations (i.e., 0°, 45°, 60°, and 90°), including Brazilian tests and CT scans, were performed to obtain a reasonably accurate description of shale. Moreover, the cyclic stress-dependent rock stiffness and strength were evaluated and compared with the monotonic test results. The method used to achieve relatively consistent test results is briefly discussed at a later stage in this paper.

MATERIALS AND METHODS

Material and Samples

The outcrop Pierre-1 shale was used in this study for the following reasons:

- The scarcity of real shale specimens.
- The difficulty in obtaining intact samples of real shale specimens due to splitting along the bedding planes during the coring process.
- Mechanically, Pierre-1 shale exhibits plastic and anisotropic behavior that could be used in borehole simulation modeling of real shale.

The outcrop shale block was preserved by wax when received. A core barrel was used to drill the core samples from a large block stored in laboratory-grade oil (Marcol™) to prevent fluid losses. Both the drilling and the subsequent end-grinding of the plugs to obtain sufficiently plane and parallel end surfaces were performed whilst surrounded in the same oil. The prepared samples were subsequently stored in Marcol™ until they were assembled in the test cell to avoid desiccation effects.

The samples were equilibrated to the brine (3.5 w% NaCl) prior to ramping up stresses to consolidation level, so as the Pierre is a outcrop material with initial low stress levels, this should assume saturation close to 1 during the tests. Porosity was determined from fluid loss by heating typically to about 105 °C.

The test samples were 38 mm in diameter and had a length of approximately two times the diameter. A summary of the test specimens and other necessary information are presented in Table 1. The mineralogical and petrophysical properties of the Pierre-1 shale used are presented in Tables 2, 3, and 4.

Table 1: Test matrix of the triaxial testing of Pierre-1 shale

Test ID	Size			Consolidation		At failure				Test time	
	Length × diameter (mm)	θ (°)	ρ (g/cc)	σ_c (MPa)	P_p (MPa)	σ_x(MPa)	σ_r (MPa)	P_p (MPa)	Strain rate (s⁻¹)	Up to consolidation (h)	Triaxial loading (h)
CIU_192_2-32	76 × 37.62	90	2.33	25	10	50.90	25	14.90	1.50E–07	38.98	29.42
CIU_192_2-33	75.92 × 37.66	90	2.33	20	10	41.14	20	13.90	1.00E–07	45.80	42.33
CIU_192_2-34	75.99 × 37.66	90	2.33	16	9.9	31.60	16	12.55	1.00E–07	40.12	39.82
CIU_192_3-6	75.9 × 37.96	0	2.33	20	10	35.97	20	15.58	1.40E–07	26.17	45.24
CIU_192_3-7	75.88 × 37.93	0	2.32	25	9.9	46.26	25	17.35	1.90E–07	28.24	33.06
CIU_192_3-8	75.88 × 37.78	0	2.32	30	9.91	55.39	30	19.5	2.00E–07	31.13	31.25
CIU_192_3-16	76.04 × 37.76	45	2.33	25	9.91	46.44	25	15.93	1.60E–07	29	33
CIU_192_3-19	75.17 × 37.71	45	2.33	18	9.93	32.08	18	14.12	1.70E–06	25	27
CIU_192_3-20	75.96 × 37.75	45	2.33	30	10.01	55.36	30	17.99	1.7E–07	32	33
CIU_192_3-11	75.65 × 37.73	60	2.33	25	10.02	46.00	25	16.20	1.00E–07	69.5	40.5
CIU_192_3-12	75.83 × 37.74	60	2.34	18	10.03	32.64	18	13.50	1.00E–07	65	63
CIU_192_3-13	76.03 × 37.71	60	2.34	30	9.98	54.17	30	17.57	1.00E–07	72	70
CID_192_2-35	75.96 × 37.72	90	2.33	17	10.07	38.60	17	10.08	2.30E–08	85	245
CID_192_3-09	75.75 × 37.77	0	2.33	17	9.97	36.80	17	9.92	2.00E–08	85	457

Table 2: Mineralogical composition of Pierre-1 shale determined by X-ray diffraction spectroscopy (with weight in terms of bulk percentage)

Quartz	k-fsp	Plag	Chl	Ka	Mi/Ill	ML	Sm	Calc	Sid	Dol/Ank	Pyr
20.1	0.7	15.7	2.2	6.8	16.6	0.3	31.5	1.8	0.7	1.8	2.0

Chl chlorite, *Ka* kaolinite, *Mi/Ill* mica and illite, *ML* mixed layer, *Sm* smectite, *Sid* siderite, *Dol* dolomite, *Ank*ankerite, *Pyr* pyrite

Table 3: Semi-quantitative X-ray diffraction analysis: fine percentage (<4 μm)

Quartz	k-fsp	Plag	Chl	Ka	Mi/Ill	ML	Sm	Calc	Sid	Dol/Ank	Pyr
7.4	0.3	1.8	7.9	8.7	15.2	0.0	57.8	0.2	0.1	0.1	0.5

Table 4: Petrophysical and physical properties of the Pierre-1 shale

Block #	Bulk density (saturated) (g/cm³)	Bulk density (dry) (g/cm³)	Water content (%)	Porosity (%)
2	2.33	2.10	9.9	23.2
3	2.34	2.12	9.6	22.4

We used samples from two neighboring blocks. All triaxial samples had densities between 2.32 and 2.34 g/cm³, with an average density of 2.33 ± 0.005 g/cm³. We also completed separate density and porosity measurements on the two blocks from which we drilled the samples. From the loss of pore fluid measurement, the densities and porosities were determined to be 2.33 g/cm³ and 23.2 % (block 2) and 2.34 g/cm³ and 22.4 % (block 3), respectively, i.e., they were fairly similar.

SAMPLE SELECTION

CT scans were performed to check for possible heterogeneities in the test samples. To illustrate the heterogeneity effect of Pierre-1 shale, two examples of CT scans (radial slices) are presented in Fig. 1a, b. Lighter areas generally correspond to denser (higher absorption) material. The very light color on one of the scans is likely calcite (Fig. 1a). Such samples were discarded from the consolidated isotropic undrained triaxial compression tests (CIU). The other image contains a less light, gray region/layer (Fig. 1b), which is not calcite but likely represents a more silty region. Such samples were used in the consolidated isotropic drained triaxial compression tests (CID).

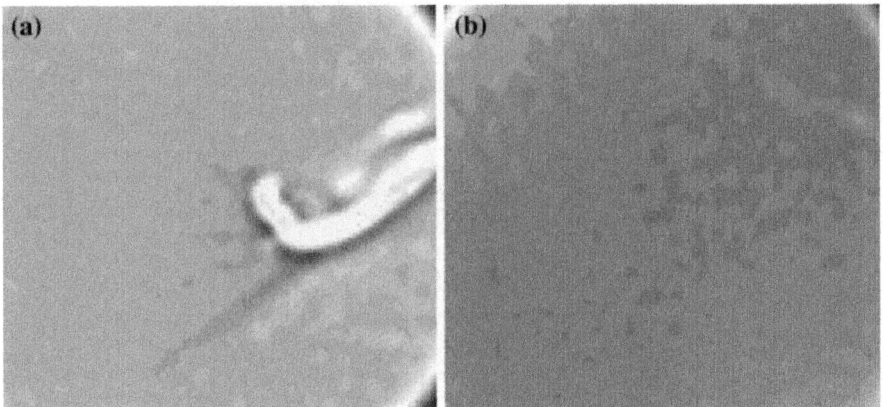

Figure. 1: CIU and CID tests for Pierre-1 shale; **a, b** heterogeneous nature of shale. The very light color on one of the scans is probably calcite (**a**). A less light, gray region/layer (Fig. 1b) is not calcite but probably a more silty region.

Bedding Plane and Loading

The bedding plane orientation in the triaxial test is denoted by the angle θ, which is measured clockwise from the loading direction relative to the bedding plane (Fig. 2). The monotonic CIU tests were performed with angles of 0°, 45°, 60°, and 90° between the load axis and the bedding plane. The same is true for the CID tests, with the angles of 0° and 90°. The postfailure samples are presented in Fig. 3. Through a postmortem analysis of the tests samples, it is observed that, in most cases, the failure plane transverses nicely through the weak bedding plane.

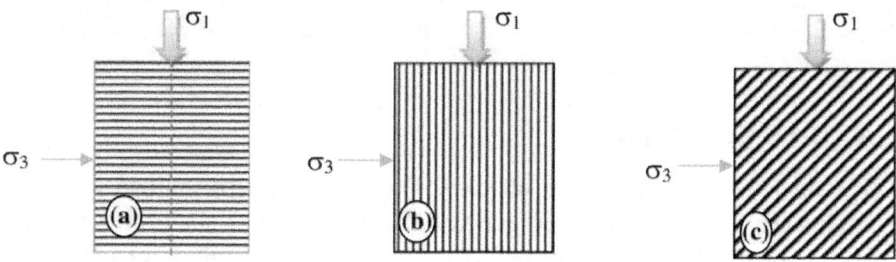

Figure. 2: Principle drawing including bedding angle and loading; **a** drilled normal to bedding ($\theta = 0°$), **b** drilled parallel to bedding ($\theta = 90°$), and **c** drilled inclined to normal to bedding (θ = inclined). The angle θ is measured clockwise from the loading direction relative to the bedding plane.

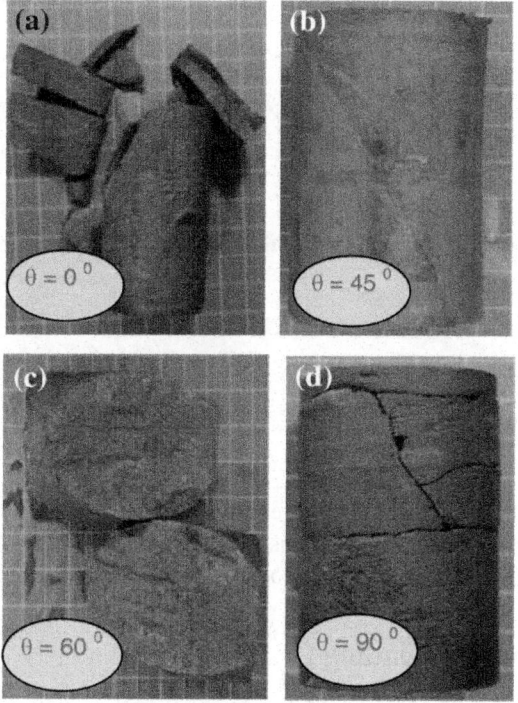

Figure. 3: Postfailure views of the Pierre-1 shale samples performed through monotonic loading; **a** loading normal to bedding ($\theta = 0°$), **b** loading 45° to the normal to

bedding ($\theta = 45°$), **c** loading $60°$ to the normal to bedding ($\theta = 60°$), and **d** loading parallel to bedding ($\theta = 90°$).

Brazilian Test

The tensile strength of rocks is among the most important parameters influencing rock deformability, rock crushing, and blasting results. To calculate the tensile strength from the indirect tensile (Brazilian) test, one must know the principal tensile stress, in particular, at the disc center, where a crack initiates. This stress can be calculated by an analytical solution given by Claesson and Bohloli (2002). A Brazilian test was conducted to estimate the indirect tensile strength at both $\theta = 90°$ and $0°$ conditions. One sample (length = 19.55 mm and diameter = 37.70 mm) with a bulk density of 2.32 g/cm^3 showed a rock tensile strength of 0.62 and 0.76 MPa at $\theta = 90°$ and $0°$, respectively. This measured tensile strength through $\theta = 0°$ is 12 % lower than the result of the standard isotropic correlation given by Claesson and Bohloli (2002). The test results also indicated that the tensile strength is higher for vertical than for horizontal bedding.

Triaxial Testing

A schematic drawing is presented in Fig. 4 to illustrate the methods of collecting data using the triaxial test.

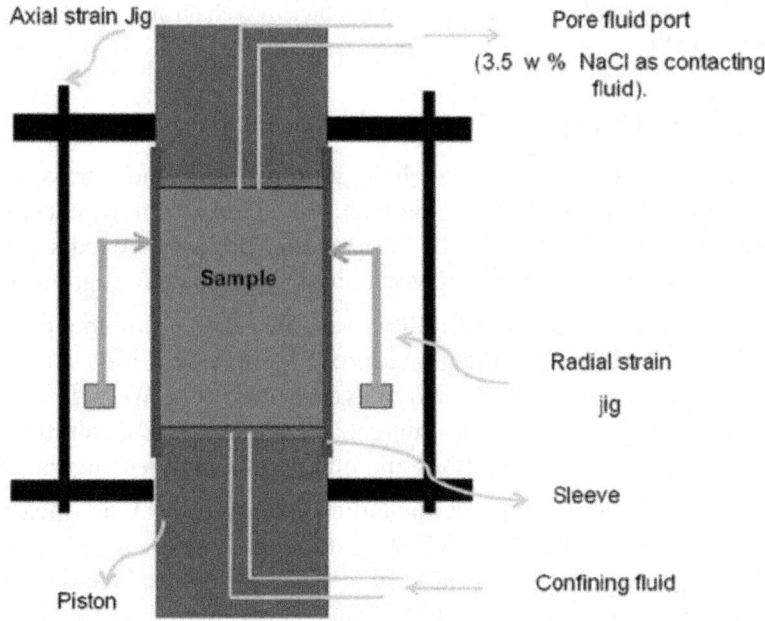

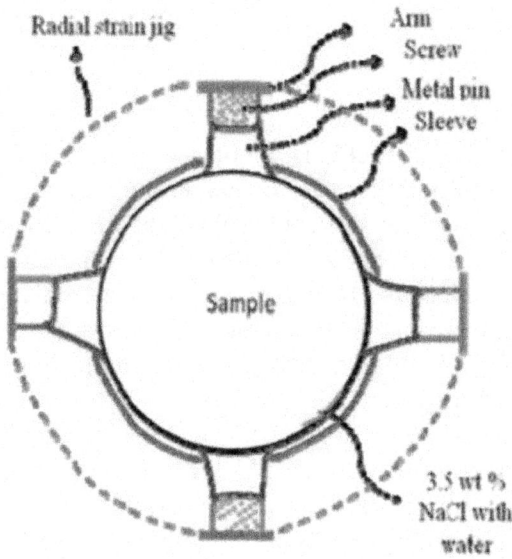

Figure. 4: Principle sketch of the interior of a triaxial cell, showing the loading piston with fluid ports, optional sintered plates for fluid distribution, a radial strain jig (suspended from the specimen) measuring two orthogonal diameters, and an axial strain jig measuring the change in sample length between the pistons. It is also possible to measure axial strain directly on the sample, to measure radial strain in more than two directions, or measure the change in circumference by a chain around the sample (after Fjær et al. 2008).

Drained and Undrained Test Mechanisms

During drilling in low permeable shale, a typical undrained stress relaxation response is observed at the borehole wall. With extended time, excess PP will dissipate and equilibrate with the surrounding pressure. A redistribution of stresses might provide a time-delayed failure in the shale formation.

In this study, fluid permeability was measured with the pulse-decay technique at a constant confining pressure (15 MPa). This is a standard pulse-decay method on samples subjected to hydrostatic stress. We used brine as the contacting pore fluid. The measurements were performed carefully to avoid effects from temperature variations and other factors influencing the results.

Due to the low shale permeability, steady-state type measurement techniques are not generally applied. Instead, techniques based on transient measurements are used (Horsrud et al. 1998). We used a technique where a thin (6-mm) circular disk (38 mm in diameter) was placed into a pressure cell, where PP and hydrostatic confining pressure could be applied separately. After loading to pre-determined levels of PP and confining pressure, ample time was given for the sample to consolidate. The PP was increased by 0.5 MPa on one side of the sample and reduced by the same amount on the other side (see Fig. 5). The decay in this pressure difference was then recorded by a differential pressure transducer, and the permeability was calculated from this time response.

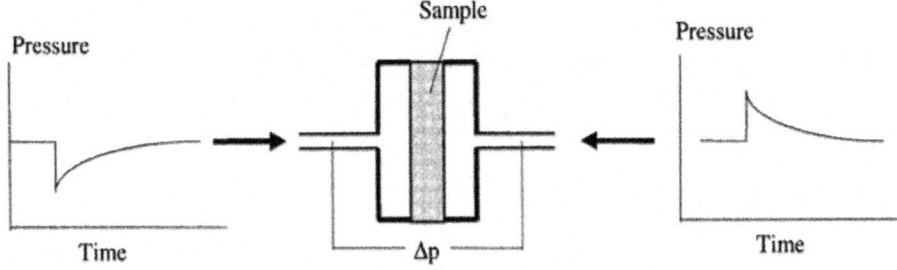

Figure. 5: Illustration of the pressure transient method used for permeability estimation (Horsrud et al. 1994).

From the permeability (k) measurements within a confining pressure of 15 MPa and a PP of 5 MPa, k was 14 and 49 ηD at $\theta = 0°$ and $90°$, respectively. Because shale is expected to be anisotropic, it is reasonable to expect the permeability to be anisotropic and to expect the largest values parallel to the bedding. Islam et al. (2009) reported that the magnitude of the consolidation time for the finite diffusive length adjacent to the borehole wall was only in the range of hours and days for samples with a permeability of 100 ηD. Therefore, drained tests were necessary for time-delayed material failure analysis. The drained stiffness was somewhat lower than the undrained stiffness, as drained stiffness accounts for effective stress.

Generally, in drained tests, the outlets through the pistons will be open such that the pore fluid pressure can be maintained at any prescribed value. During the hold periods, the PP equilibration is expected to take place inside the sample. During the testing of shale, the drained condition means that the test must run sufficiently slowly to avoid unacceptable pressure build-up when the sample deforms. In this case, a strain rate of 2×10^{-8} s^{-1} was maintained until failure. How long the drained test will take is fairly difficult to estimate and basically depends on the strain rate used during the triaxial loading of the specific sample. The standard procedure is to estimate this strain rate from the consolidation behavior in the initial hydrostatic phase, which depends on the specific sample, confining stress, and PP, as well as how effective the PP drainage is working for a given sample. The desired strain rate for the drained condition is often 10 times slower than that for the undrained condition. For example, based on an assumed strain rate of 2×10^{-7} s^{-1} (and a 10-mm axial strain at failure) and 24 h initial consolidation, 68 h was required to complete an undrained test (sample drilled parallel to the bedding; $\theta = 90°$), whereas for the drained test, it would be approximately 10 times longer. A brief discussion is included on this topic in the next section. From a theoretical point of view, drained conditions mean that the PP in the sample is an independent variable, while for undrained tests, it is a dependent variable. In real-life borehole stability simulations, both drained and undrained rock mechanical properties are necessary because they depend on the time scale. In the case of a short borehole simulation, undrained rock properties can be used, whereas for long simulations, the drained rock properties are applied.

Triaxial Testing of Shales

The triaxial tests were run as consolidated drained and undrained tests in a servo-controlled load frame. Consolidated undrained (CIU) tests were selected for the testing of shales (Steiger and Leung 1992; Nakken et al. 1989; Horsrud et al. 1998). The procedure was adopted from soil testing (e.g., Head 1984). This test consisted of three distinct phases:

- Loading to the predetermined level of confining pressure and PP.
- Consolidating, maintaining a constant confining pressure, and allowing drainage of the pore fluid against a constant PP.
- Undrained axial loading under a constant axial displacement rate beyond failure of the sample.

The internal instrumentation of the test sample is shown in Fig. 4. In addition to the measurements of the external load, pressure, and deformations, the PP on both ends of the sample were recorded, and acoustic wave trends in both the axial and radial directions were also recorded.

CIU tests were performed at angles of 0°, 45°, 60°, and 90° between the sample axis and the bedding plane. In addition, two CID tests were performed to obtain the effective stress parameters. For triaxial testing, all of the samples were consolidated up to 10 MPa of PP. The confining pressure was ramped up to 20, 25, and 30 MPa to simulate a closer range of downhole effective stress conditions.

The samples were left for 24 h to establish consolidation. We monitored the strain (deformation) so as to evaluate when consolidation would be sufficient. In general, 24 h of consolidation for our sample geometries is considered to be a sufficient amount of time. The time dependency of the consolidation phase was also used to determine the loading rate during the main triaxial phases (which were run in a constant strain mode). In some undrained tests, the strain rate was $1-2 \times 10^{-7}$ s^{-1}. For the drained tests, the rate was approximately 2×10^{-8} s^{-1}. In addition, the size of the test sample and the time required to perform a test depended highly on the permeability of the shale (Nes et al. 1998; Islam et al. 2009; Loret et al. 2001) and the fluid viscosity.

A typical response from the triaxial loading of the consolidation type is shown in Fig. 6. The drained test was extremely time consuming up to sample failure (see Fig. 6e). For a drained test, approximately 20 days was required to reach yielding with a strain rate of 2×10^{-8} s^{-1}. This is the main reason for running the triaxial loading part in the undrained mode. Concepts from soil mechanical testing can be applied to assist in determining when consolidation is complete and also when the appropriate displacement rate is achieved in the undrained part of the test (Katsube et al. 1991). Determination of the strain rate in the undrained part was also based on the consolidation response of the sample to ensure that PP equilibrium was achieved throughout the sample. To reduce the test time as much as possible, both axial and radial drainage of the sample were included.

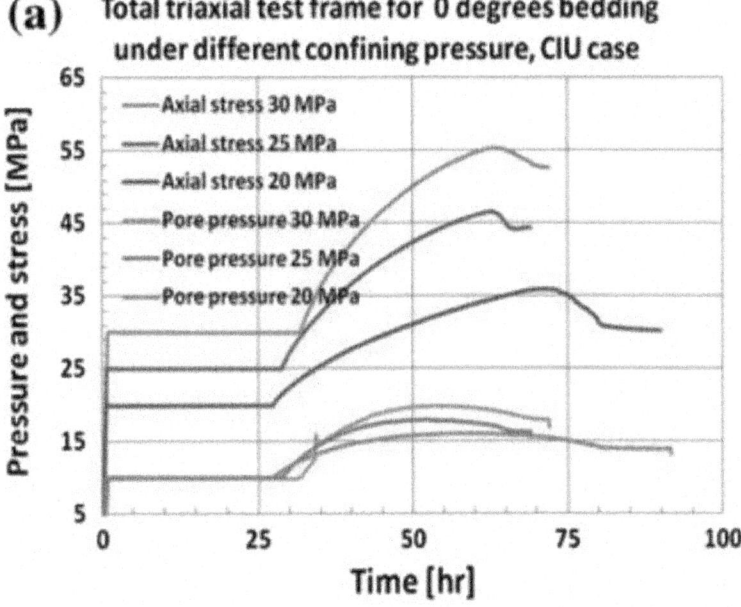

(a) Total triaxial test frame for 0 degrees bedding under different confining pressure, CIU case

(b) Triaxial loading part under at an initial confining pressure of 25 MPa, CIU case

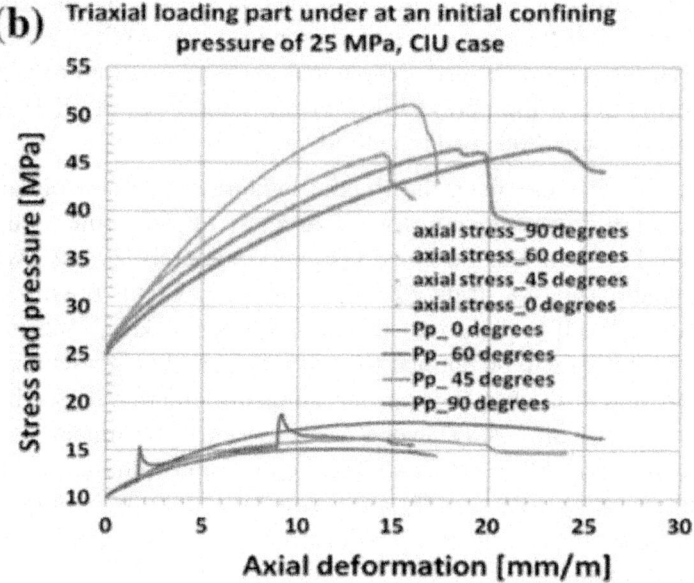

Axial deformation [mm/m]

(c) Total triaxial test frame for 90 degrees bedding under different confining pressure, CIU case

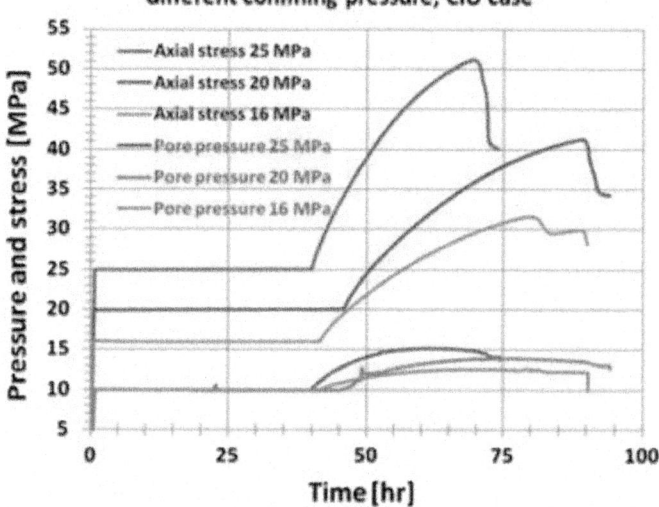

Time [hr]

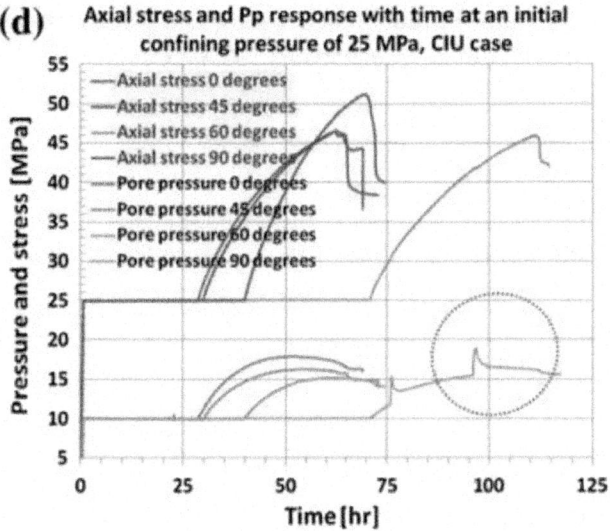

(d) Axial stress and Pp response with time at an initial confining pressure of 25 MPa, CIU case

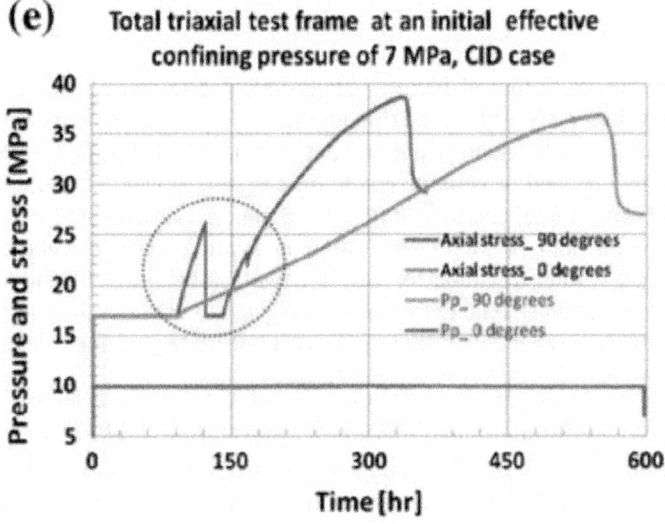

(e) Total triaxial test frame at an initial effective confining pressure of 7 MPa, CID case

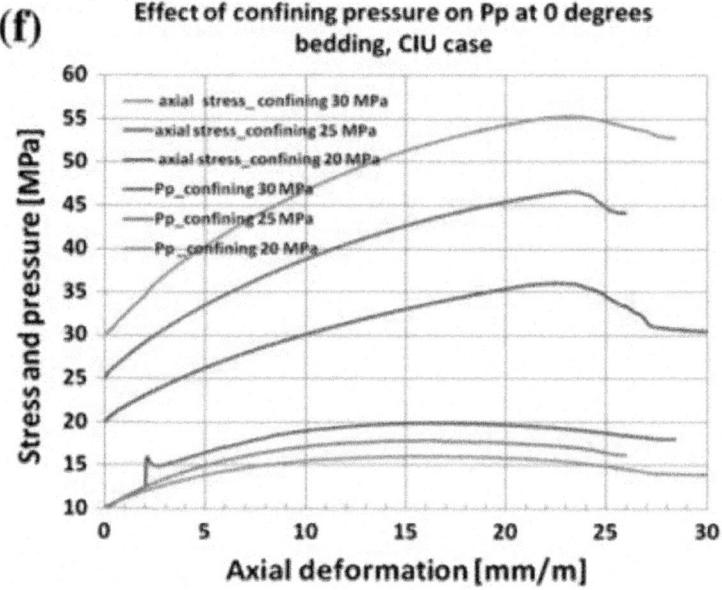

Figure. 6: Triaxial testing of Pierre-1 shale subjected to consolidation, triaxial load-ing, and development of PP; **a** the timeline shows the development of PP and failure load with variable confinement pressure, at sample θ = 0°,**b** effect of bedding plane on PP and failure load, **c** variable confining pressure to estimate failure load and to develop PP against total test time at θ = 90°, **d** material failure and PP response under constant confinement pressure with different bedding plane orientations, **e** failure response against total test time for CID for two samples, θ = 0° and 90°, **f** PP and failure response against axial deformation subjected to variable confinement pressure. The two *dotted circles* indicate abnormal strain rate and unusual PP development.

In the case of CIU tests conditions, a total test time of 2–4 days, on average, was required, including consolidation of approximately 20–25 h to failure. The bedding plane sample at 60° took a maximum of 5 days to fail (Fig. 6d). From the material stiffness analysis, the strongest sample was determined to be the sample drilled parallel to the bedding and would normally take a longer time to fail.

The PP increased up to a certain level with increasing axial stress, after which the PP tended to decrease until the sample failed. The response of PP during triaxial testing is presented in Fig. 6. Referring to Fig. 6a, with 30 MPa of confining pressure, the net PP increased 9 MPa, but at the failure position, the PP dropped 7 MPa and decreased further. This decrease indicates that the sample dilates before failure. In general, induced cracks or microfractures were created before the material reached failure. Basically, induced cracks or microfractures increase the pore volume (Fjær et al. 2008). The increased pore volume led to material dilatancy, consequently reducing the PP. In general, the failure position is controlled by increased pore volume, which reduces the PP. The development of PP under different bedding planes and confining pressures is presented in Fig. 6a–f. From this analysis, PP changed due to the change of the bedding plane. The confining pressure was quantified. The PP development versus confining pressure response is shown in Fig. 6f. This analysis indicates that the PP development reached a maximum at 0° bedding and was the lowest for 90° bedding. For the CID condition, PP was constant (see Fig. 6e). Thus, the actual effective stress-related parameters can be obtained through such tests.

PP development is linked with the effective stress to define material failures. According to the Mohr–Coulomb failure envelope, a material with 0° or 45° bedding to normal will reach the failure line prior to other orientations. One anomaly is indicated by a dotted circle at 90° bedding in Fig. 6e. The anomaly may be due to a contribution in the pore-pressure fluid volume or uncertainities related to existing cracks. The stiffness analysis trends are presented in Fig. 6b, which shows that samples drilled parallel to the bedding are the strongest. A summary of the rock failure-related data applicable at different sample orientations is presented in Table 1. The summary implies that net PP developments during CIU tests are not similar. The bedding plane and material heterogeneity are two key concerns to control the material failure and the development of PP.

ELASTIC MODULI OF SHALE—THEORY

The elastic moduli of a material are primarily described by its Young's modulus (E) and Poisson's ratio (v). These two parameters are interpreted in this study using the triaxial test results through a stress–strain trend analysis. Several simplifications have been introduced to calculate the elastic properties of shale from the triaxial tests. One key simplification is that the elastic stress–strain response is assumed to be linear. This is generally not the case for both the hydrostatic and triaxial phases. Thus, when presenting the elastic moduli, it is essential to define how the interpretation has been made. The following alternatives are applicable to estimate the Young's modulus (Wood 1990; Fjær et al. 2008):

- Initial modulus, given as the initial slope of the stress–strain curve.
- Secant modulus, measured at a fixed percentage of the peak stress.
- Tangent modulus, given at a specific percentage of the peak stress.

- Average modulus, given within a specific maximum and minimum stress level.

Of these alternatives, the tangent modulus at 50 % of the peak stress value was chosen in this study simply because it is the most widely used method.

The Young's modulus was determined as the tangent modulus measured at 50 % of the peak stress (see Fig. 7a) and is, therefore, denoted as E_{50}. The Poisson's ratio was determined from the inclination of a straight line that passes from the origin to the point of the curve that corresponds to 50 % of the peak stress in an ε_{vol} versus ε_1 diagram (see Fig. 7b), and is, therefore, denoted as v_{50}.

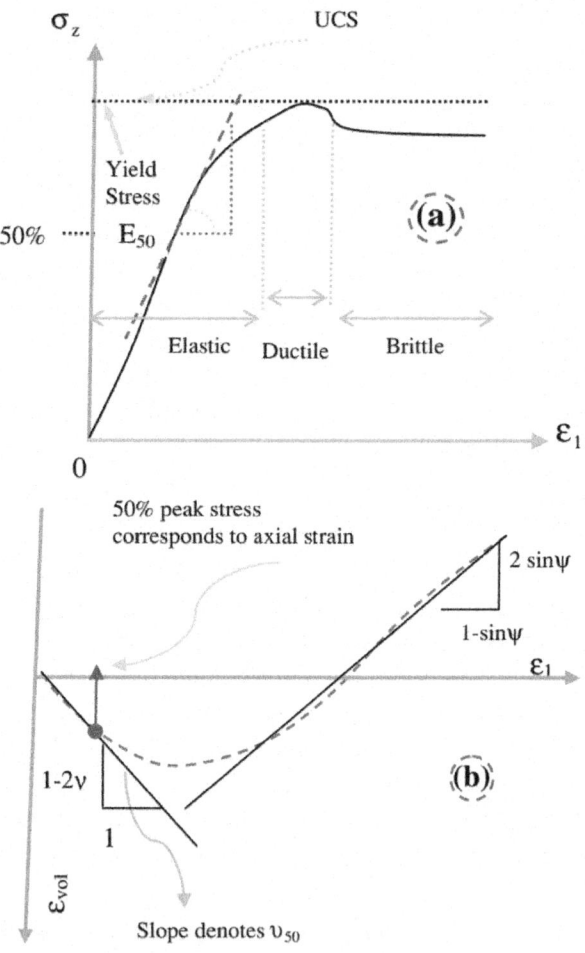

Figure. 7: Elastic moduli estimation techniques of shale; **a** Young's modulus, **b** Poisson's ratio. The symbol ψ is the dilatancy angle and v is the Poisson's ratio. In practice, the ductile region may be very small.

According to the American Society for Testing and Materials (ASTM) and International Society for Rock Mechanics (ISRM) standards, it is recommended to use the deformation rates at 50 % of the peak stress level for the determination of v. However, if the curve is strongly non-linear, complete information can only be given if the entire curve is presented. Shale is an anisotropic material, thus, the Poisson's ratio and the Young's modulus are not the proper parameters to describe the mechanical behavior. The elastic stiffness tensor, defined by Fjær et al. (2008), is required for an accurate description of the shale.

ELASTIC MODULI—RESULTS AND DISCUSSION

Strain Measurements

We have carefully leveled the strike and slip directions of the test samples. One arm of the cantilever was placed in the strike direction (along the bedding), and another arm was placed in the slip direction (perpendicular to the bedding). Radial deformation was measured by two pairs of calipers (pairs #1 and #2) mounted orthogonal to one another. One of the pairs always measured deformation along the strike direction.

Because we performed several tests along different bedding inclinations, a summary of sample orientations and caliper positions to measure the radial deformations are presented below:

- Perpendicular to bedding ($\theta = 0°$): pair #1 measures deformation along the strike direction.
- Angle to bed ($\theta = 45°$): pair #1 measures deformation along the dip direction.
- Angle to bed ($\theta = 60°$): pair #1 measures deformation along the dip direction.
- Parallel to bedding ($\theta = 90°$): pair #1 measures deformation along the slip direction.

Elastic Moduli of Shale—the CIU Case

During the CIU testing of shale, the confining stress was generally kept constant on a total stress basis. However, the effective confining stress is affected by the PP response. Consequently, the anisotropic effective stress parameters were difficult to obtain because the radial effective stress changed. In the CIU test conditions, the stiffness of the water was also taken into account and, because there was no volumetric change (because water is incompressible), the Young's modulus was expected to be greater than for the CID case. In general, for partially undrained conditions, the stiffness is higher and the volumetric strain is reduced (Holt et al. 2011; Rozhko 2011).

As mentioned earlier, the Young's modulus normally appeared to be greater for the undrained conditions. In addition, the higher the confining stresses, the higher the Young's modulus for the same type of test and the same angle to the bedding. For similar confining pressures, the Young's modulus was greater (the material is stiffer) for loading parallel to the bedding and smaller for loading perpendicular to the bedding. The stress–strain curves are presented in Fig. 8 to show how the material stiffness varied due to the effects of confining pressure and of the bedding plane. The Young's moduli and the Poisson's ratios were calculated from Fig. 8, in both drained and undrained test conditions. The determined elastic parameters, including both E_{50} and v_{50}, together with the bulk modulus, shear modulus, and Lamé modulus, are summarized in Table 5. The effect of the temperature on these elastic moduli is not included in this paper. Similar studies (Horsrud et al.1998; Søreide et al. 2009) indicate that, at a downhole temperature of 120 °C, a significant reduction in both stiffness (approx. −25 %) and strength (approx. −35 %) of the shale is observed. According to Søreide et al. (2009), the downhole undrained stiffness for North Sea shale at 140 °C is 57 % of the stiffness at room temperature. One possible countermeasure of the temperature increase is to improve the saturation of the sample, weaken the rock frame, and/or increase the fluid modulus, which would reduce capillary effects.

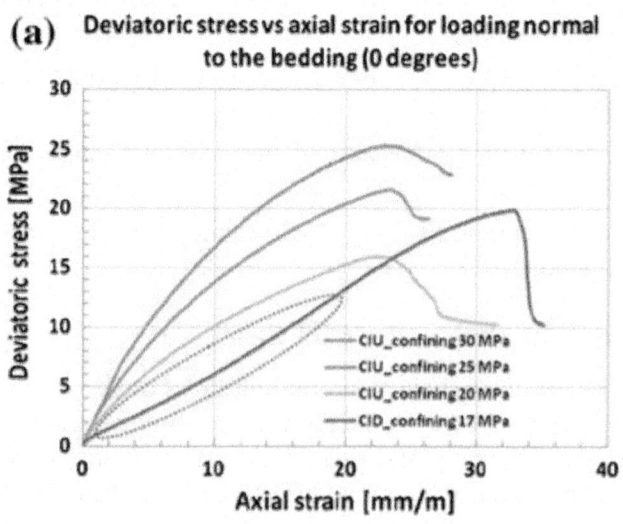

(a) **Deviatoric stress vs axial strain for loading normal to the bedding (0 degrees)**

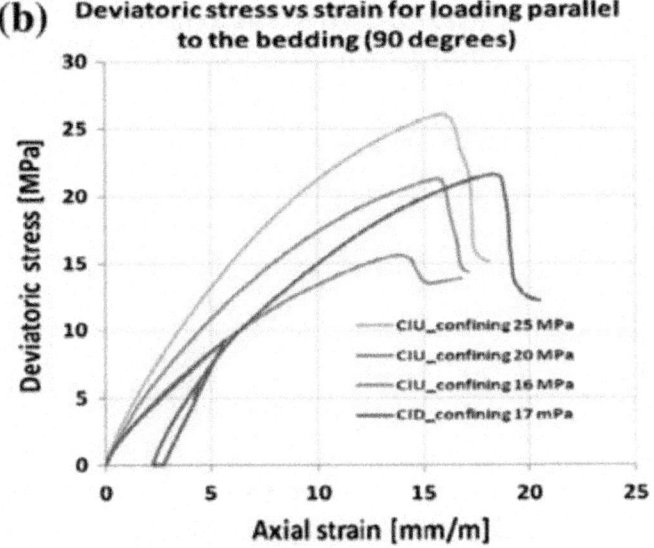

(b) Deviatoric stress vs strain for loading parallel to the bedding (90 degrees)

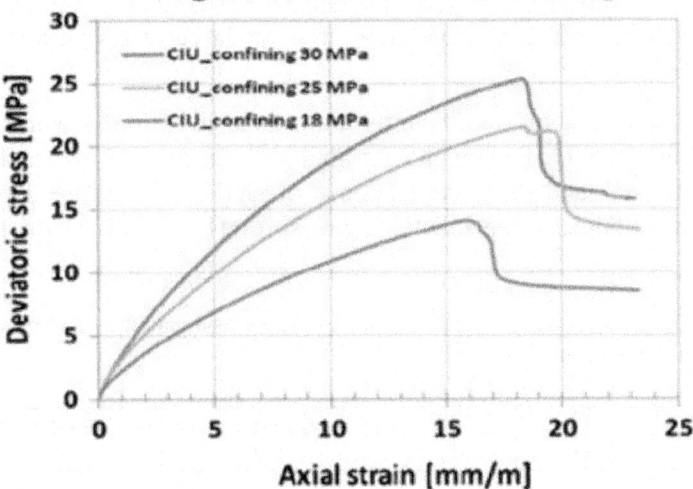

(c) Deviatoric stress vs axial strain for loading 45 degrees to normal to the bedding

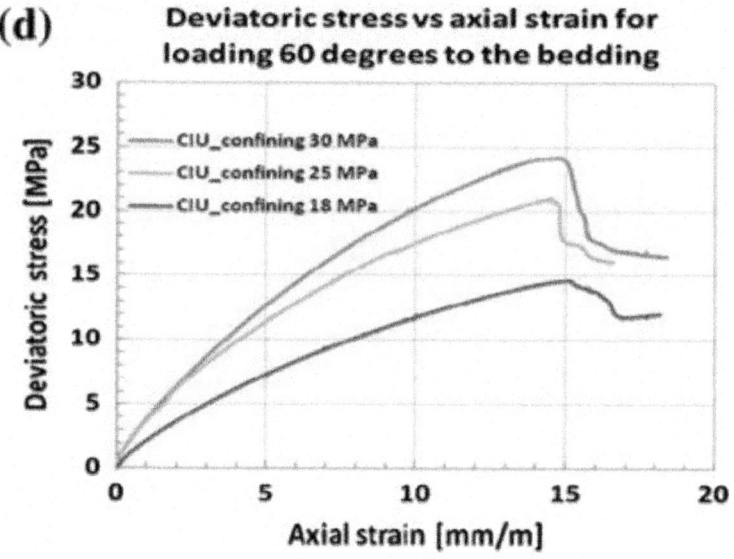

Figure. 8: Stress–strain response subjected to different confining pressures and bedding angles; **a** θ = 0°, **b** θ = 90°, **c** θ = 45°, and **d** θ = 60°. At low confinement pressure, natural microfractures cause non-linear stress–strain curve response (marked with the *dotted ellipse* in **a**).

This can contribute significantly to the strength of rocks with small pores (Schmitt et al. 1994; Papamichos et al. 1997). In general, it was found that, compared with sandstone, the response of shales were more dependent on temperature and less dependent on pressure (Holt et al. 1996; Horsrud et al. 1998).

Table 5: Estimated elastic moduli of Pierre-1 shale associated with different bedding planes and confinement pressures

θ (°)	σ_c (MPa)	E_{50} (GPa)	v_1	v_2	G_1(GPa)	G_2(GPa)	K_1(GPa)	K_2(GPa)	λ_1(GPa)	λ_1(GPa)
CIU										
0	20	1.360	0.44	0.54	0.47	0.44	3.78	5.67	3.46	5.96
0	25	1.550	0.52	0.53	0.51	0.51	12.92	8.61	13.26	8.95
0	30	1.900	0.41	0.56	0.67	0.61	3.52	5.28	3.07	5.68
45	18	1.150	0.61	0.34	0.36	0.43	1.74	1.20	1.98	0.91
45	25	1.900	0.46	0.35	0.65	0.70	7.92	2.11	7.48	1.64
45	30	2.300	0.58	0.38	0.73	0.83	4.79	3.19	5.28	2.64

60	18	1.450	0.62	0.33	0.45	0.55	2.01	1.42	2.31	1.06
60	25	2.400	0.72	0.38	0.70	0.87	1.82	3.33	2.28	2.75
60	30	2.600	0.56	0.30	0.83	1.00	7.22	2.17	7.78	1.50
90	16	1.730	0.70	0.31	0.51	0.66	1.44	1.52	1.78	1.08
90	20	2.200	0.75	0.32	0.63	0.83	1.47	2.04	1.89	1.48
90	25	2.650	0.65	0.29	0.80	1.03	2.94	2.10	3.48	1.42
CID										
0	17	0.640	0.1	0.1	0.29	0.29	0.27	0.27	0.07	0.07
90	17	1.560	0.19	0.095	0.66	0.71	0.84	0.64	0.40	0.17

θ = angle between loading and normal of the bedding planes, σ_c = confinement pressure, E_{50} = Young's modulus, v_1 and v_2 = Poisson's ratios, G_1 and G_2 = shear moduli, K_1 and K_2 = bulk moduli, λ_1 and λ_2 = elastic moduli (Lamé parameters)

Figure 8a represents the stress–strain response at $\theta = 0°$ bedding subjected to the effects of confinement. The material stiffness under CID and CIU tests differed significantly. Under CIU tests, the stiffness curves increased non-linearly up to a certain level and then decreased. A possible reason for this behavior is that the PP develops asymptotically up to a certain level close to the peak stress. The CIU test subjected to drilling normal to bedding would, thus, provide a higher development response of the PP. The developed PP under this test did not account for the effective stress because the volumetric strain was constant. However, for the CID test, the effective stresses play a role in determining the rock strength and stiffness. The initial non-linear trend in Fig. 8a (marked by a dotted elipse) may be due to the presence of induced cracks or natural microfractures. Through the postfailure analysis (Fig. 3) of these samples, it was inferred that the additional induced cracks or microfractures can further contribute to increased pore volume prior to failure. The pore fluid may penetrate through the microfracture network and further reduce the PP, resulting in reduced rock strength. Moreover, the natural properties of shale in terms of crack or microfracture orientation may also produce various non-linear behaviors. Depending on the loading direction, the fracture network/crack can close or open. This variance leads to a decrease or increase in pore volume in response to the PP. A similar observation was found at 90° bedding (Fig. 8b). The concave-type CIU test results were mostly defined by the PP increase through the triaxial phase. The volumetric-axial strain curve (Fig. 9) was used to calculate the Poisson's ratio. The results are presented in Table 5. For the undrained tests, the value of the Poisson's ratios ranged between 0.34 and 0.75, while for the drained case, it decreased to 0.2. This behavior is illustrated and discussed later in this paper.

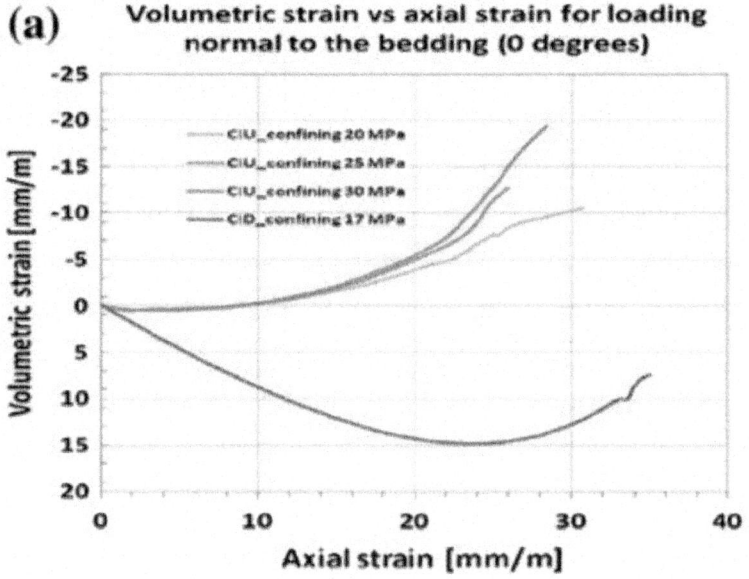

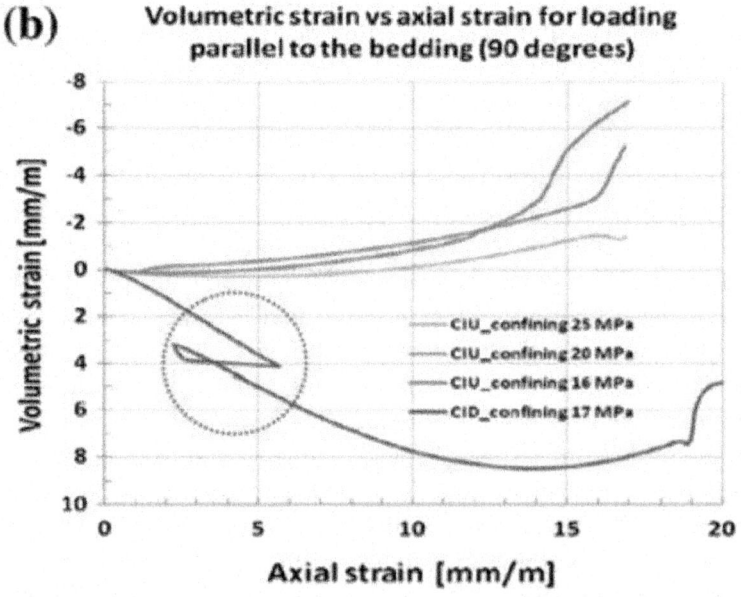

Figure. 9: Poisson's ratio from axial strain versus volumetric strain through CID tests; **a** θ = 0°, **b** θ = 90°. The *dotted circle* indicates the strain rate effect due to a temporary stop in the experiment.

Elastic Moduli of Shale—the CID Case

The drained test for low permeability rocks is essential to obtain the effective stress parameters. These effective stress parameters are important for borehole stability modeling, especially in shale. Directional stiffness parameters (E_t and E_p) are essential for the shale anisotropy borehole stability model. The normal or transversal stiffness parameter (E_t) was determined at sample $\theta = 0°$ (load normal or transversal to the bedding), and E_p was estimated when $\theta = 90°$ (loading parallel to the bedding). Søreide et al. (2009) showed analytically that the stiffness for the samples with loading parallel to the bedding (E_p) is higher than for loading perpendicular to the bedding (E_t). In our case, the estimated values of E_t and E_p were 0.65 and 1.55 GPa, respectively (see Fig. 10a). The estimated mean relative difference of the E between these two samples (in absolute value) was approximately 58 %. A significant difference between these two moduli indicated a strong heterogeneous nature.

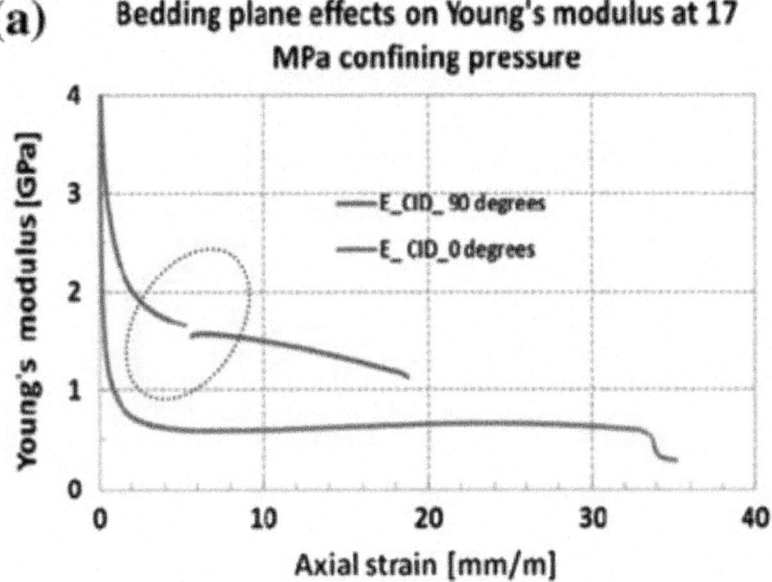

(a) Bedding plane effects on Young's modulus at 17 MPa confining pressure

(b)   Bedding plane effects on Young's modulus at 17 MPa confining pressure

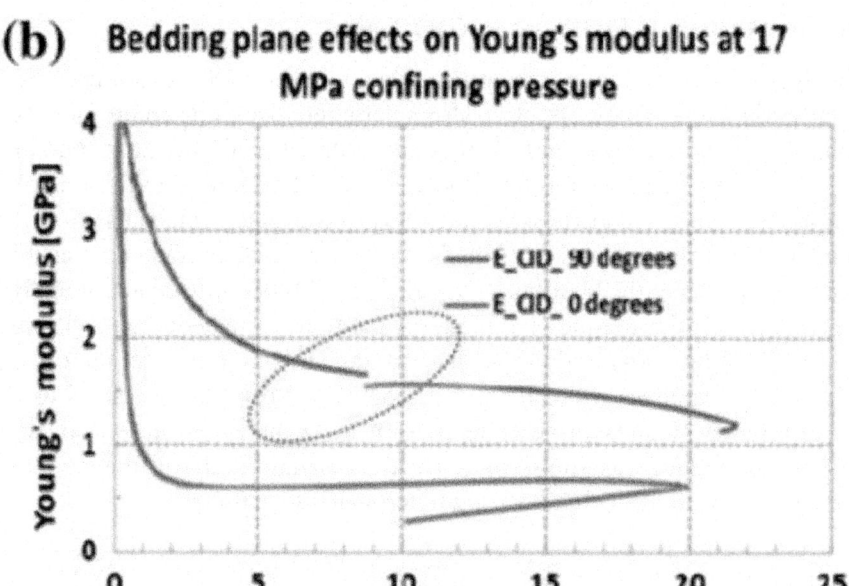

(c)   Bedding plane effects on Shear Modulus at 17 MPa confining pressure

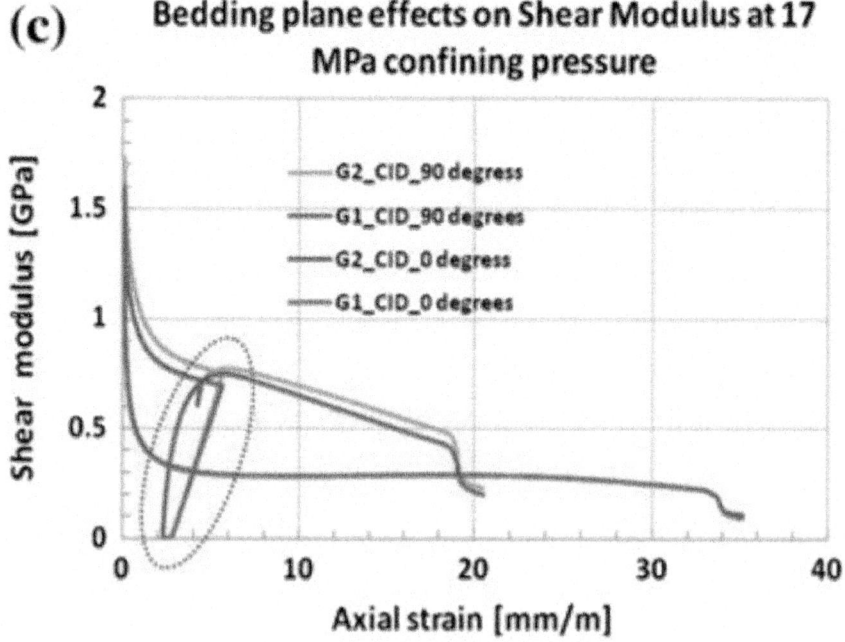

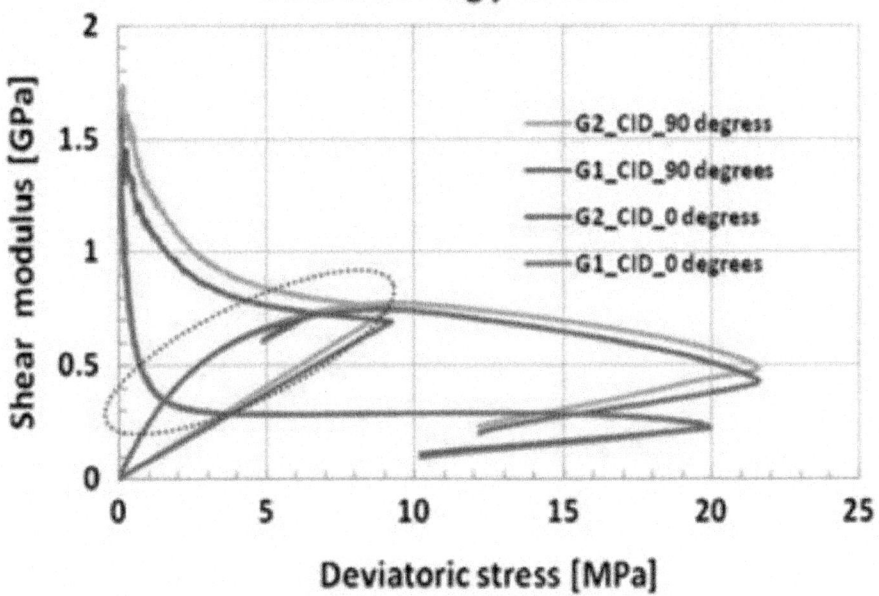

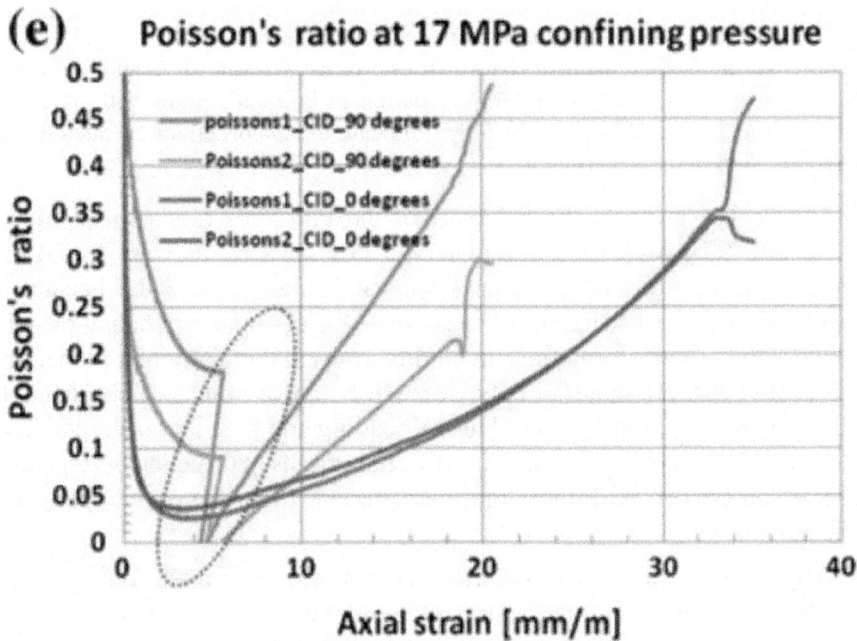

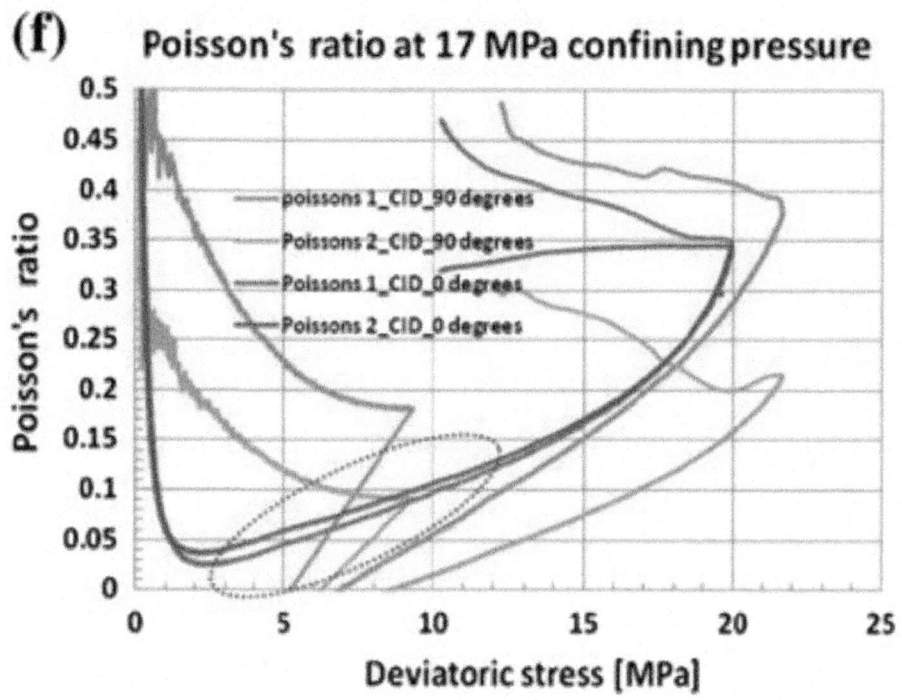

(f)   Poisson's ratio at 17 MPa confining pressure

Legend:
- poissons 1_CID_90 degrees
- Poissons 2_CID_90 degrees
- Poissons 1_CID_0 degrees
- Poissons 2_CID_0 degrees

Figure. 10: Shale elastic moduli at drained test conditions; **a** variation of E with axial strain, **b** variation of E with deviatoric stress, **c, d** variation of G with axial strain and deviatoric stress, and, finally, **e, f** variation of Poisson's ratios with axial strain and deviatoric stress. The *dotted ellipses* indicate the abnormal strain rates due to a temporary stop in the experiment.

Figure 10e presents the behavior of the Poisson's ratios for samples at $\theta = 0°$ and $90°$. At $\theta = 0°$; the observed Poisson's ratios were found to be similar. The two measured radial strains will always be equal because the sample plane acts as a ring and, thus, carries similar material behavior, which produces the same deformations in all directions in that particular plane. In the case of $\theta = 90°$, the material behavior will be anisotropic, which means that one radial deformation is parallel to the bedding and the other is perpendicular to the bedding. These two measured radial strains will never be equal, and the Poisson's ratios will, therefore, be different. Figure 10e, f supports these conclusions.

The shear modulus is calculated based on the estimated Young's moduli and Poisson's ratios, and the resulting calculations are presented in Fig. 10c, d. The lower limit of the shear modulus for $\theta = 0°$ was 0.24 GPa, but for $\theta = 90°$, it was 0.62 GPa (on average). In Fig. 10a, b, there is an irregularity, marked by dotted ellipses, caused by stopping the test at a high strain rate. This irregularity may affect the quality of the results. One point that should be carefully addressed is how to select the measured data points. The trend of the Young's modulus at the horizontal bedding is almost linear, but it is non-linear at $\theta = 90°$.

Elastic Moduli of Shale—the CIU Case with Constant Confining Pressure

At constant confining pressure (i.e., 25 MPa), the E and the v were calculated and are presented in Fig. 11. The response of the E and v decreased exponentially with increasing axial strain or deviatoric stress. For example, in Fig. 11a, for 10 % axial strain, E for the 90°, 60°, 45°, and 0° samples are 2.1, 1.7, 1.5, and 1.4 GPa, respectively. The samples showed the highest stiffness at $\theta = 90°$ and the lowest stiffness at $\theta = 0°$. It is recommended to choose the E data point based on the expected axial strain or deviatoric stress condition.

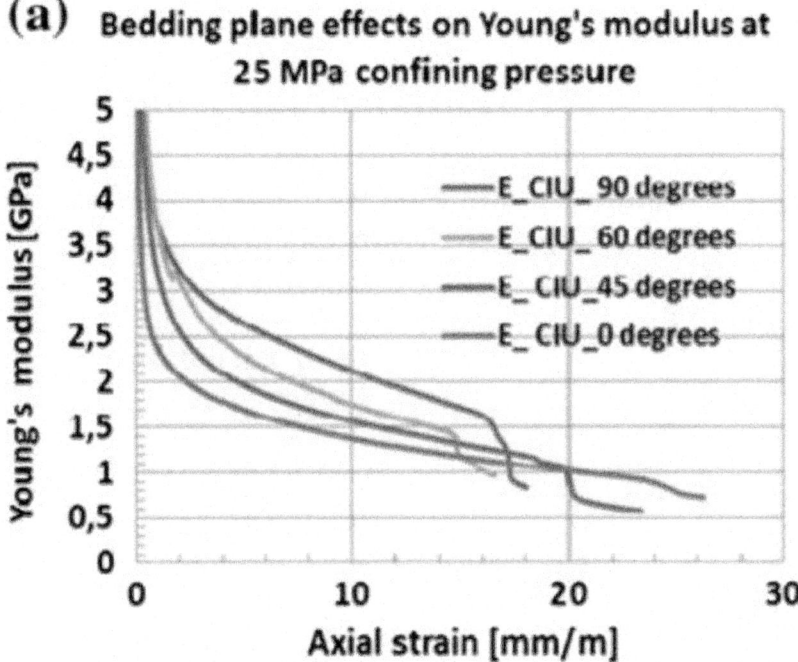

(a) Bedding plane effects on Young's modulus at 25 MPa confining pressure

(b) **Bedding plane effects on Young's modulus at 25 MPa confining pressure**

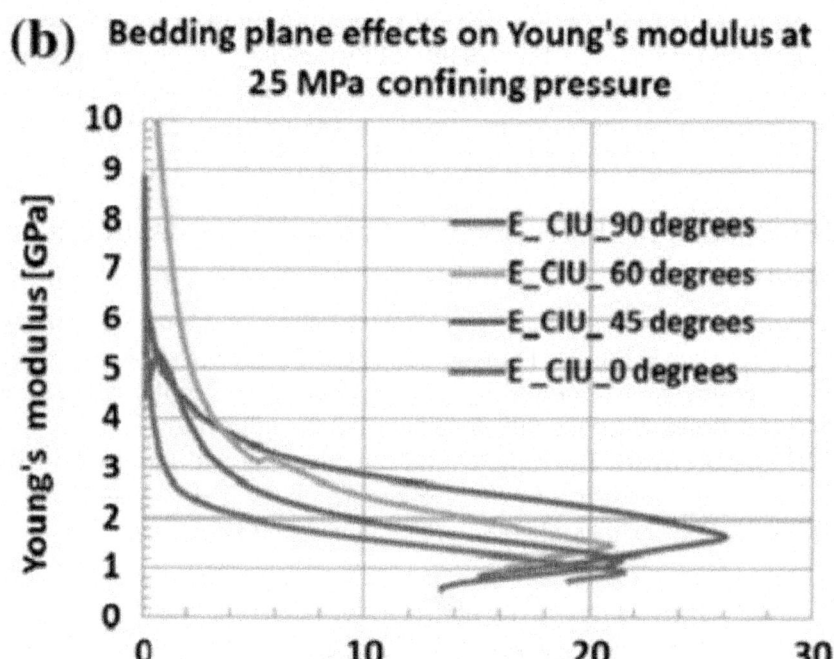

(c) **Bedding plane effects on Shear modulus at 25 MPa confining pressure**

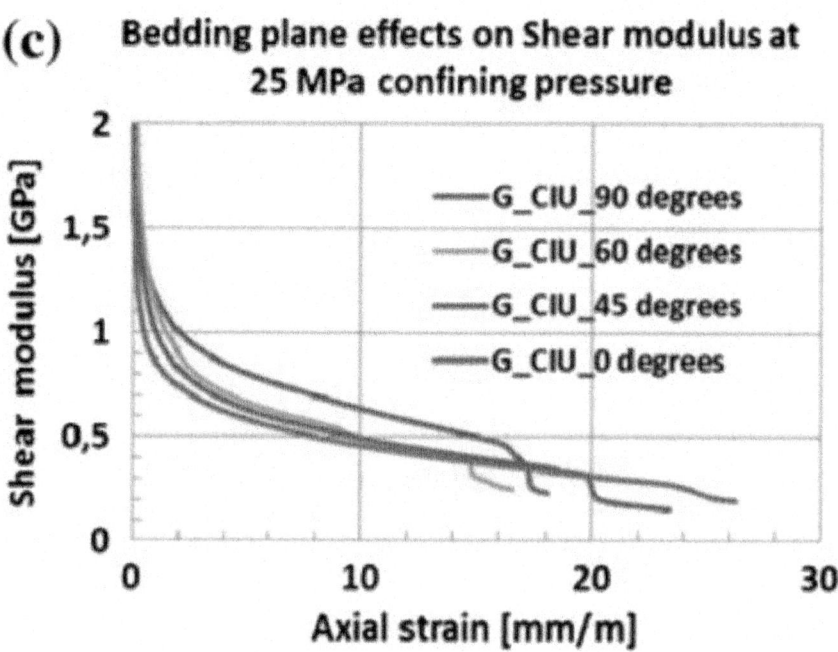

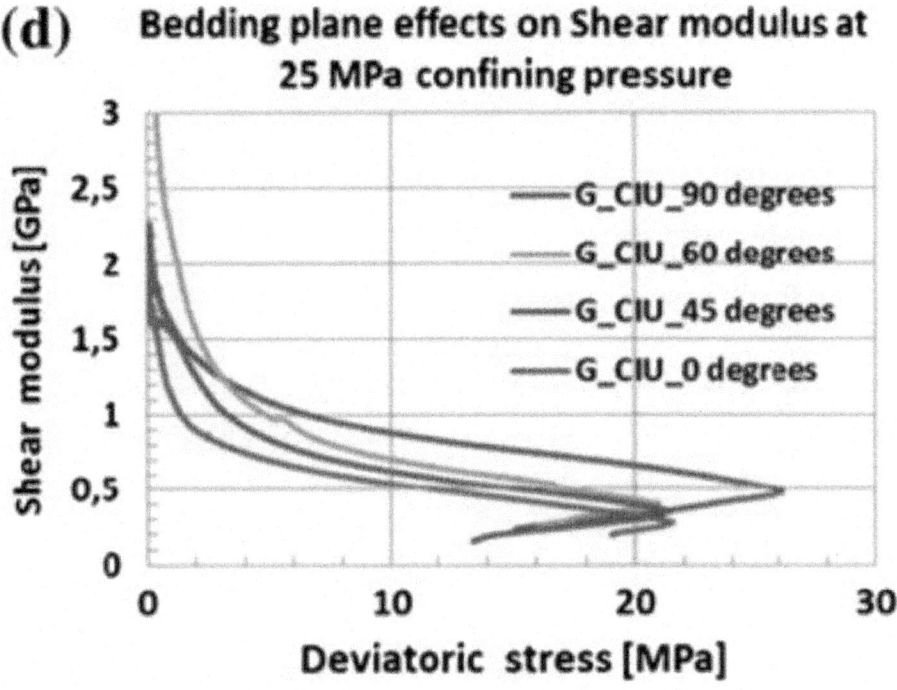

Figure. 11: Estimation of Young's modulus in CIU tests at different constant confinement pressures.

Two Poisson's ratios were defined, v_1 and v_2. The values of v_1 vary significantly with respect to all bedding plane samples. These values are increasing exponentially but, in some cases, also increasing gradually versus changing deviatoric stress (Fig. 12a, b). For $\theta = 0°$, the magnitudes of v_1 and v_2 followed a non-linear trend and varied from 0.4 to 0.7. At other sample orientations (except $\theta = 0°$), v_1 showed extremely high values between 0.6 and 0.8. However, a significant reduction occurred in the interpreted v_2, with values varying between 0.25 and 0.4. At $\theta = 0°$ and $\theta = 90°$, the Poisson's ratios $v_1 = 0.52$ and 0.68, respectively, while v_2 was calculated to be 0.53 and 0.35, respectively.

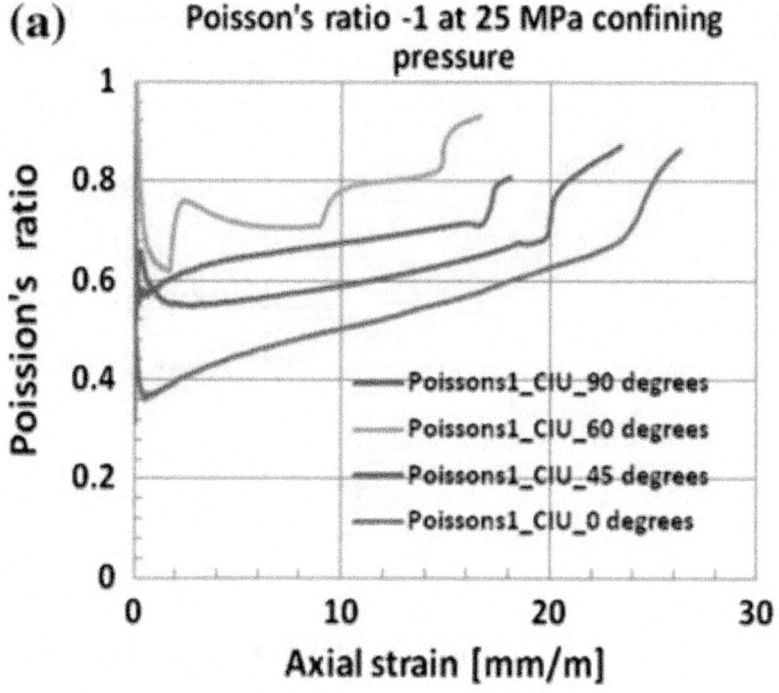

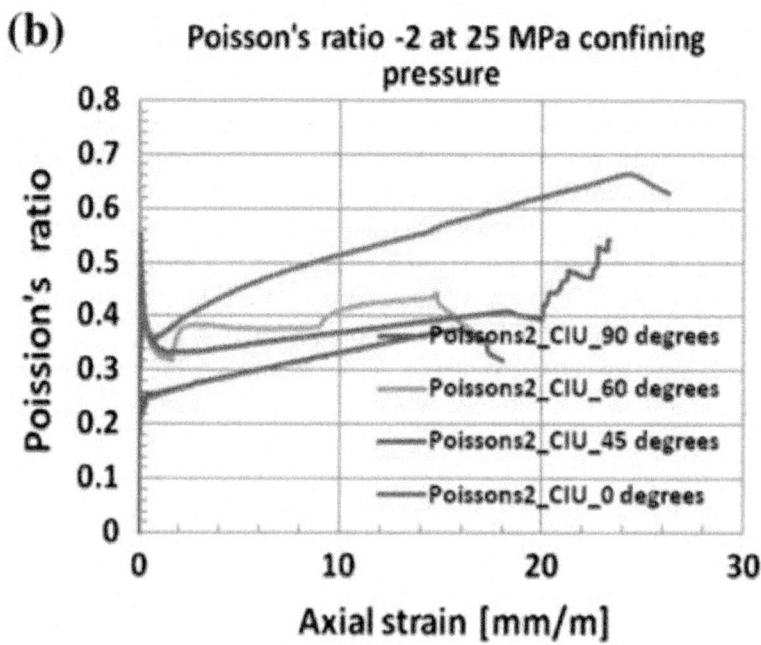

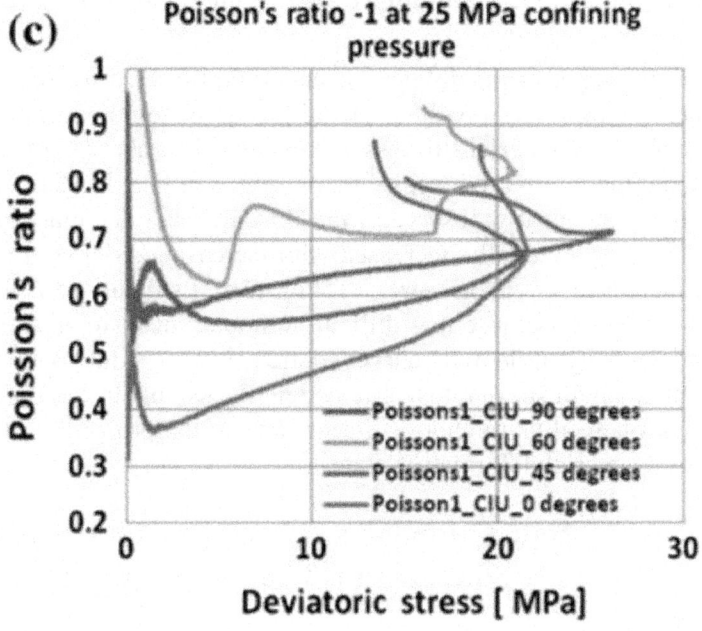

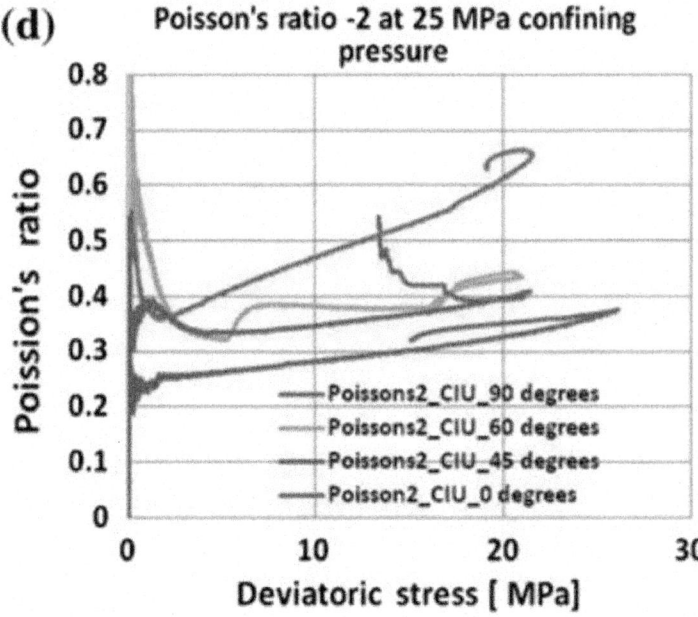

Figure. 12: Estimation of Poisson's ratios in CIU tests at different constant confinement pressures.

Effects of Variable Confining Pressure on *E* (for Both CIU and CID Tests)

The confining pressure influences material stiffness and shear strength. The higher the confining pressure, the stiffer the material and the higher the shear strength. The material stiffness was evaluated by changing the confining pressure, and the interpreted results are presented in Fig. 13. The variation of rock stiffness was clearly observed from this analysis. The material stiffness for the CIU cases gradually decreased with increasing axial strain (Fig. 13a) and with increasing deviatoric stresses (Fig. 13b). Estimates of *E* for the CIU cases are confusing because it is difficult to find a linear trend, as in the CID cases. The anomaly of the CID curves (dotted circles) is due to a pause in the experiment at a high strain rate and reloading the sample to run the test again. This analysis helped us to choose *E* according to the projection of confining pressure and material deformation state.

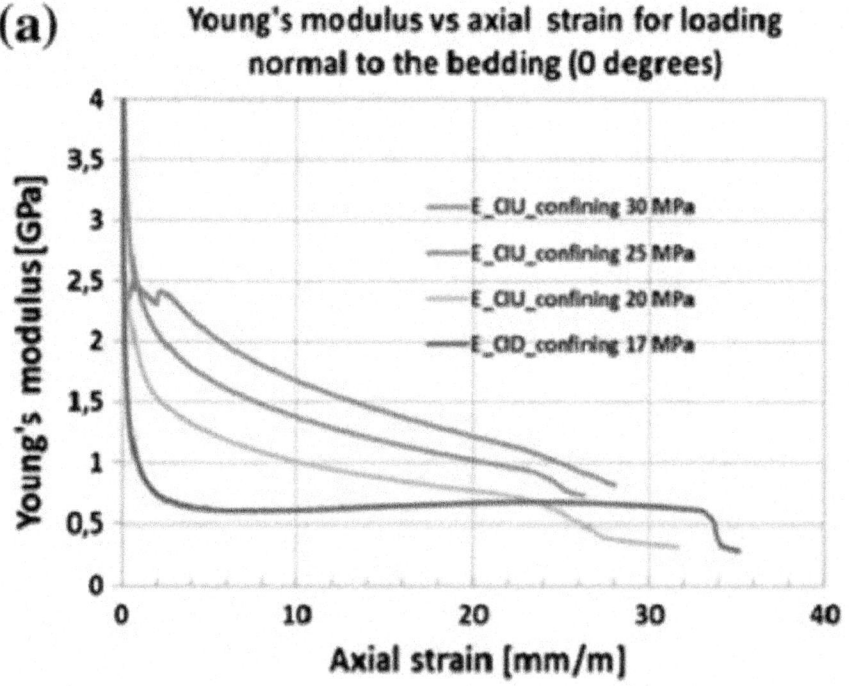

(a) **Young's modulus vs axial strain for loading normal to the bedding (0 degrees)**

(b) Young's modulus vs deviatoric stress for loading normal to the bedding (0 degrees)

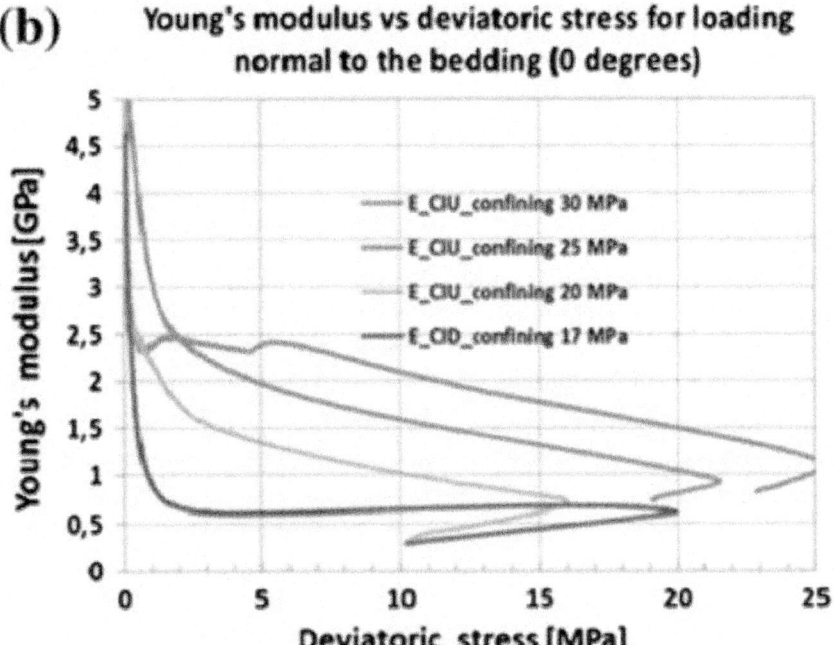

(c) Young's modulus vs axial strain for loading 45 degrees to normal to the bedding

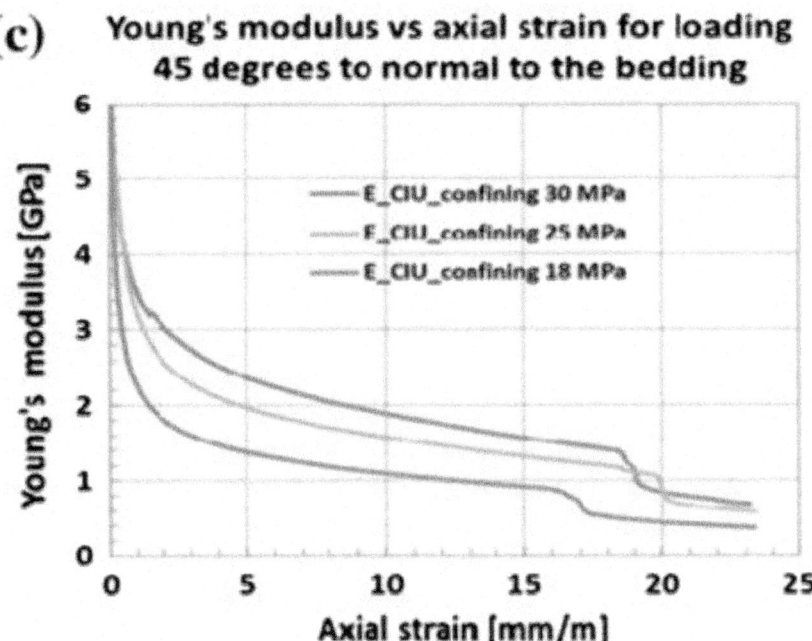

(d)

Young's modulus vs deviatoric stress for loading 45 degrees to normal to the bedding

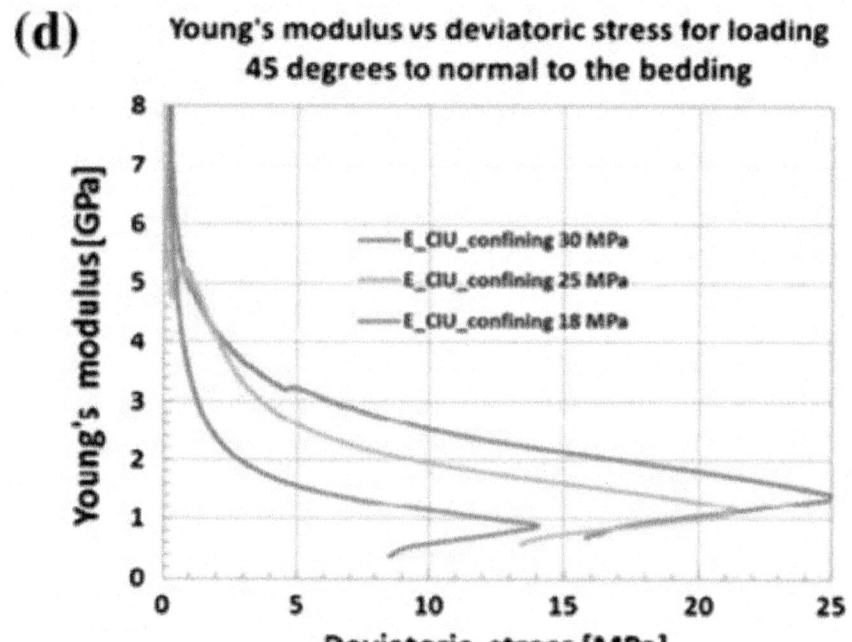

(e)

Young's modulus vs axial strain for loading 60 degrees to the bedding

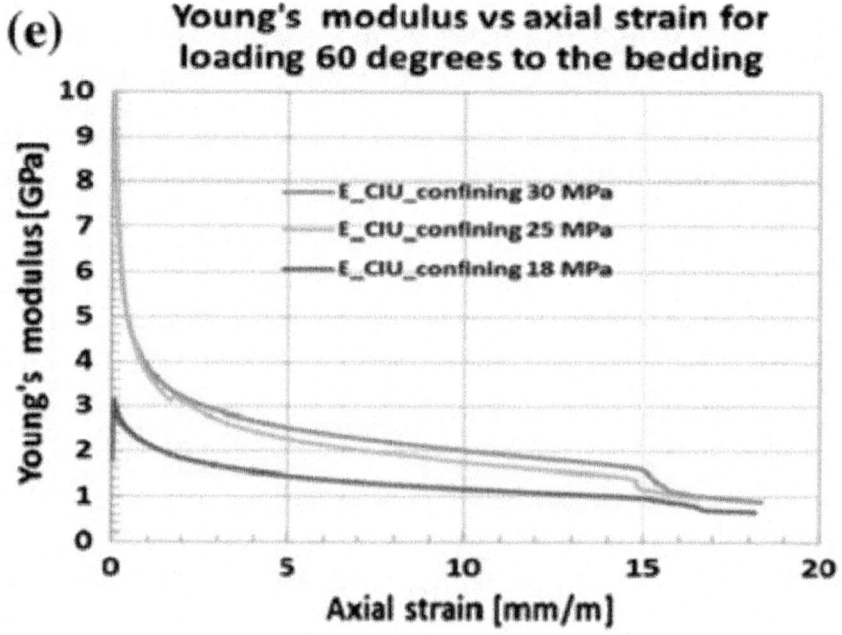

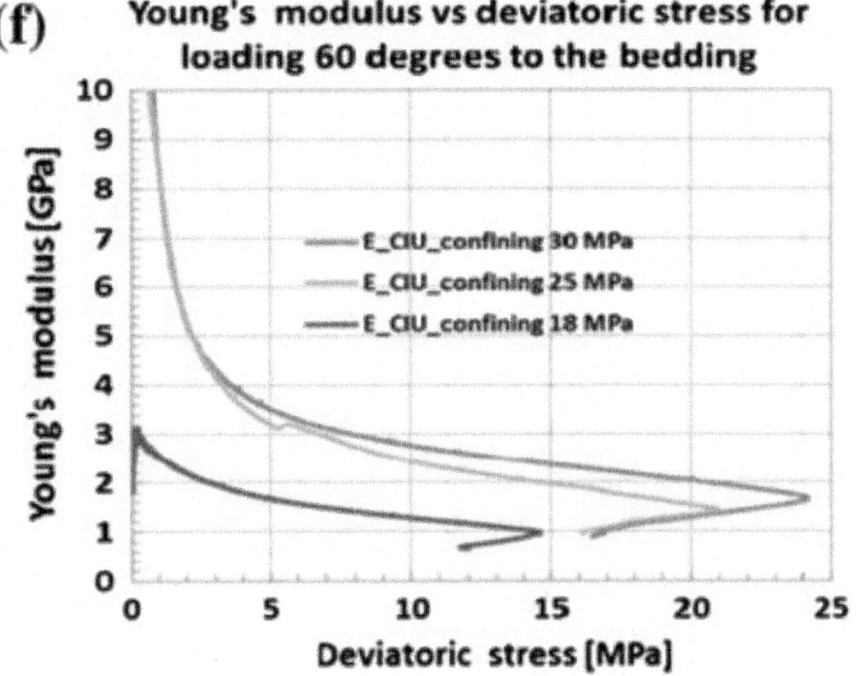

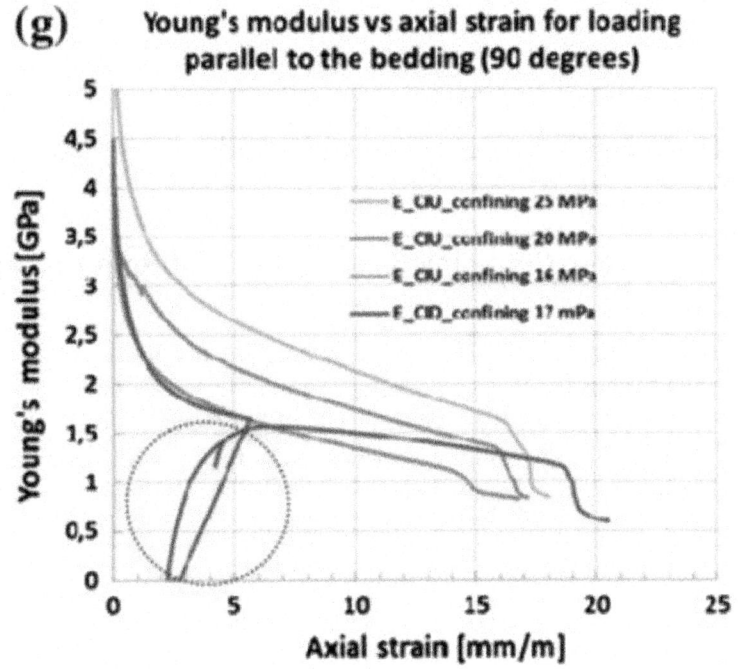

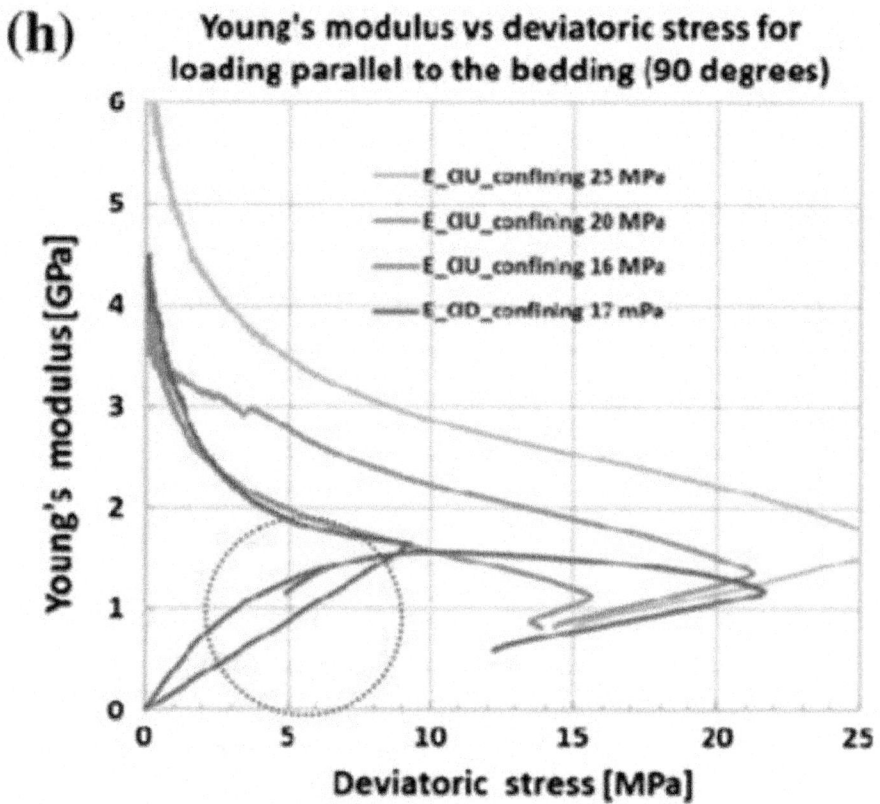

(h)   Young's modulus vs deviatoric stress for loading parallel to the bedding (90 degrees)

Figure. 13: **a–h** Variation of the Young's modulus subjected to variable confinement pressures.

Effects of Variable Confining Pressure on   (for Both CIU and CID Tests)

For homogeneous materials, it is generally accepted that the higher the confining pressure, the higher the Poisson's ratio. Figure 14a shows that, for samples at $\theta = 0°$, both v_1 and v_2 increased non-linearly with increasing confining pressure. The changing values are minor compared with other bedding orientations.

(a)

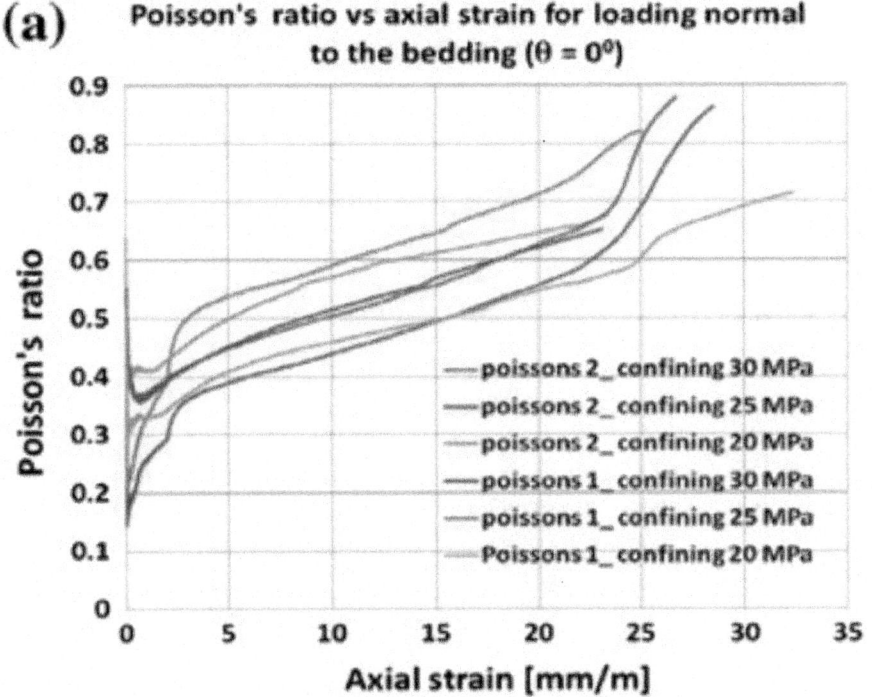

Poisson's ratio vs axial strain for loading normal to the bedding (θ = 0⁰)

(b)

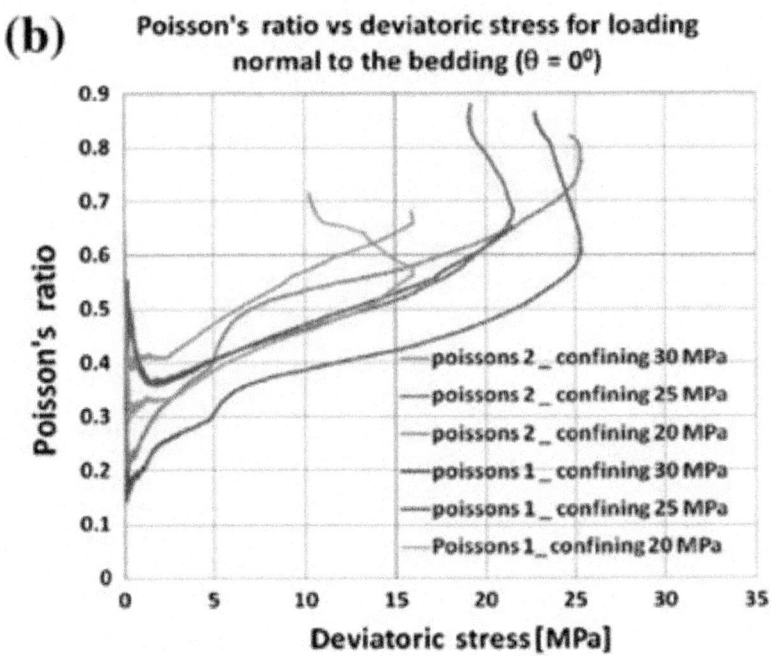

Poisson's ratio vs deviatoric stress for loading normal to the bedding (θ = 0⁰)

(c)

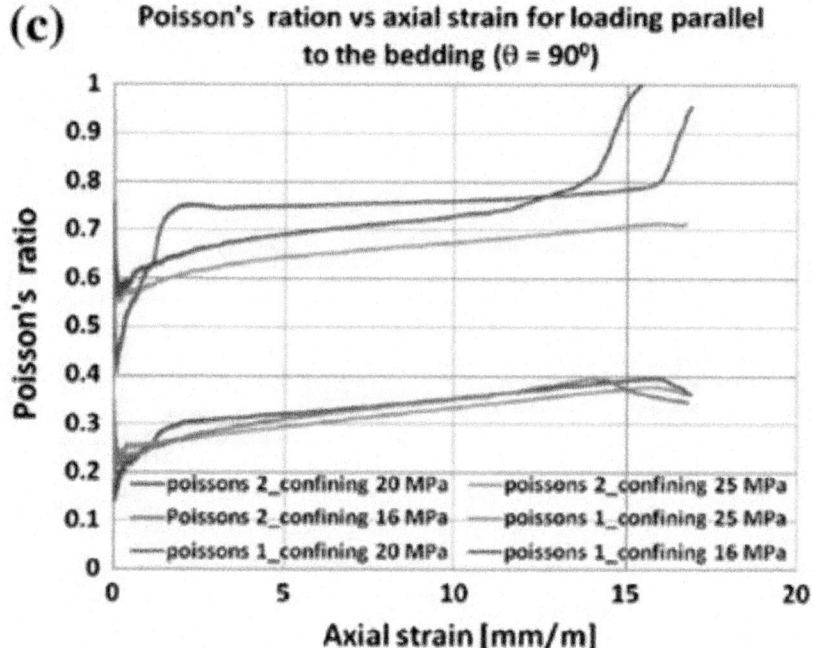

(d)

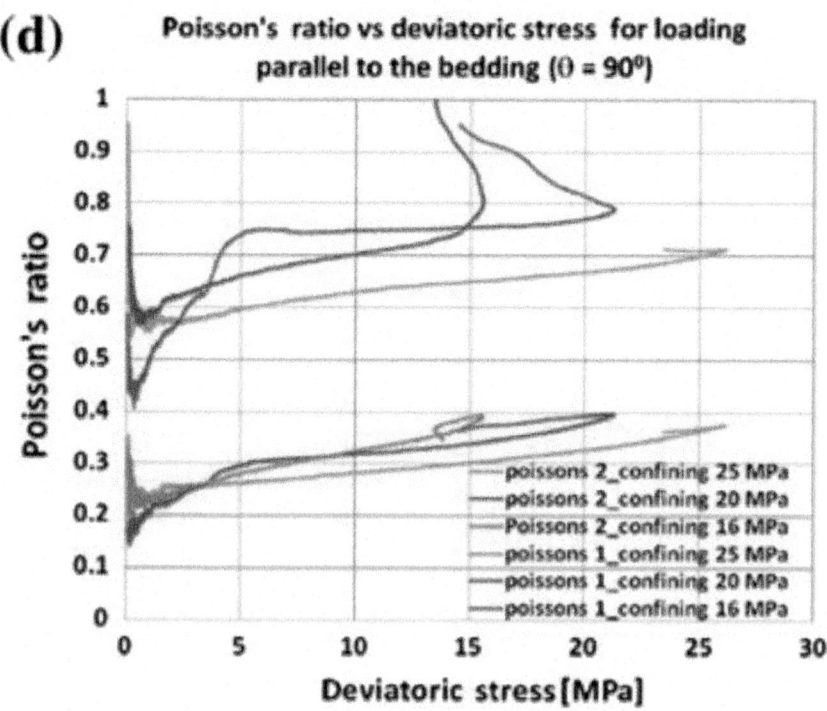

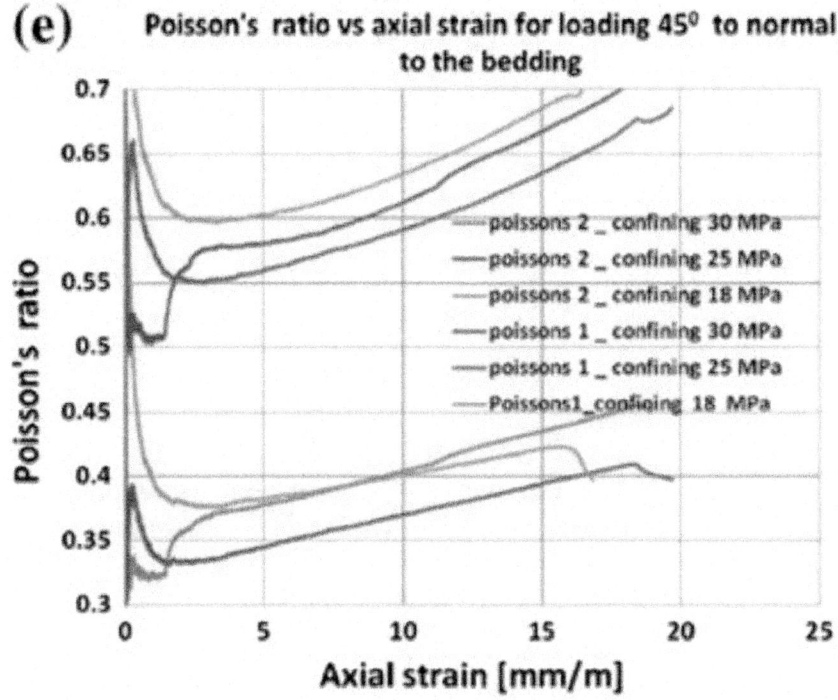

(e) Poisson's ratio vs axial strain for loading 45⁰ to normal to the bedding

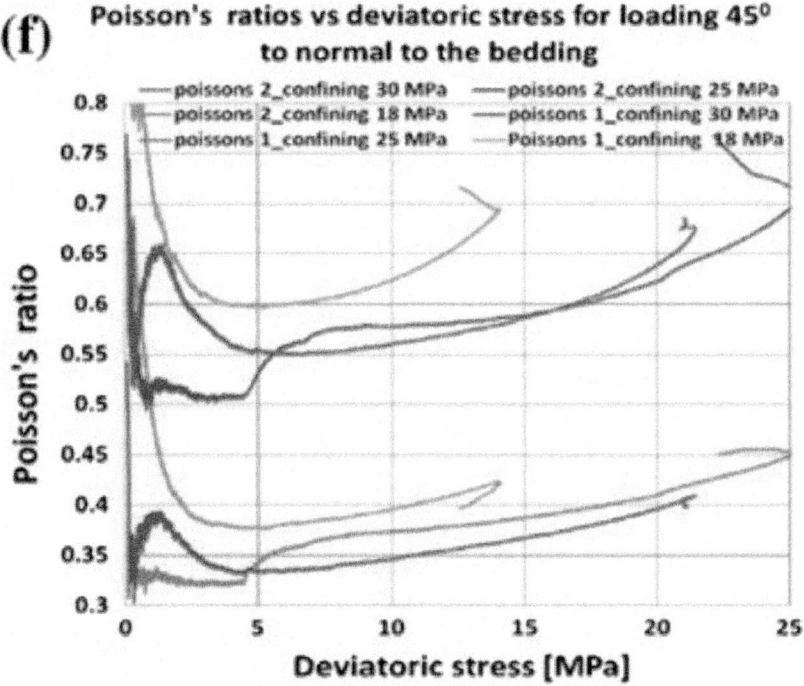

(f) Poisson's ratios vs deviatoric stress for loading 45⁰ to normal to the bedding

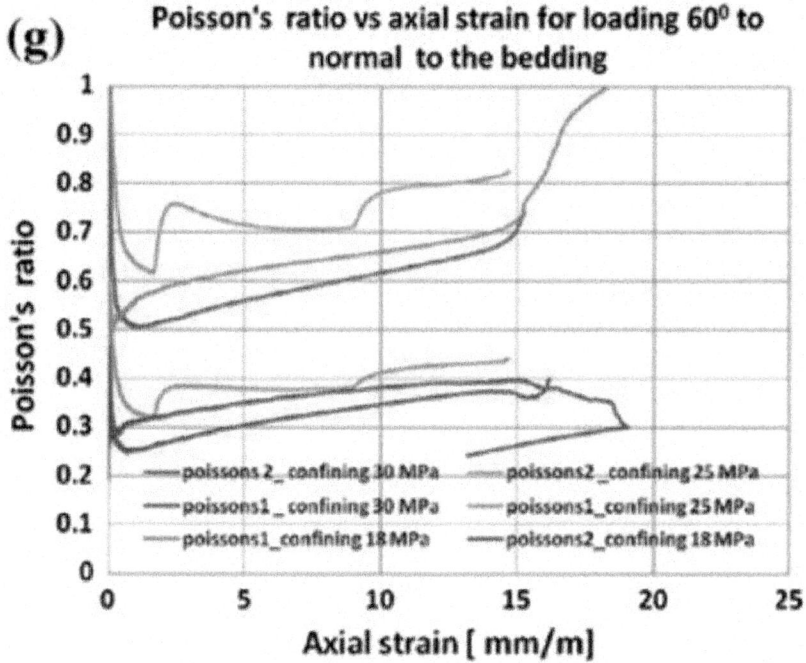

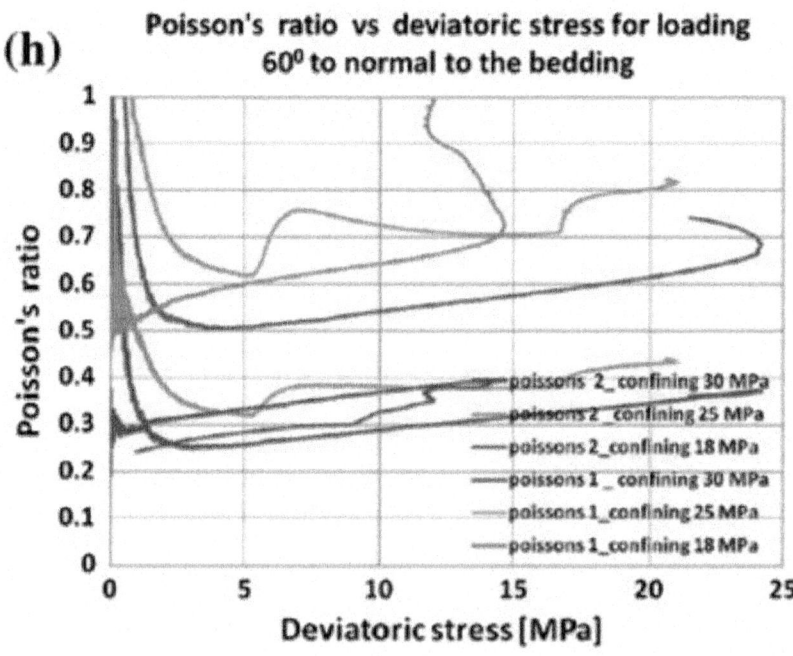

Figure. 14: a–h Quantifying the effects of confining pressure on the Poisson's ratio.

On the other hand, a large gap between v_1 and v_2 was found for the sample drilled at $\theta = 90°$ (Fig. 14c, d), and similar trends were observed for $\theta = 45°$ and 60°. At $\theta = 90°$, values of 0.3 and 0.7 were obtained for v_1 and v_2, respectively, and this difference is significant. In the undrained test, the Poisson's ratios observed in the tested shale rocks tended to be larger than 0.5. This observation is supported by Aadnøy and Chenevert (1987), who showed that, for laminated materials, the Poisson's ratio can be higher than 0.5 and is, in fact, dependent on the elasticity modulus ratio (a ratio between the vertical and horizontal moduli).

CYCLIC VERSUS MONOTONIC TEST EFFECTS ON ROCK STRENGTH

The CIU test was performed under cyclic loading–reloading, followed by a 4-MPa cyclic amplitude on a sample drilled perpendicular to the bedding ($\theta = 0°$). The idea was to see more of the elastic response during such a small cycle due to non-linearity (Fig. 15c). The test sample for the cyclic test was taken from the same core block used in the previous monotonic samples tested. The cyclic test included 3–4 unloading–reloading cycles. The cycle for each step was 5 MPa during the triaxial phase, with a cycling amplitude of 4 MPa. The PP during consolidation was 10 MPa, and the confining stress was 25 MPa. In the triaxial phase, the strain rate was set to 2×10^{-7} s^{-1}. The test required 4 days to reach yielding.

(a) 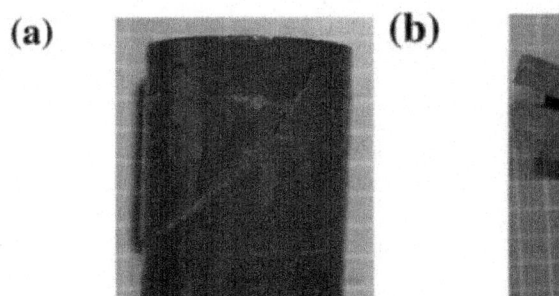   **(b)**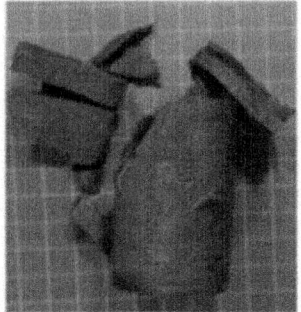

(c) Cyclic and monotonic triaxial test at 25 MPa confining pressure, samples normal to bedding

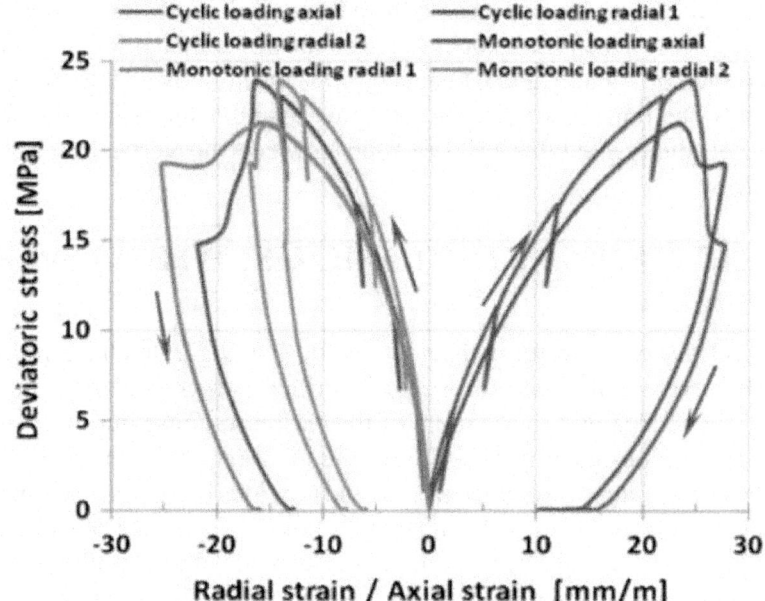

(d) Cyclic and static triaxial test at 25 MPa confining pressure

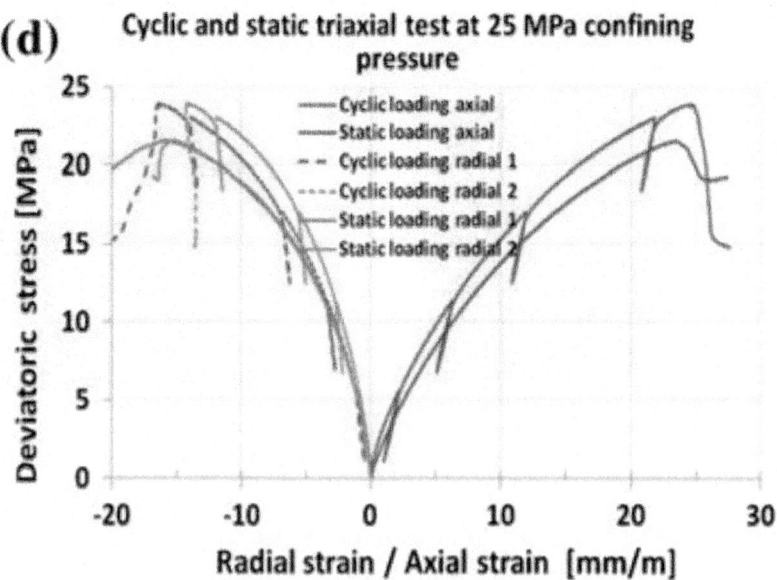

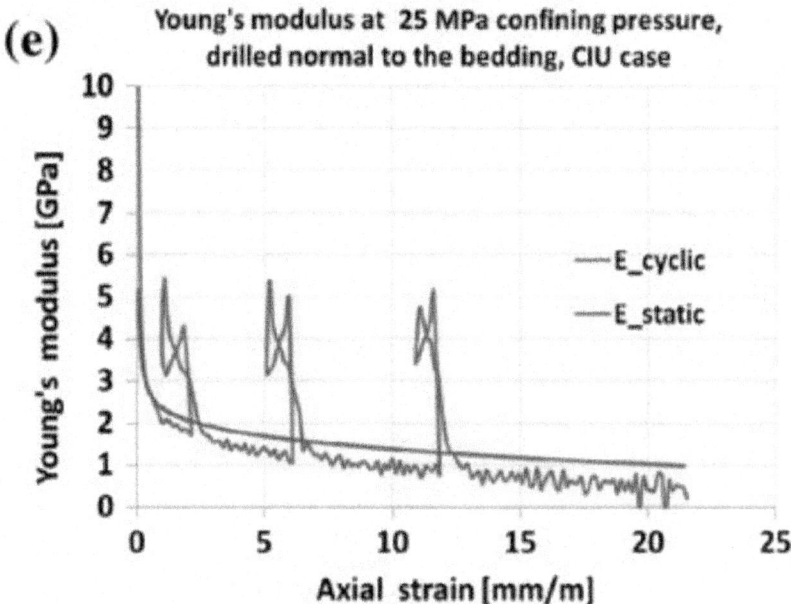

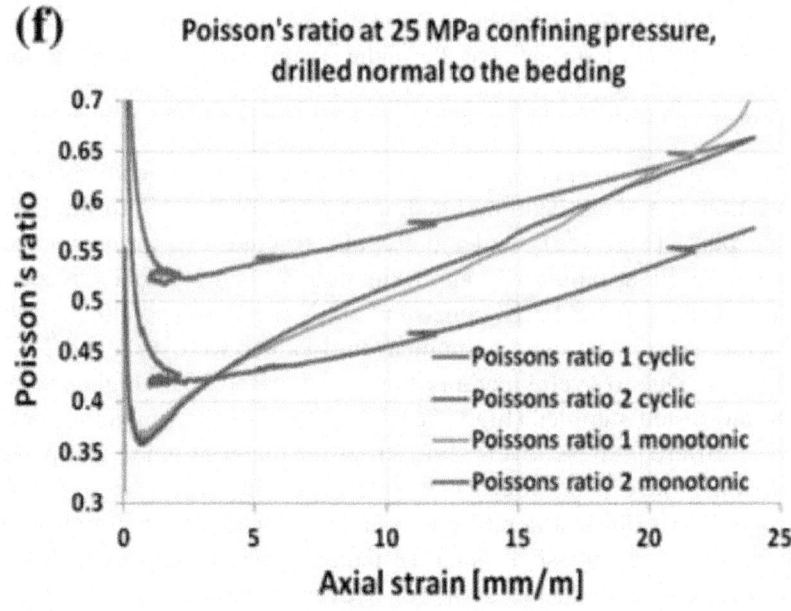

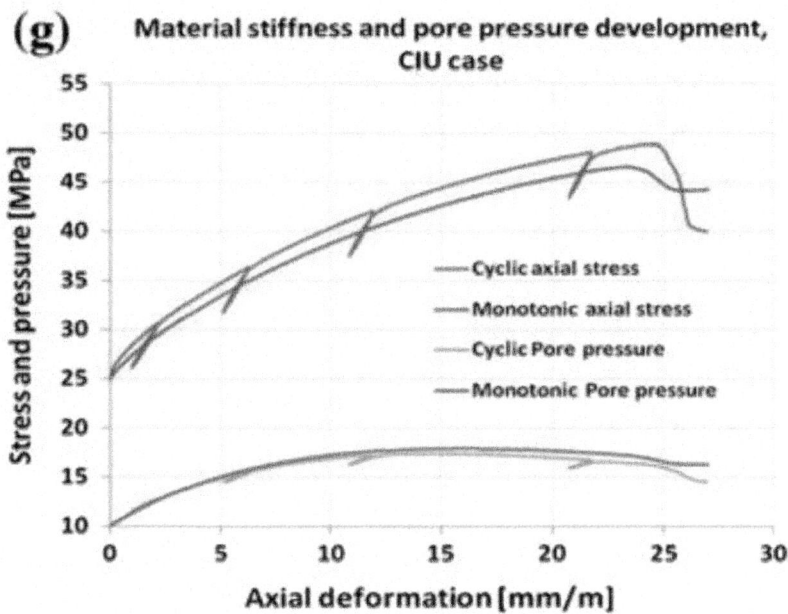

Figure. 15: Comparative studies to evaluate the mechanical properties of shale under cyclic and monotonic test conditions. Postfailure samples **a** cyclic test and **b** monotonic test, **c** stiffness curves, **d**, **e** *E* calculation, **f** variations of Poisson's ratio, and **g** PP development and stiffness. All samples were drilled perpendicular to bedding ($\theta = 0°$).

The postmortem analysis of the sample showed a localization of the deformation in a shear band inclined at an angle of $\theta = 45°$ to the horizontal bedding plane (Fig. 15a). Under the cyclic triaxial test, the estimated axial stress and the PP at failure was approximately 49 and 16.3 MPa, respectively, which appeared to be fairly consistent with the corresponding established monotonic triaxial test, where the measured values were 46.5 and 17.2 MPa, respectively. Due to cyclic loading, the PP increased 10 % higher than in the monotonic tested samples (Fig. 15g). The slope of the unloading–reloading cycles at different stress levels showed stiffer material when compared with only loading once (Fig. 15e, g). In this particular case, the shale stiffness under cyclic triaxial test conditions was approximately 50 % higher than in the monotonic triaxial test. This increase may be due to an irreversible change in the microstructure of the rock. An irreversible strain (plastic strain) was observed. However, the degradation effects are negligible here because we performed only one cycle and were still well below the peak strength. We cycled at approximately 50 % of the peak stress.

In the cyclic triaxial test condition, an exception was observed in the data for calculating the Poisson's ratio. The two Poisson's ratios were similar under the monotonic triaxial testing, whereas they were completely different in cyclic triaxial test conditions (Fig. 15f). For this particular case, v_1 and v_2 were determined to be 0.6 and 0.46, respectively, for the cyclic test. The possible reasons for such dissimilarity may be the shale heterogeneity or the material deformation state under the cyclic stress state (see Fig. 15c, d). For the static test, the Poisson's ratios were 0.52 and 0.53, respectively. It was also noticed that the calculated PP under the cyclic triaxial test was lower than for the monotonic triaxial test (Fig. 15g). This study confirmed that the PP plays an adverse role in the determination of the lower stiffness in monotonic tests compared with the cyclic tests. The conclusion is that the PP development is a critical parameter under the CIU tests that can specifically control material stiffness in clay-dominant samples.

There are many factors, i.e., induced cracks and their orientation, partial saturation, material heterogeneity and anisotropy, plasticity, magnitudes of the loading–reloading cycles, strain rate, etc., that could all influence the geomechanical elastic properties of shale. A detailed discussion of the elastic response due to cyclic loading has been made by Fjær et al. (2008, p. 267), (2011), Holt et al. (2011), and Niandou et al. (1997). From their analysis, the non-elastic behavior during loading is largely dependent on the stress history of the rock.

ROCK DEFORMATION VERSUS BEDDING PLANE

A comparative study of the material deformations under shearing on different bedding samples is presented in Fig. 16. This analysis is interpreted as the peak deviation with respect to axial or radial deformation. Basically, such an analysis helps provide a clear picture of weak bedding in terms of the stiffness. It is observed that the variation of the peak stress varies significantly with the bedding plane orientation. Under the CIU test matrix, the peak deviation is largest for samples drilled parallel to the bedding. The same analysis indicated larger radial deformations but relatively lower peak deviations for samples at 0° (18 %) and 45° (8 %) bedding and, thus, showed lower stiffness. In the case of the CID tests, the maximum axial deformation (35 %) was observed for samples drilled perpendicular to the bedding, whereas it was 20 % for samples drilled parallel to the bedding. Lower stiffness was measured for samples tested under the CID conditions compared with samples tested under the CIU conditions.

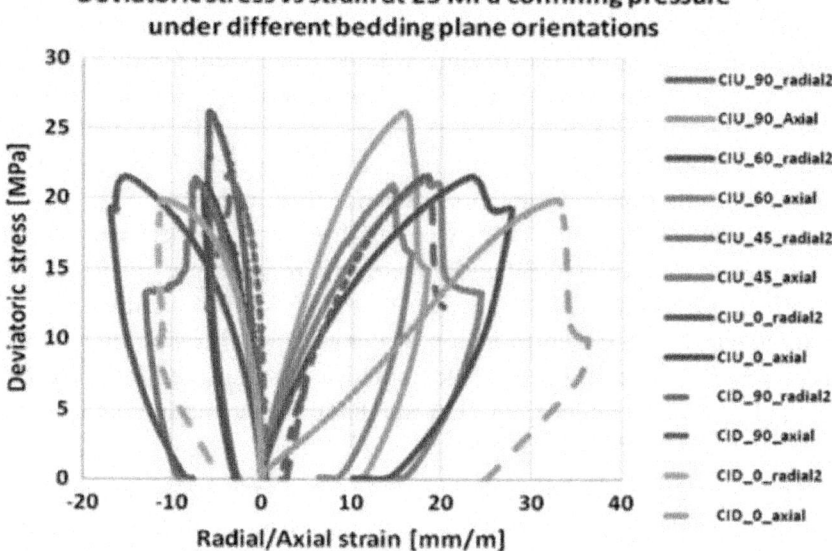

Figure. 16: Stress–strain relationship under different bedding planes.

SHALE ANISOTROPY

For an ideal undrained test condition, the effective stress path should be vertical because there is no change in the mean effective stress (zero change in the total volume). The total stress path will then be inclined, and the horizontal change in the mean total stress will be equal to the PP change. However, for the drained case, the PP was constant and the total stress path inclined 3–1 in the p'–q diagram (Fig. 17b). This resulted in an inclination of 3–1 for the effective stress path. The shear strength for samples at $\theta = 90°$ was greater than that for samples at $\theta = 0°$. For the CIU tests, the total stress path was different from the effective stress path due to the building of the PP. Because the PP was constant for the CID tests, both the total and effective stress paths were identical.

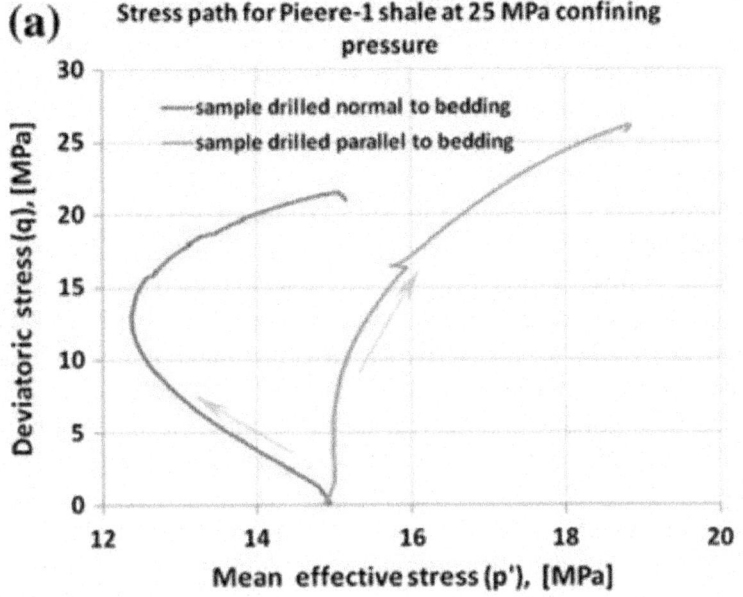

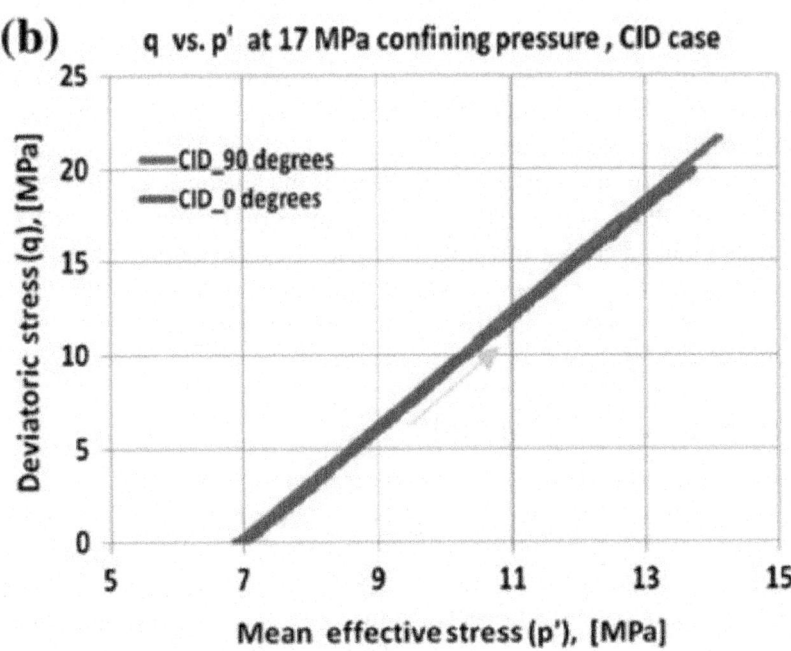

Figure. 17: Stress path and material anisotropy effects under **a** CIU response and **b** CID response.

Due to the inherent anisotropic nature of clay platelets (both texturally and mechanically), one would intuitively expect shale to be anisotropic. Figure 17a shows the stress paths for two samples at the same effective confining pressure. One of the samples was drilled with the sample axis along the bedding and the other was drilled with the sample axis normal to the bedding. The sample that was drilled parallel to the bedding was both stiffer and stronger than that drilled normal to the bedding (Fig. 17a). With increasing deviatoric stress, the mean effective stress decreased more for the sample at $\theta = 0°$ due to material contraction. At this position, the pore volumes decreased and led to an increase in the PP with increasing shearing. However, when shearing passed a certain level, i.e., at 15 MPa for this particular case, the material then started to dilate or create cracks or microfractures. These attributes accelerated and caused increasing pore volume and decreasing PP. Samples at 0° bedding had both contraction and dilation tendencies, whereas for the 90° bedding, only dilation was observed (Islam 2011). The material drilled parallel to the bedding ($\theta = 90°$) was much stronger, more brittle, and also tilted more to the right side compared to when drilled normal to the bedding ($\theta = 0°$). Therefore, wells drilled at high deviation, i.e., closer to horizontal, and also at larger depths will, thus, be more susceptible to potential stability problems due to the anisotropy.

There are several factors that may affect the degree of anisotropy. One would expect that the larger the clay content, the larger the degree of anisotropy, but this also depends on the mineral type. Both porosity and depth are important. Anisotropy effects were not, however, the main subject of this study. The need for a complete description of the anisotropy of the mechanical properties of shale will depend on the application. Generally, the acoustic properties of shale are significantly influenced by the heterogeneous nature of shale. Many authors have already implicitly analyzed this issue (Horsrud et al. 1994, 1998; Holt et al. 1996; Lockner and Stanchits 2002; Fjær et al. 2008; Sarout et al. 2007).

PORE PRESSURE RESPONSE

The non-elastic behavior of tested shale through the unloading–reloading cycle is presented in Sect. 5. The stress path behavior is observed to be different between the cyclic and static tests, and the elastic moduli vary significantly. However, the non-elastic behavior of the tested shale under the undrained situation is governed by the PP development. The PP response during the testing of shales may indicate whether the sample is fully saturated. To quantify the PP response (ΔP_f), the Skempton parameters A and B can be defined, as suggested by Skempton (1954):

$$\Delta P_f = B[\Delta\sigma_3 + A(\Delta\sigma_1 - \Delta\sigma_3)] \qquad (1)$$

For triaxial loading conditions, the mean effective stress can be expressed as

$$\Delta\sigma'_m = \frac{1}{3}\frac{\Delta P_f}{AB} - \Delta P_f \tag{2}$$

with the change in the mean effective stress during triaxial loading yielding σ'm=ΔP'.σm'=ΔP'.

For the poroelastic case where $k_f \ll k_s$, AB can be expressed by elastic constants (Fjær et al. 2008), as suggested by

$$AB = \frac{1}{3}\left[\frac{1}{1 + \frac{\varphi k_{fr}\left(1 - \frac{k_f}{k_s}\right)}{\alpha k_f}}\right] \tag{3}$$

where φ is the fractional porosity, α is the Biot coefficient (= $1 - k_{fr}/k_s$), k_{fr} is the bulk modulus of the framework, and k_f is the fluid modulus.

The analytical expression (Eq. 3) can be used to explain the non-elastic response of the tested shale because this equation can now be related to the stress paths of the p'–q plots (Figs. 17a, 18). With $\varphi k_{fr} \ll \alpha k_p$, $AB = 1/3$ from Eq. 2 and $\Delta P' = 0$; thus, the curve in the undrained triaxial test is vertical. This result is commonly referred to as the 'weak frame limit', which is the common assumption for any soil. According to Skempton (1954), for a soil in the elastic case, it can be shown that $A = 1/3$ and $B = 1$. In this study, no vertical stress path was found, the weak frame assumption did not hold, and $AB \neq 1/3$.

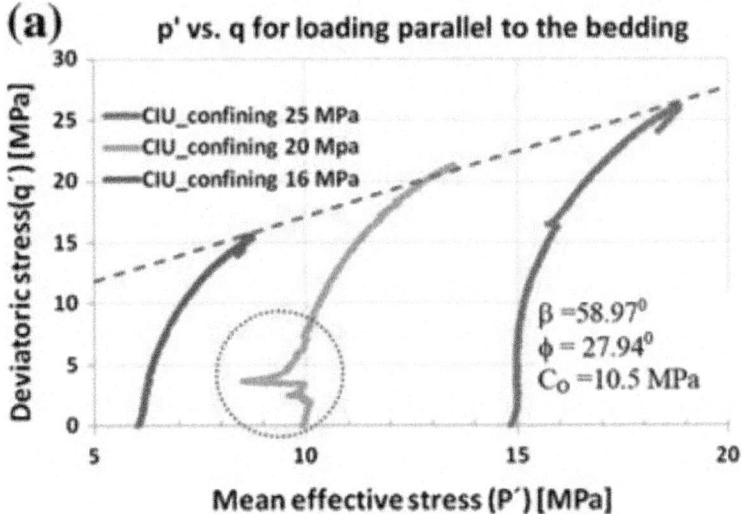

(a)   **p' vs. q for loading parallel to the bedding**

Deviatoric stress(q´) [MPa]

30

25 ━CIU_confining 25 MPa

━CIU_confining 20 Mpa

20 ━CIU_confining 16 MPa

15

10

5

0

β =58.97°
ɸ = 27.94°
C₀ =10.5 MPa

5   10   15   20

Mean effective stress (P´) [MPa]

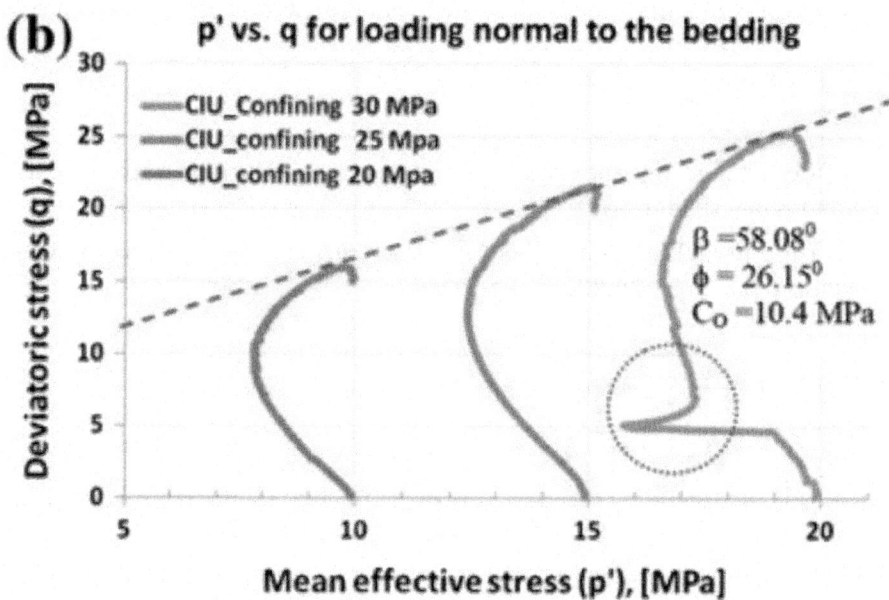

(b) p' vs. q for loading normal to the bedding

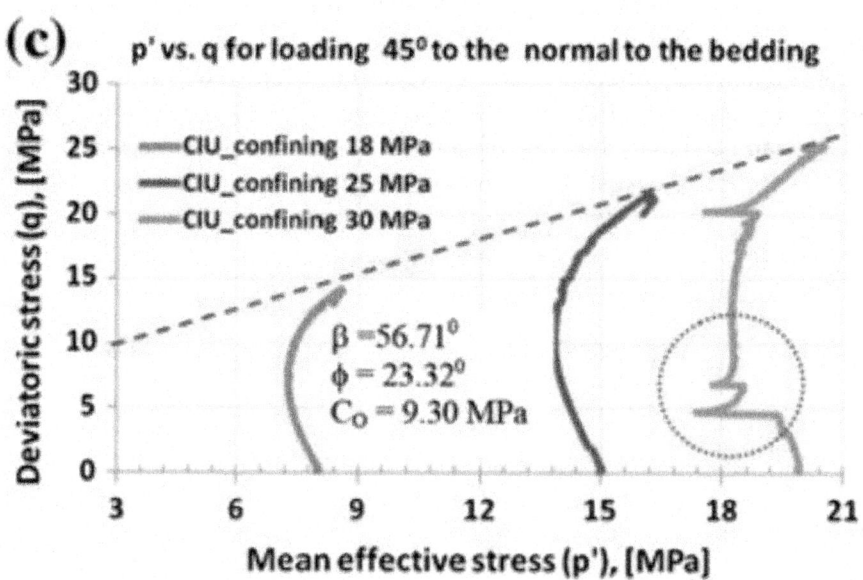

(c) p' vs. q for loading 45° to the normal to the bedding

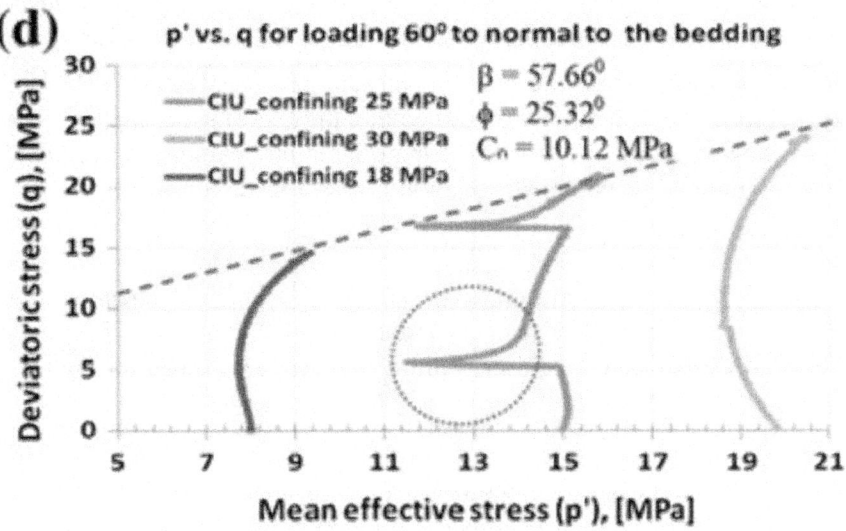

Figure. 18: CIU test of Pierre-1 shale, showing the Mohr–Coulomb failure lines from laboratory tests and estimating Mohr–Coulomb failure parameters; **a** loading parallel to bedding ($\theta = 90°$), **b** loading perpendicular to bedding ($\theta = 0°$), **c** loading 45° to the normal to bedding ($\theta = 45°$), and **d** loading 60° to the normal to bedding ($\theta = 60°$). The anomalies marked by the *dotted circles* were caused by the test performances.

If $AB < 1/3$, from Eq. 2, $\Delta P' = (-m)\Delta P_f$, where m is a multiplier factor and is less than 1. As a result, the stress path curves will tilt to the right, as shown in Fig. 18a (only loading parallel to bedding). In most cases, the stress path tilts slightly (i.e., at bedding angles of 45° and 60°) or more inclined (i.e., 0°) to the left (see Fig. 18). This behavior can be justified if $\Delta P' < 0$.

Examining Eq. 3, the only solution to this equation is if $k_f > k_s$, which is not realistic. This type of behavior, thus, indicates that the rock is no longer elastic. This is, for instance, the case for a normally consolidated material such as shale. Several authors also presented the PP response in sandstone-based rock strength in different ways (Skempton 1954; Horsrud et al. 1994, 1998; Lockner and Stanchits 2002; Fjær et al. 2008). However, in the end, the same conclusion was noted.

INTERPRETATION OF THE MOHR–COULOMB FAILURE PARAMETERS FROM THE TRIAXIAL TESTS

The Mohr–Coulomb failure criterion satisfies linear elastic condition. It is simpler than other models but, at the same time, the most widely used criterion in the oil industry. The material failure parameter, i.e., cohesion and uniaxial compressive strength (UCS), can be interpreted from the triaxial tests with existing mathematical expressions. For example, the τ–p' space interprets cohesion directly, while the $\beta = \frac{\pi}{4} + \frac{\varphi}{2}$ space provides the UCS. Sometimes, the q–p' space is used, but it provides neither the cohesion nor the UCS directly. To use the q–p' space to evaluate the material failure parameters for the Mohr–Coulomb model, the parameter relation is rearranged:

$$q = 2S_0 + 2 \tan \phi \sigma'_1 \qquad (4)$$

Here, q is the deviatoric stress, S_0 is the inherent shear strength or cohesion, ϕ is the internal friction angle, and σ' is the mean effective stress. The parameter UCS (C_0) and the failure angle (β) are related through

$$C_0 = 2S_0 \tan\beta \qquad (5)$$

These equations determine the Mohr–Coulomb model input parameters (ϕ, β, C_0). The peak stress values in Fig. 18 falls essentially on a straight line. In soil mechanical terms, this is a projection of the Hvorslev surface onto the p'–q plane (for a specific given volume). A uniform clay sample obeying the critical state theory would follow the Hvorslev surface up to the critical state line. Over consolidated and cemented rocks will eventually behave in a non-uniform manner. This was clearly the case for the samples shown in Fig. 18. When approaching the peak stress value, localization took place, and shear bands developed, which eventually formed a macroscopic shear plane through the sample. The behavior after the peak was, thus, in this case, more dependent on the characteristics typical of a rock (e.g., cementation) rather than a soil. In addition, the dotted circles on the stress paths indicate rapid PP development, which are due to strain rate effects or to existing cracking. The straight lines in Fig. 18 could be translated into a Mohr–Coulomb failure criterion, providing an extrapolated UCS, failure angle, and friction angle. A set of Mohr–Coulomb failure model data is shown in Fig. 18. The friction angle appears to be high for the CIU tests for the relatively soft shale. Similar analysis was performed for different bedding plane orientations and at different confining pressures. The observed results are presented in Fig. 18. Depending on the bedding plane, Fig. 18 implies that the UCS varied between 9.3 and 10.5 MPa, and the failure angle varied between 23.4° and 27.9°. In the case of North Sea shale, studied by Horsrud et al. (1994), the UCS varied between 6 and 77.5 MPa, and the failure angle (β) ranged between 48° and 60°. The Young's modulus (E) correlated with a UCS of $6.55E$ ($R^2 = 0.99$). Aadnøy et al. (2009) reported that E for green river shale in Canada varied between 60 and 160 GPa.

It is difficult to explain the exact reasons behind the high frictional angles obtained in the CIU tests. In the CIU tests, we worked with Pierre-1 shale, which has high heterogeneity. We believe that we obtained a higher frictional angle due to the frictional behavior. However, partial saturation could also lead to a high frictional angle (Sønstebø and Horsrud 1996; Schmitt et al. 1994). We cannot guarantee that the test samples for this outcrop achieved 100 % saturation. We carefully attempted to obtain good saturation with brine as a contacting pore fluid, as reported in Sect. 2.1. The sample should also improve its saturation during the mechanical loading. Some of the apparent plastic behavior may be related to the low permeability and PP development during loading.

The Mohr–Coulomb failure parameters and the necessary correlations developed through this study are presented in Fig. 19. Both the CIU and CID test results are presented. The 90° bedding samples seem to have the strongest correlation, whereas the 45° bedding samples have the weakest correlation.

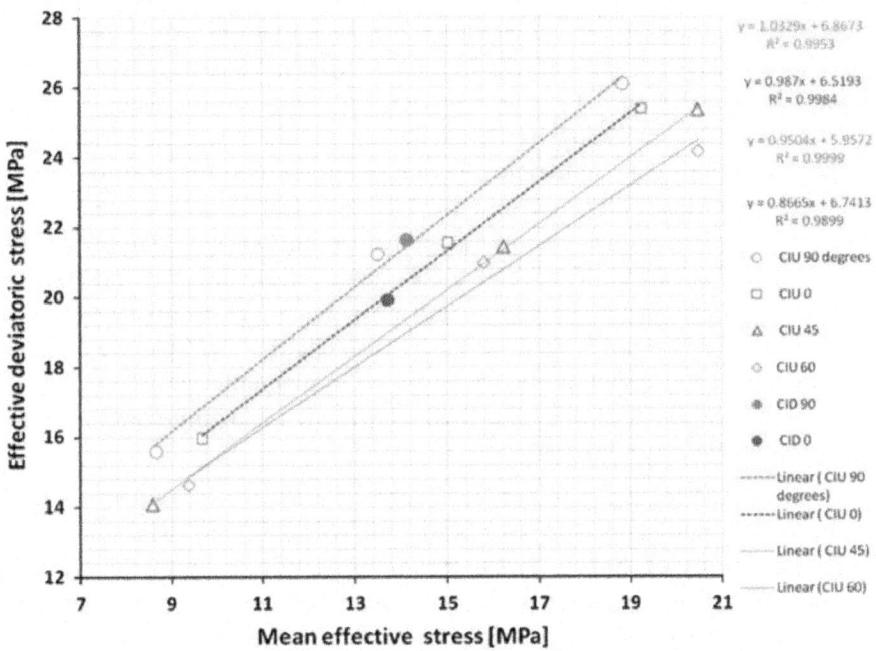

Figure. 19: Mohr–Coulomb failure lines connected for different bedding planes. The different colors are indications of the respective failure trend lines of the different bedding plane samples.

POTENTIAL APPLICATIONS

The rock strength behavior of the shale tested under drained and undrained conditions is considered to be a challenging and costly task. The most obvious application of this study is to supply data sets for numerical borehole stability modeling in shale. The mechanical properties of shale are demanding parameters not only for drilling engineering purposes but also in the geomechanical field. For example, the Poisson's ratio is often used within geophysics and sanding prediction. Experimental results and theoretical considerations have shown that the Poisson's ratio is not a single-valued, well-defined parameter for a given rock. The same observations were observed for measurements of the Young's modulus. Young's moduli for drained/undrained conditions varied largely with the stress level and with the amplitude and duration of the applied stress changes. Because the Poisson's ratio represents the relationship between the P and S waves in the identification of lithology from the seismic data and the AVO analysis, finding a suitable value of this parameter is vital, particularly at interbedded formations.

In recent years, the shale-oil and shale-gas potential represents new unconventional assets where shale-related information could be valuable for other researchers working in this field. The same aspect is true for the underground storage of CO_2. Our analysis will be useful for such issues. As a result, rock failure analysis was included in this work. The contribution relating to caprock failure data and shale characterization could be used for updating existing borehole stability analysis models. Obviously, improved shale property information under drained and undrained test conditions would improve model efficiency and accuracy. Very few studies have considered such extensive test matrices in shale. Moreover, this study provides some correlations between the material friction angle and the UCS. Because the UCS can be calculated from the sonic logs, such correlation can be implemented more readily. This correlation may be useful in the field of petroleum engineering with more testing under field conditions.

A dedicated testing program for Pierre-1 shale has provided a valuable database for shale properties, especially for weaker shale, which may cause borehole-stability problems. The correlations developed in this study can be used as an engineering tool to provide more reliable and more continuous estimates of the mechanical properties of shale, keeping in mind that the validity of the correlations should be verified when used in other geological and geophysical areas. Other sources of uncertainty also exist (e.g., core damage effects, temperature effects, etc.) and should be a focus in further work.

CONCLUSIONS

This study supplied data sets to implement an anisotropic material model that was used to simulate shale rocks. An isotropic model cannot simulate the real material behavior and, thus, should not be considered adequate. The anisotropic behavior is primarily due to the bedding planes that are formed, and because a borehole can be made at any angle to those planes, one should be able to simulate the real anisotropic behavior at any angle to the bedding plane. Thus, care should be taken when simulating such rock masses. The elastic moduli of such masses should never be considered as constant in all directions; instead, as the tests showed, we must expect them to vary considerably. This study also showed the dependency of the material on the confining stress. Obviously, the correct confining stress must be derived and used to simulate the real behavior as closely as possible.

Transverse isotropic shale has generally been studied experimentally using triaxial tests subjected to globally axisymmetric loading states, although the true stress states will not be axisymmetric due to the bedding plane inclination, inhomogeneities in the specimens, and end-effects. This study showed that the planes of weakness in bedded rocks could lead to severe borehole collapse problems.

The key findings of this study were elaborately presented in every section. However, the most crucial observations are listed here:

- Cyclic triaxial tests provided approximately 6 % higher rock strength, 50 % higher stiffness, and 10 % higher PP development than the only monotonic triaxial loading test. PP controls the rock stiffness under the CIU tests.

- Poisson's ratio effects in the samples drilled parallel to bedding are more vital than at any other sample orientation. A large variation in Poisson's ratios was found for drained and undrained samples drilled at the same bedding angle in transverse isotropic material. For the undrained fluid flow condition, the Poisson's ratios were observed in the tested shale rocks to be larger than 0.5 and varied between 0.3 and 0.75. However, for the drained case, the maximum limit of Poisson's ratios was 0.2. Therefore, in the case of shale, merely assuming a constant Poisson's ratio is a risky approach.

- The elastic moduli are non-linear functions of the confining pressure but are also dependent on the effective stress. E_p (loading parallel to bedding) was higher than E_t (loading perpendicular to bedding). The estimated values of E_t and E_p were 0.65 and 1.55 GPa, respectively. The estimated mean relative difference of the E for these two samples (in absolute value) was approximately 58 %. A significant difference between these two moduli indicates a strong heterogeneous nature.

- The elastic moduli for drained and undrained test conditions varied largely. The drained Young's modulus was approximately 48 % of the undrained value. The drained Poisson's ratios were, on average,

40 % or lower than the undrained value. These mechanical properties were significantly impacted by the bedding plane orientation and the confinement pressure.

- The bedding plane, anisotropy, and material heterogeneity are three prime factors to achieve accurate estimation of the material failure and development of the PP.

- The dilation behavior is significant under the drained test conditions. At low confinement pressure, the failure is brittle, with a sudden loss of strength and a transition from compressive to dilatants volumetric strain during failure and postfailure.

- The time dependency of parameters is significant due to the low permeability of shale.

ACKNOWLEDGMENTS

The authors thank the Department of Petroleum Engineering and Applied Geophysics, Norwegian University of Science and Technology, for their support and providing the permission to write this paper. We would like to express our appreciation to Ole Kristian Søreide and Per Horsrud of STATOIL, and Jørn Stenebråten, Erling Fjær, and Olav-Magner Nes of SINTEF Petroleum Research for their contribution to the discussions of critical issues in this work. We especially thank Konstantinos Kalomoiris. His help greatly improved our work. A thorough peer review and constructive comments by the reviewers are also appreciated and acknowledged. It further helped us to improve the quality of the paper. We thank STATOIL for providing funds for the experimental investigation of borehole stability. In addition, we appreciate and acknowledge the extensive laboratory work that was performed and partially funded by SINTEF Petroleum Research.

REFERENCES

1. Aadnøy BS, Chenevert ME (1987) Stability of highly inclined boreholes. In: Proceedings of the IADC/SPE drilling conference, New Orleans, Louisiana, 15–18 March 1987, SPE 16052

2. Aadnøy BS, Hareland G, Kustamsi A (2009) Borehole failure related to bedding plane. In: Proceedings of the ARMA conference, Asheville, North Carolina, 28 June–1 July 2009

3. Bradley WB (1979) Failure of inclined boreholes. J Energy Resour Technol 101:232–239

4. Claesson J, Bohloli B (2002) Brazilian test: stress field and tensile strength of anisotropic rocks using an analytical solution. Int J Rock

Mech Min Sci 39:991–1004

5. Closmann PJ, Bradley WB (1979) The effect of temperature on tensile and compressive strengths and Young's modulus of oil shale. In: Proceedings of the SPE-AIME 52nd annual fall technical conference and exhibition, Denver, Colorado, 9–12 October 1977, SPE 6734

6. Crook AJL, Yu J-G, Willson SM (2002) Development of an orthotropic 3D elastoplastic material model for shale. In: Presented at the SPE/ISRM rock mechanics conference, San Antonio, Texas, 20–23 October 2002

7. Fjær E, Holt RM, Horsrud P, Raaen AM, Risnes R (2008) Petroleum related rock mechanics, 2nd edn. Elsevier, Amsterdam, p 267

8. Fjær E, Holt RM, Nes O-M (2011) The transition from elastic to non-elastic behavior. In: Presented at the 45th US rock mechanics/ geomechanics symposium, San Francisco, CA, 26–29 June 2011

9. Head KH (1984) Soil laboratory testing, vol I–III. ELE Int., London

10. Hoang SK, Abousleiman YN (2010) Openhole stability and solids production simulation of emerging gas shales using anisotropic thick wall cylinders. In: Proceedings of the IADC/SPE Asia Pacific drilling technology conference and exhibition, Ho Chi Minh City, Vietnam, 1–3 November 2010, IADC/SPE 135865

11. Horsrud P (1998) Estimating mechanical properties of shale from empirical correlations. J SPE 16(2):68–73

12. Horsrud P, Holt RM, Sonstebo EF, Svano G, Bostrom B (1994) Time dependent borehole stability: laboratory studies and numerical simulation of different mechanisms in shale. In: Presented at the Eurock SPE/ ISRM rock mechanics in petroleum engineering conference, Delft, The Netherlands, 29–31 August 1994

13. Horsrud P, Sønstebø EF, Bøe R (1998) Mechanical and petrophysical properties of North Sea shales. Int J Rock Mech Min Sci 35(8):1009–1020

14. Holt RM, Sønstebø EF, Horsrud P (1996) Acoustic velocities of North Sea shales. In: Proceedings of the 58th EAGE conference and technical exhibition, Amsterdam, The Netherlands, 3–7 June 1996

15. Holt RM, Fjær E, Nes O-M, Alassi HT (2011) A shaly look at brittleness. In: Presented at the 45th US rock mechanics/geomechanics symposium, San Francisco, California, 26–29 June 2011

16. Hofmann CE, Johnson RK (2006) Estimation of formation pressures from log-derived shale properties. J Petrol Technol 17:717–722

17. Islam MdA (2011) Modeling and prediction of borehole collapse pressure

during underbalanced drilling in shale. Ph.D. thesis at NTNU, Norway. ISBN 978-82-471-2599-1

18. Islam MdA, Skalle P, Faruk ABM, Pierre B (2009) Analytical and numerical study of consolidation effect on time delayed borehole stability during underbalanced drilling in shale. In: Proceedings of the KIPCE 09/SPE drilling conference, Kuwait City, Kuwait, 14–16 December 2009, SPE 127554

19. Jaeger JC, Cook NGW (1979) Fundamentals of rock mechanics, 3rd edn. Chapman and Hall, New York

20. Johnston DH (1987) Physical properties of shale at temperature and pressure. Geophysics 52:1391–1401

21. Katsube TJ, Mudford BS, Best ME (1991) Petrophysical characteristics of shales from the Scotian shelf. Geophysics 56:1681–1689

22. Lockner DA, Stanchits SA (2002) Undrained poroelastic response of sandstones to deviatoric stress change. J Geophys Res 107(B12):2353. doi:10.1029/2001JB001460

23. Loret B, Rizzi E, Zerfa Z (2001) Relations between drained and undrained moduli in anisotropic poroelasticity. J Mech Phys Solids 40:2593–2619

24. Maury V, Santarelli FJ (1988) Borehole stability: a new challenge for an old problem. In: Cundall PA, Sterling RL, Starfield AM (eds) Key questions in rock mechanics. Balkema, Rotterdam, pp 453–460. ISBN 90 6191 8359

25. Nakken SJ, Christensen TL, Marsden R, Holt RM (1989) Mechanical behaviour of clays at high stress levels for well bore stability applications. In: Maury V, Fourmaintraux D (eds) Rock at great depth. Balkema, Rotterdam, pp 141–148

26. Nes O-M, Sønstebø EF, Holt RM (1998) Dynamic and static measurements on mm-size shale samples. In: Proceedings of Eurock 98, Trondheim, Norway, Trondheim, 8–10 July 1998, pp 23–32, vol 2, SPE/ISRM 47200

27. Niandou H, Shao JF, Henry JP, Fourmaintraux D (1997) Laboratory investigation of the mechanical behavior of tournemire shale. Int J Rock Mech Min Sci 34(1):3–16

28. Papamichos E, Brignoli M, Santarelli FJ (1997) An experimental and theoretical study of a partially saturated collapsible rock. Mech Cohes Frict Mater 2:251–278

29. Rozhko AY (2011) Capillary phenomena in partially-saturated rocks: theory of effective stress. In: Presented at the 45th US rock mechanics/geomechanics symposium, San Francisco, California, 26–29 June 2011

30. Sarout J, Molez L, Guéguen Y, Hoteit N (2007) Shale dynamic properties and anisotropy under triaxial loading: experimental and theoretical investigations. Phys Chem Earth 32:896–906

31. Schmitt L, Forsans T, Santarelli FJ (1994) Shale testing and capillary phenomena. Int J Rock Mech Min Sci Geomech Abstr 31:411–427

32. Skempton AW (1954) The pore-pressure coefficients A and B. Geotechnique 4:143–147

33. Sønstebø EF, Horsrud P (1996) Effects of brines on mechanical properties of shales under different test conditions. In: Proceedings of Eurock 96, Turin, Italy, 2–5 September 1996. Balkema, Rotterdam, pp 91–98

34. Søreide OK, Bostrøm B, Horsrud P (2009) Borehole stability simulations of an HPHT field using anisotropic shale modeling. In: Proceedings of the ARMA conference, Asheville, North Carolina, June 28–July 1 2008

35. Spaar JR, Ledgerwood LW, Hughes C, Goodman H, Graff RL, Moo TJ (1995) Formation compressive strength estimates for predicting drillability and PDC bit selection. In: Presented at the SPE/IADC drilling conference, Amsterdam, The Netherlands, 28 February–2 March 1995, SPE/IADC 29397

36. Steiger RP, Leung PK (1992) Quantitative determination of the mechanical properties of shales. SPE Drill Eng 7:181–185

37. Wood DM (1990) Soil behaviour and critical state soil mechanics. Cambridge University Press, Cambridge

Chapter 4

ROCK MAGNETIC PROPERTIES OF SEDIMENTARY ROCKS IN CENTRAL HOKKAIDO — INSIGHTS INTO SEDIMENTARY AND TECTONIC PROCESSES ON AN ACTIVE MARGIN

Yasuto Itoh[1], Machiko Tamaki[2], and Osamu Takano[3]

[1]Graduate School of Science, Osaka Prefecture University, Osaka, Japan

[2]Japan Oil Engineering Co. Ltd., Tokyo, Japan

[3]JAPEX Research Center, Japan Petroleum Exploration Co. Ltd., Chiba, Japan

INTRODUCTION

Reflecting a complicated subduction and collision history on the eastern Eurasian margin, central Hokkaido has been a site of various types of basin formation. Thick piles of the Cretaceous and Paleogene sediments (Figure 1; [1]) buried a regional forearc basin subducted by the Izanagi/Kula and Pacific Plates. Paleomagnetic studies of the Cretaceous Yezo Supergroup [2,3] showed that the present forearc is divided into some basins developed in different areas. Sedimentary system and forearc basin architecture in the Paleogene was studied in detail by Takano and Waseda [4] and Takano et al. [5].

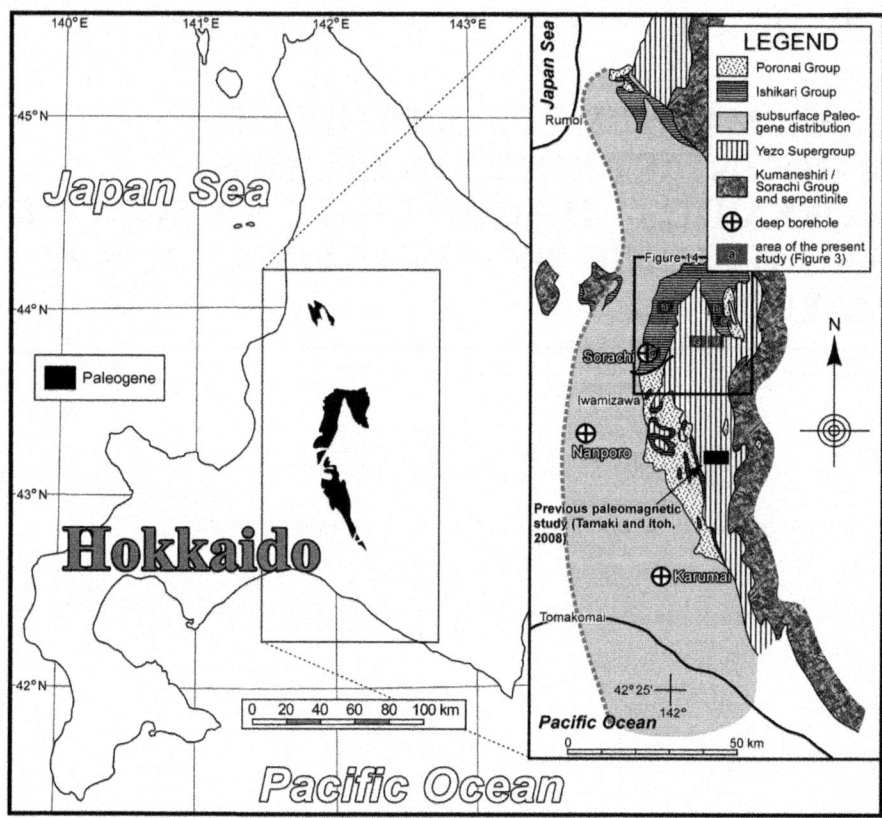

Figure 1. Index map of the study area of the Cretaceous and Paleogene strata. Geologic map is after Editorial Committee of Hokkaido, Regional Geology of Japan [1]

Under the influence of arc-arc collision on the Pacific convergent margin, vigorous mountain building and formation of foreland basins became active since the late Cenozoic. The Ishikari-Teshio belt (see Figure 2) is underlain with thick middle Miocene clastic strata. These are the Kawabata and its correlative formations, derived from the longitudinal mountainous ranges that were uplifted and eroded during that time [6]. It is generally regarded as a typical foreland setting, and the burial history of turbidites and associated coarse clastics of the Kawabata Formation has previously been studied from a sedimentological viewpoint (e.g., [7]). The process through which the Miocene basin developed in central Hokkaido is not only governed by compressive stress in the collision zone, but also by coeval tectonic events like back-arc spreading in the Japan Sea (e.g., [8]) and dextral transcurrent faulting along the Eurasian margin (e.g., [9]).

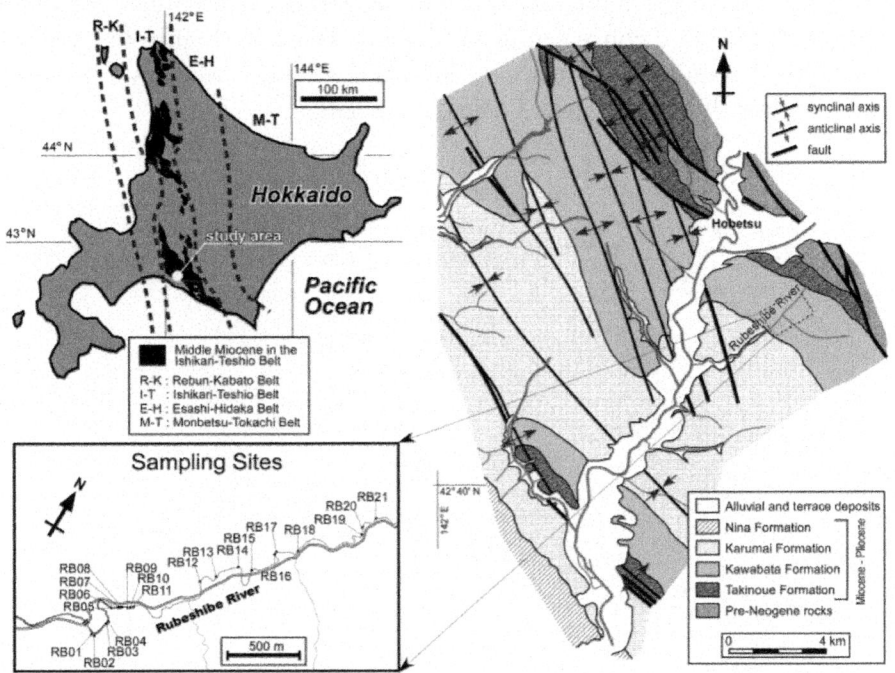

Figure 2. Cenozoic tectonic context of Hokkaido, geology of the study area of the Neogene strata (simplified from Kawakami et al. [7]), and locations of rock magnetic samples.

In this paper, we present preliminary results of rock magnetic analyses of the Cretaceous Yezo Supergroup, the Eocene Ishikari Group and the Miocene Kawabata Formation in order to detect tectonic movements around the basin and to describe the microfabric of sedimentary rocks related to the tectonic regime and sedimentation processes in the mobile zone. This study is an attempt to apply magnetic properties to tectono-sedimentology.

GEOLOGY

Background

The Yezo Supergroup deposited on the Cretaceous forearc and consists of monotonous mudstone intercalated by coarse clastics and ash layers. After a stagnant subsidence stage at the beginning of the Cenozoic, fluvial sediments of the Ishikari Group and its correlative units began to bury depressions on the forearc. As a result of strong deformation and continued sedimentation on the active margin, surface distribution of the Eocene Ishikari Group is rather restricted. However, numerous exploration drilling clarified that voluminous Paleogene units are concealed under the alluvial plain (Figure 1). Paleogene depositional sequence and facies classification were described by Takano et al. [5]. They are shown in Figure 3 using abbreviations.

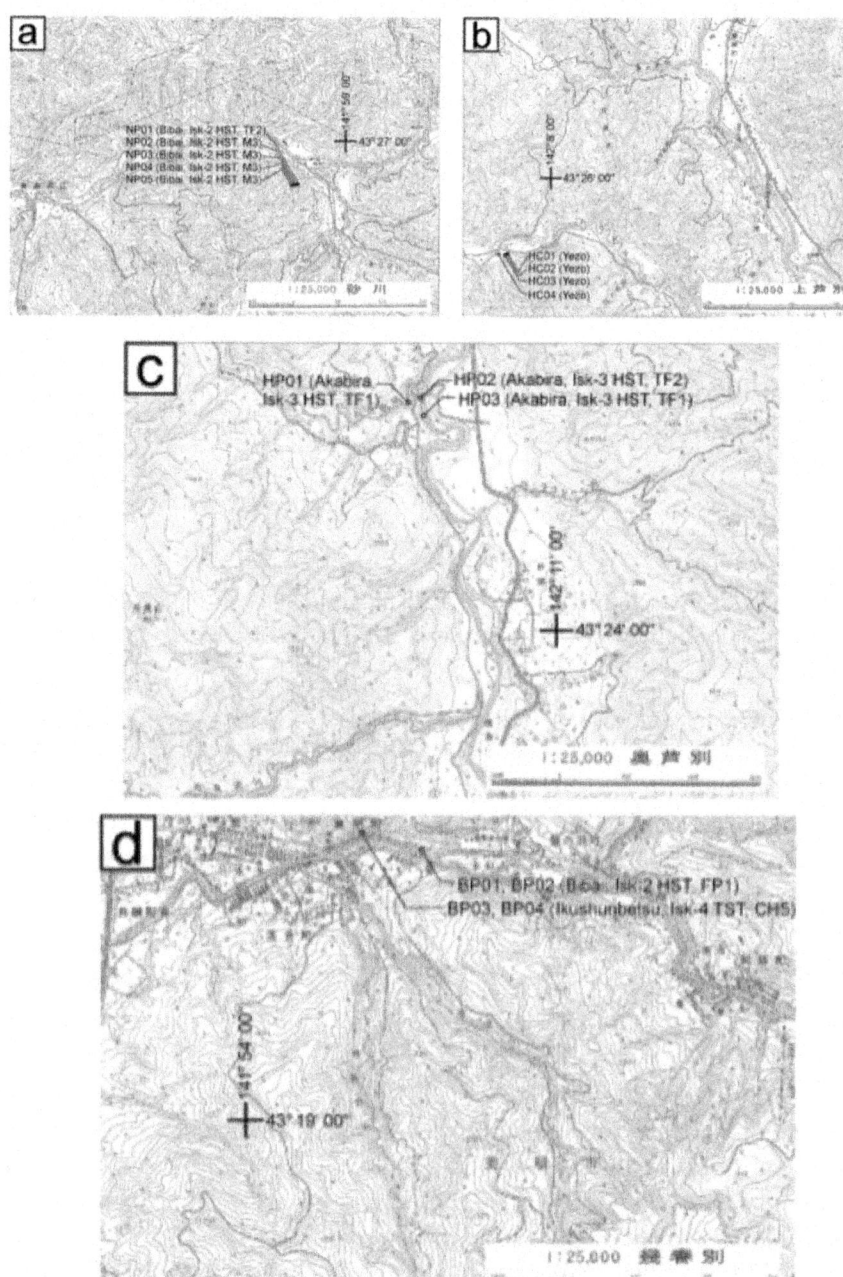

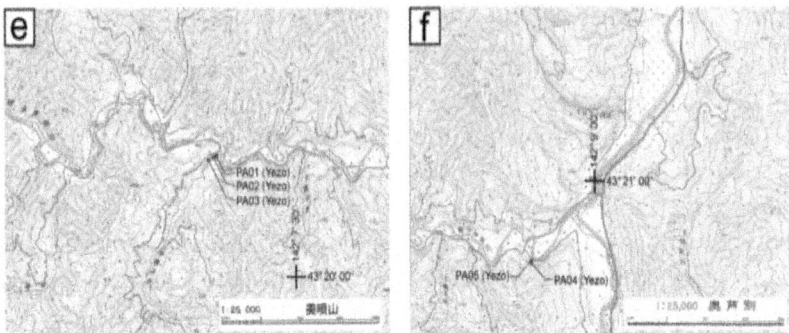

Figure 3. Sampling localities for rock magnetic analyses of the Cretaceous and Paleo-gene strata. The base maps are parts of the "Sunagawa", "Kamiashibetsu", "Okuashi-betsu", "Ikushunbetsu" and "Bibaiyama" 1:25,000 topographic maps published by the Geographical Survey Institute. As for the Paleogene sites (a, c and d), geologic units (Yezo, Yezo Supergroup; Bibai, Bibai Formation; Akabira, Akabira Formation; Ikush-unbetsu, Ikushunbetsu Formation), depositional sequence and facies classification are shown in parentheses after Takano et al. [5]

The study area of the Kawabata Formation is located in the southern part of the middle Miocene basins of the Ishikari-Teshio belt. Folded sedimentary units are distributed with a NNW-SSE trend, and are cut by numerous faults (Figure 2). The area is divided into the following formations in ascending order [7]: the Takinoue Formation, the Kawabata Formation, the Karumai Formation, and the Nina Formation (Figure 4). They represent the sequence by which an elongate N-S foreland basin was filled. The middle Miocene Kawabata Formation comprises mainly turbidites and associated coarse clastic rocks derived from the eastern hinterland [7].

Age		Unit	Lithology	Description
(Ma)	L. Mio.	Nina Formation	not exposed	Turbidites and related coarse clastics, showing fining-upward succession at the basal part
10	Middle Miocene	Karumai Formation		Upper Member : coarse sediment (gravity flow deposits) Lower Member : sandy siltstone
		Kawabata Formation		Mainly consists of turbidites, and intercalates coarse sediment (gravity flow deposits)
15		Takinoue Formation		Muddy sandstone, fining-upward to massive mudstone, intercalated with thin-bedded conglomerate

Conglomerate		Alternating beds of sandstone & mudstone	
Muddy sandstone		Mudstone	

Legend

Figure 4. Neogene stratigraphy of the study area of the Kawabata Formation

Sedimentary Facies of the Miocene Unit

This study conducted sedimentary facies analysis for the Kawabata Formation along the Rubeshibe River (Figure 2). The analysis revealed that the turbidites of the Kawabata Formation mainly consisted of sheet-flow turbidite facies association and channel-levee facies association (Figure 5). The sheet-flow turbidite facies association comprises aggradational stacking of rhythmic alternating beds of turbidite sandstone and mudstone with rare upward thickening or thinning successions, and is interpreted to be sheet-like turbidites with minor occurrences of depositional lobes, which occupied major part of the trough-like foreland basin fill [7,10]. The channel-levee facies association is composed of thick amalgamated sandstone facies with slump blocks and thinly bedded alternating beds of sandstone and mudstone. These two facies appearing coupled is indicative of an elongated channel-levee system made of the main channel with levees on both sides. These two facies associations are believed to have been deposited in an elongated trough-like foredeep in the foreland basin [7]. The turbidites of the Kawabata Formation commonly contain sedimentary structures indicating paleocurrent directions; e.g., sole marks (mostly flute marks) at the bottom of individual turbidite bed, and current ripples in Bouma Tc division [11].

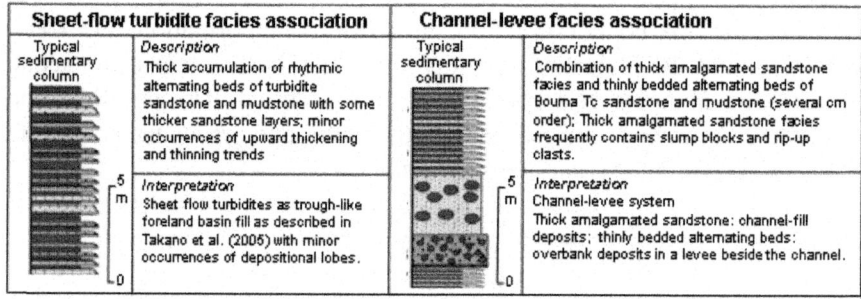

Figure 5. Facies association classification of the Kawabata Formation along the Rubeshibe River.

ROCK MAGNETISM

We obtained samples for rock magnetic analyses exclusively from fine-grained parts of the target sedimentary units, since fine sedimentary rocks generally preserve stable detrital remanent magnetization (DRM). Few visible markers of the sedimentation process accompany such sediments, so we attempted to measure their microscopic magnetic fabric, which may be related to paleocurrent directions (e.g., [12]).

Basic Measurements

The Cretaceous and Eocene samples were taken from outcrops along the streambed in central Hokkaido (Figure 3) using an engine or electric drill at 21 sites. Samples of the Kawabata Formation were collected with a battery-powered electric drill at 21 sites along the Rubeshibe River (Figure 2). The bedding attitudes were measured on outcrops to allow us to compensate for tectonic tilting later. Between seven and sixteen independently oriented cores 25 mm in diameter were obtained at each site using a magnetic compass. Cylindrical specimens 22 mm in length were cut from each core and the natural remanent magnetization (NRM) of each specimen was measured using a cryogenic magnetometer (model 760-R SRM, 2-G Enterprises). Low-field magnetic susceptibility was measured on a Bartington MS2 susceptibility meter, and the anisotropy of magnetic susceptibility (AMS) was measured using an AGICO KappaBridge KLY-3 S magnetic susceptibility meter. After the basic measurements, pilot specimens with average NRM intensities, directions and susceptibility levels were selected from each site for subsequent demagnetization tests.

Demagnetization Tests

In order to isolate stable components of the remanent magnetization, progressive alternating field demagnetization (PAFD) and progressive thermal demagnetization (PThD) tests were carried out on two pilot specimens per site that had average NRM directions. The PAFD test loading ranged from 0 to 80 mT using a three-axis tumbling system with specimens contained in a μ-metal envelope. The PThD test was performed using an electric furnace, with a residual magnetic field less than 10 nT, beginning at 100 °C and continuing until the specimen was either fully demagnetized and a characteristic remanent magnetization (ChRM) component was isolated, or until the thermal treatment provoked erratic behavior of the magnetic direction. Specimens' low-field bulk magnetic susceptibilities were measured using a susceptibility meter after each PThD step in order to monitor chemical changes in ferromagnetic minerals.

Figure 6 presents typical PThD and PAFD results for the Yezo Supergroup and Ishikari Group. It is obvious that the ChRM direction was not isolated because of unstable behavior in thermal treatment (Figure 6a), overlapping spectra of primary and secondary magnetization (Figure 6b) and partial remagnetization within a site (Figure 6c,d). Therefore further analyses for magnetic granulometry were not applied on the Cretaceous and Eocene samples. On the other hand, PThD treatment was effective for isolating stable ChRM in the sedimentary rocks of the Kawabata Formation. Figure 7 shows typical results of the progressive demagnetization tests.

a) NP04-31 (Ishikari Group, Bibai Formation) PThD *in situ*

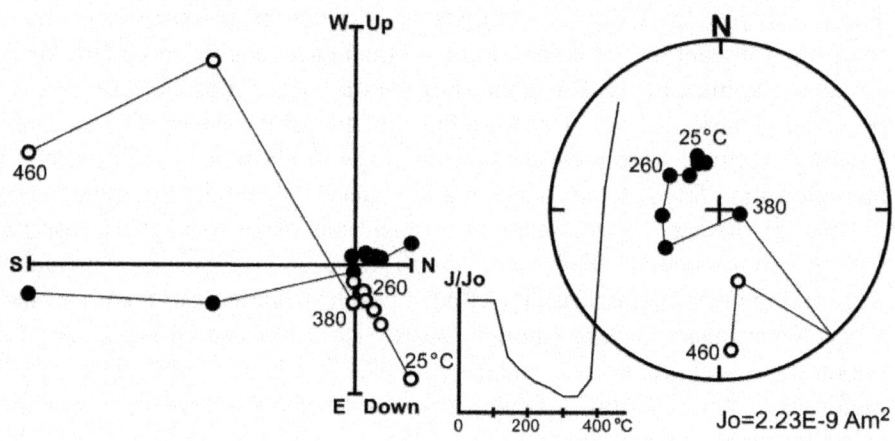

Jo=2.23E-9 Am^2

b) HP02-31 (Ishikari Group, Akabira Formation) PThD *in situ*

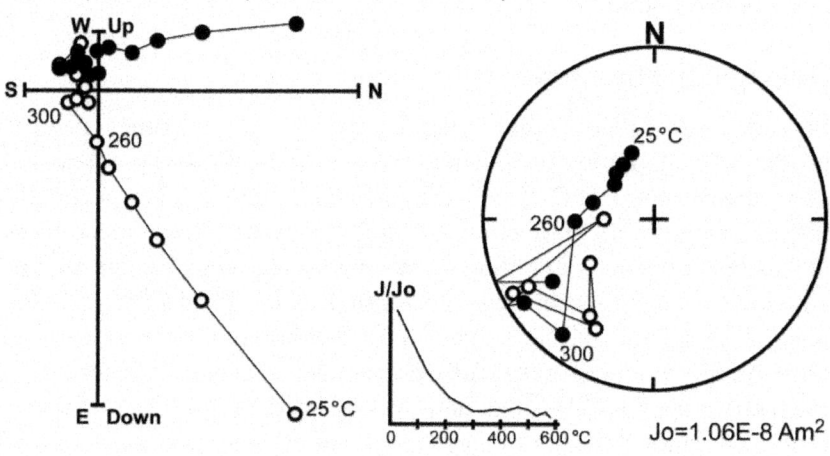

Jo=1.06E-8 Am^2

c) HC03-61 (Yezo Supergroup) PAFD *in situ*

d) HC03-21 (Yezo Supergroup) PAFD *in situ*

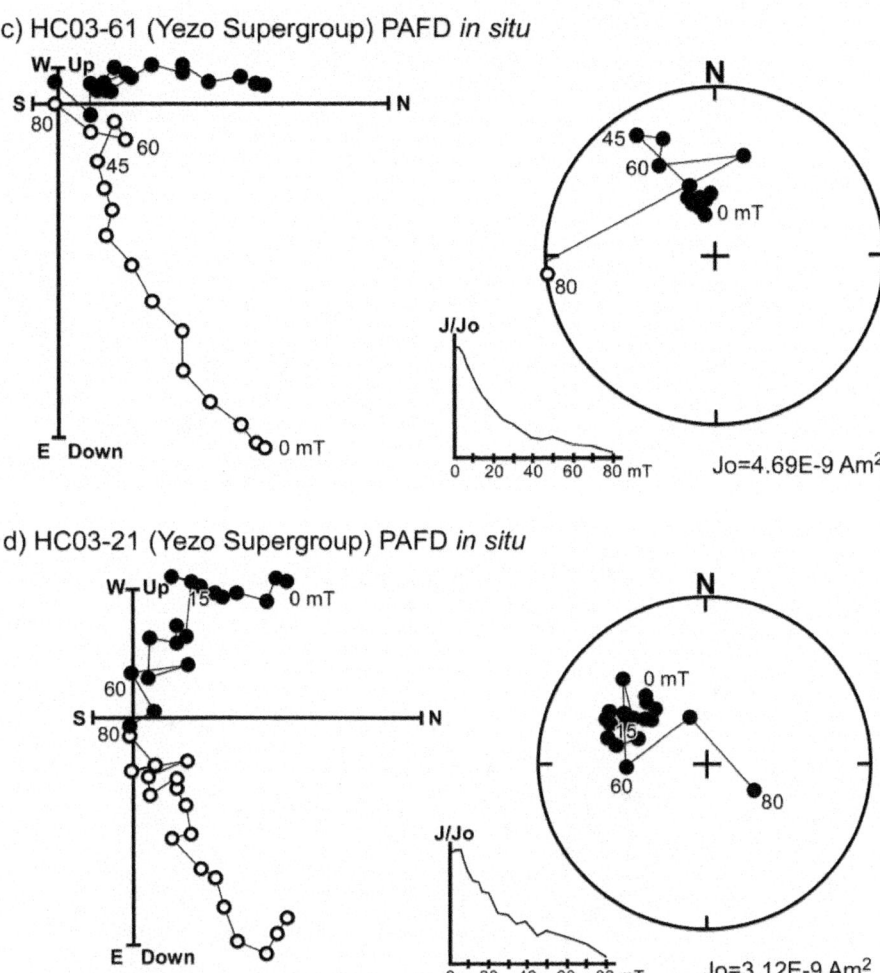

Figure 6. Typical results of progressive thermal demagnetization (PThD) and progressive alternating field demagnetization (PAFD) in geographic coordinates for the Paleogene Ishikari Group (a,b) and the Cretaceous Yezo Supergroup (c,d). On the vector-demagnetization diagrams, solid (open) circles are projection of vector end-points on horizontal (N-S vertical) plane. Equal-area projection and normalized intensity decay curve are shown on the right-side of each vector diagram. Solid (open) circles in equal-area nets are projections on the lower (upper) hemisphere. Numbers attached on data points are demagnetization levels in °C or mT

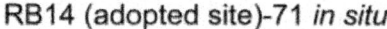

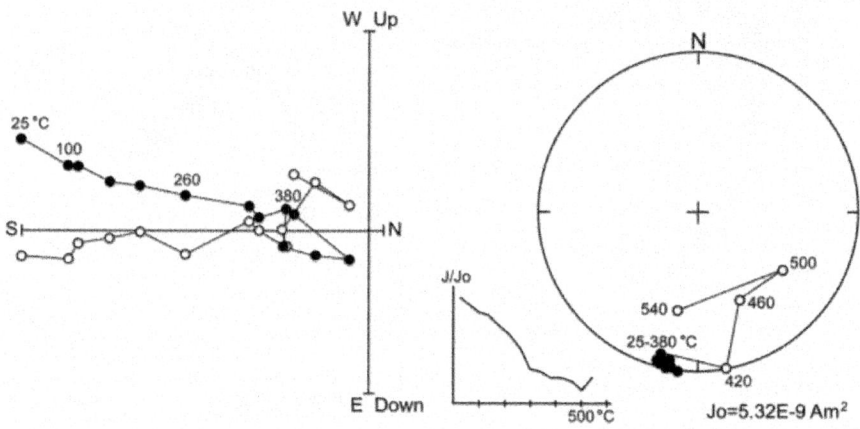

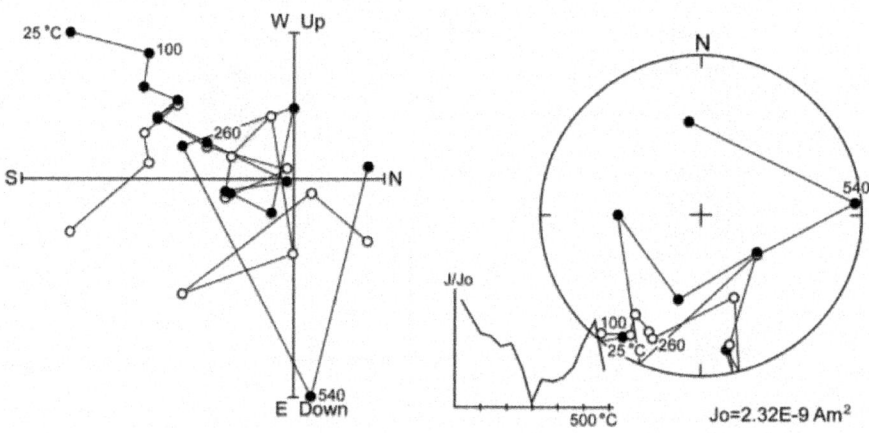

Figure 7. Results of progressive thermal demagnetization for samples of the Neogene Kawabata Formation with stable (upper) and unstable (lower) magnetization. All co-ordinates are geographic (*in situ*). Units are bulk remanent intensity. The solid and open circles in the vector-demagnetization diagrams (left) are projections of vector end-points on the horizontal and north-south vertical planes, respectively. The solid and open circles in the equal-area Schmidt nets (right) are projections on the lower and upper hemispheres, respectively.

Hysteresis Properties

Hysteresis parameters were determined for the Kawabata samples with an alternating gradient magnetometer (Princeton Measurements Corporation, MicroMag 2900). Ten sample chips up to 1 mm in size were randomly selected from site RB16, where stable ChRM has been successfully isolated.Figure 8 displays typical hysteresis of the Kawabata mudstones. The raw diagram seems to suggest the absence of ferromagnetic material. After correcting the linear gradient of paramagnetism, a weak ferromagnetic behavior signature can be recognized. Saturation magnetization (Js), saturation remanence (Jrs) and coercive force (Hc) values were determined for all samples from their hysteresis loops. Their relatively low Hc (~ 100 mT) implies that magnetite is the dominant remanence carrier. After acquiring coercivity of remanence (Hcr) values through backfield demagnetization experiments, we constructed a correlation plot of Jrs/Js versus Hcr/Hc [13] as shown in Figure 9. All the data are plotted in the pseudo-single domain (PSD) region of magnetite.

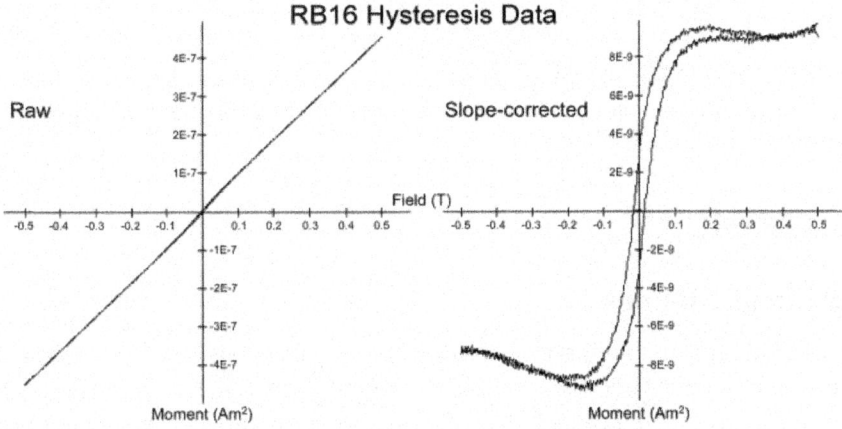

Figure 8. An example of hysteresis loop for a sample of the Kawabata Formation from site RB16 (Left: raw data, Right: data corrected for slope of paramagnetism).

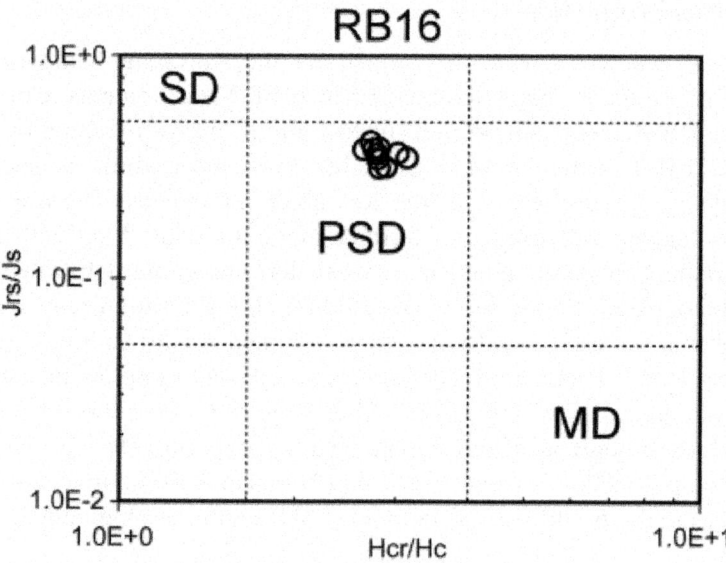

Figure 9. Logarithmic plot of hysteresis parameters [13] of ten samples of the Kawabata Formation from site RB16. Abbreviations: SD, single domain; PSD, pseudo-single domain; MD, multi-domain.

DISCUSSION

Rotational Motions

We found stable magnetic components at three sites of the Kawabata Formation. Their directions were determined with a three-dimensional least squares analysis technique [14]. Figure 10 and Table 1 present site-mean ChRM directions obtained from the Kawabata Formation. They exhibit antipodal directions, and precision parameter (κ) improves after tilt correction. Although the number of data points is minimal for tectonic discussion, we can interpret the site-mean directions as a record of the Earth's dipole magnetic field, acquired before the strata tilted. The declination of the formation mean exhibits a significant westerly deflection, which suggests counterclockwise rotation of the study area.

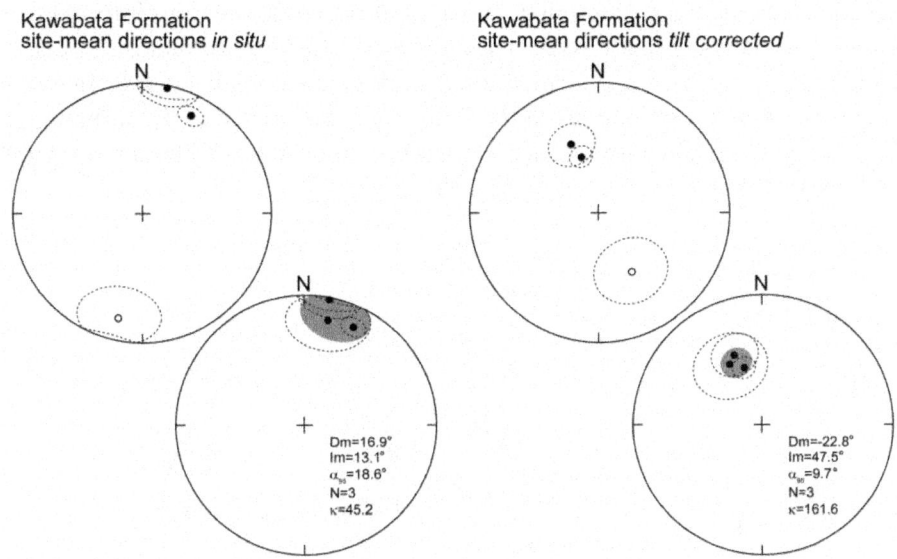

Kawabata Formation
site-mean directions *in situ*

Kawabata Formation
site-mean directions *tilt corrected*

Dm=16.9°
Im=13.1°
α₉₅=18.6°
N=3
κ=45.2

Dm=-22.8°
Im=47.5°
α₉₅=9.7°
N=3
κ=161.6

Figure 10. Site-mean ChRM directions of the Kawabata Formation in the study area. The solid and open circles in all the equal-area nets are projections on the lower and upper hemispheres, respectively. Dotted ovals show 95 % confidence limits. Lower diagrams are polarity-converted for calculating formation mean directions and Fisher's precision parameters as annotated in the diagrams (Shaded ovals depict 95 % confidence for the formation means).

Table 1. Paleomagnetic directions of the Kawabata Formation

Site	Latitude	Longitude	D	I	Dc	Ic	α95	κ	N	φ	λ
RB14	42.7361	142.1771	-167.1	-18.7	151.2	-47.0	21.9	13.1	5	62.5	29.5
RB16	42.7379	142.1793	11.4	2.8	-21.8	42.5	14.0	14.4	9	64.5	13.9
RB17	42.7381	142.1793	26.8	17.4	-17.4	52.7	6.8	66.5	8	73.3	23.1

[i] - D and I, *in situ* site-mean declination and inclination before tilt correction in degrees, respectively; Dc and Ic, site-mean declination and inclination after tilt correction in degrees, respectively; α95, radius of 95% confidence circle in degrees; κ, precision parameter; N, number of specimens; φ and λ, latitude (N) and longitude (E) of north-seeking virtual geomagnetic pole for untilted site-mean direction in degrees, respectively.

A previous study [15] suggested a clockwise tectonic rotation around central Hokkaido based on a paleomagnetic study of the Kawabata Formation. Takeuchi et al. [16] proposed a coherent rotational model with 'domino-style' rigid crustal blocks. However, Tamaki et al. [17] criticized the block rotation scheme as being overly simplistic based on differential rotations inferred from Oligocene paleomagnetic data. They restored crustal deformation in central Hokkaido using dislocation modeling, and found complicated vertical-axis rotations around terminations of the faults that contributed to the formation of N-S elongate sedimentary basins. Figure 11 demonstrates differential rotation in central Hokkaido since the middle Miocene.

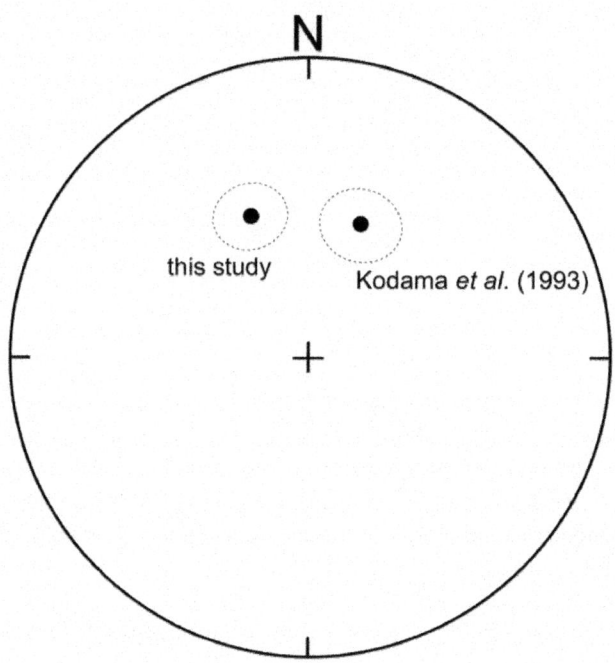

Figure 11. Comparison of the mean paleomagnetic directions of the Kawabata Formation in central Hokkaido between this study and [15]. Data are plotted on the lower hemisphere of the equal-area projection. Dotted ovals represent 95 % confidence limits.

Sedimentation Process Inferred From AMS Fabric

We found that the AMS fabric (orientation of principal axes) were precisely determined at all the sampled localities. Tables 2 and 3 show the AMS parameters for the Cretaceous/Eocene units and the Miocene unit, respectively. Figure 12 delineates typical AMS fabric obtained from the Ishikari (left) and Yezo (right) samples. After tilt-correction, the maximum (K_1) and intermediate (K_2) axes of AMS are bound to the horizontal plane with a subtle imbrication suggestive of hydrodynamic forcing.

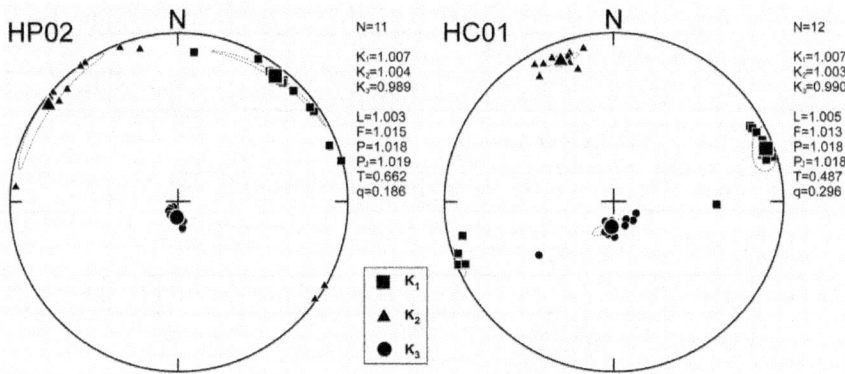

Figure 12. Anisotropy of magnetic susceptibility (AMS) fabric (principal susceptibility axes) for all specimens of typical sites of the Ishikari Group (HP02) and Yezo
Supergroup (HC01) plotted on the lower hemisphere of equal-area projections. Data
are shown in stratigraphic coordinates. Ovals surrounding mean directions of three
axes (shown by larger symbols) are 95% confidence regions. See Table 2 for all the
AMS parameters.

Table 2. Site-mean AMS parameters of the Paleogene and Cretaceous units in central
Hokkaido

Site	N	K_1Str.	K_2Str.	K_3Str.	L	F	P	P^J	T	q	Unit / Sequence
		(D, I)	(D, I)	(D, I)	(K_1/K_2)	(K_2/K_3)	(K_1/K_3)				
Paleogene											
BP01	11	1.010	1.005	0.984	1.005	1.021	1.026	1.028	0.619	0.213	Bb / Isk-2HST
		(207, 4)	(297, 1)	(35, 86)							
BP02	11	1.009	1.008	0.982	1.001	1.026	1.027	1.031	0.917	0.043	Bb / Isk-2HST
		(226, 3)	(136, 4)	(0, 85)							
BP03	11	1.010	1.005	0.985	1.005	1.020	1.025	1.027	0.593	0.229	Ik / Isk-4TST
		(233, 9)	(141, 7)	(16, 78)							
BP04	13	1.010	1.002	0.988	1.008	1.014	1.022	1.022	0.259	0.458	Ik / Isk-4TST
		(225, 6)	(134, 3)	(15, 83)							
HP01	18	1.010	1.004	0.987	1.006	1.017	1.023	1.024	0.479	0.303	Ak / Isk-3HST
		(271, 4)	(180, 8)	(29, 81)							
HP02	11	1.007	1.004	0.989	1.003	1.015	1.018	1.019	0.662	0.186	Ak / Isk-3HST
		(38, 6)	(307, 4)	(187, 83)							

NP01	15	1.009	1.006	0.985	1.003	1.022	1.025	1.027	0.762	0.128	Bb / Isk-2HST
		(264, 2)	(174, 7)	(11, 83)							
NP02	9	1.013	1.006	0.981	1.006	1.026	1.032	1.034	0.597	0.227	Bb / Isk-2HST
		(252, 10)	(162, 2)	(59, 80)							
NP03	10	1.006	1.004	0.989	1.002	1.016	1.018	1.019	0.779	0.118	Bb / Isk-2HST
		(253, 1)	(163, 7)	(349, 83)							
NP04	7	1.009	1.006	0.985	1.002	1.022	1.024	1.027	0.791	0.111	Bb / Isk-2HST
		(80, 10)	(170, 4)	(282, 79)							
NP05	9	1.007	1.003	0.990	1.004	1.013	1.017	1.018	0.536	0.264	Bb / Isk-2HST
		(39, 28)	(141, 21)	(261, 54)							
Cretaceous											
HC01	12	1.007	1.003	0.990	1.005	1.013	1.018	1.018	0.487	0.296	
		(71, 5)	(340, 11)	(187, 78)							
HC02	10	1.011	1.005	0.984	1.006	1.021	1.027	1.028	0.541	0.262	
		(66, 2)	(156, 4)	(315, 86)							
HC03	11	1.010	1.002	0.988	1.008	1.014	1.022	1.022	0.274	0.447	
		(51, 4)	(321, 11)	(163, 78)							
PA01	13	1.006	1.003	0.991	1.003	1.012	1.015	1.016	0.650	0.193	
		(336, 11)	(70, 19)	(217, 68)							
PA04	14	1.017	1.014	0.969	1.004	1.046	1.049	1.055	0.852	0.078	
		(8, 9)	(99, 11)	(240, 76)							

[i] - N denotes the number of specimens. Directions of AMS principal axes are in stratigraphic coordinates. Abbreviations for the Paleogene geologic units: Ak, Akabira Formation; Bb, Bibai Formation; Ik, Ikushunbetsu Formation. Depositional sequence is after Takano & Waseda [4] and Takano et al. [5].

Figure 13 delineates typical AMS fabrics of the Kawabata Formation. Site RB08 typifies an elongate (prolate) fabric reflecting aligned detrital grains. Site RB14 has highly oblate fabric, as shown by a positive T parameter near unity. This fabric is essentially confined to the bedding plane under gravitational force. As the hysteresis study showed a negligible amount of ferromagnetic material in the Kawabata samples (Figure 8), we consider the AMS fabric as being governed simply by the shape anisotropy of paramagnetic minerals, i.e. alignments of elongate or platy grains such as amphibole or mica.

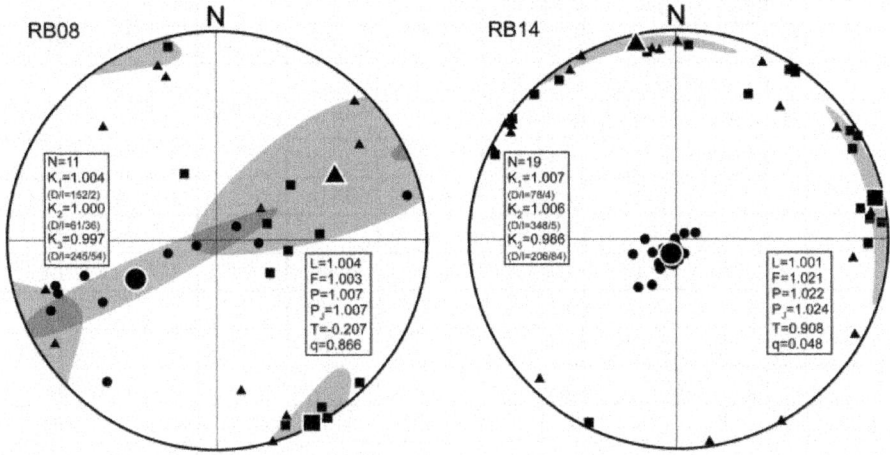

Figure 13. Typical tilt-corrected AMS fabric for the Kawabata Formation muddy samples. Prolate (left) and oblate (right) fabrics are numerically described by negative and positive T parameters, respectively, posted on the equal-area diagrams. All the data are plotted on the lower hemisphere. Square, triangular and circular symbols represent orthogonal maximum (K_1), intermediate (K_2), and minimum (K_3) AMS principal axes, respectively, and larger symbols show their mean directions. Shaded areas are 95 % confidence limits based upon Bingham statistics.

Table 3. Site-mean AMS parameters of the Kawabata Formation

Site	N	K^1	K^2	K^3	L (K_1/K_2)	F (K_2/K_3)	P (K_1/K_3)	P^J	T	q
		(D, I)	(D, I)	(D, I)						
RB01	11	1.013	1.006	0.981	1.007	1.025	1.033	1.034	0.546	0.260
		(16, 6)	(107, 7)	(242, 81)						
RB02	11	1.007	1.003	0.991	1.004	1.012	1.016	1.017	0.508	0.282
		(3, 20)	(98, 11)	(215, 66)						
RB03	12	1.006	1.002	0.992	1.004	1.010	1.014	1.015	0.475	0.304
		(326, 11)	(58, 11)	(190, 74)						
RB04	13	1.008	1.002	0.990	1.005	1.013	1.018	1.019	0.406	0.352
		(159, 2)	(249, 6)	(52, 83)						
RB05	15	1.005	1.001	0.993	1.004	1.008	1.012	1.012	0.330	0.404

		(4, 19)	(101, 19)	(232, 62)						
RB06	14	1.007	1.001	0.992	1.005	1.009	1.014	1.015	0.256	0.459
		(336, 5)	(66, 7)	(208, 82)						
RB07	12	1.005	0.999	0.996	1.007	1.002	1.009	1.009	-0.459	1.151
		(135, 24)	(41, 9)	(291, 64)						
RB08	11	1.004	1.000	0.997	1.004	1.003	1.007	1.007	-0.207	0.866
		(152, 2)	(61, 36)	(245, 54)						
RB09	17	1.005	1.001	0.994	1.004	1.007	1.011	1.011	0.313	0.416
		(121, 1)	(31, 7)	(218, 83)						
RB10	10	1.010	1.007	0.983	1.003	1.024	1.027	1.030	0.777	0.120
		(343, 2)	(252, 12)	(84, 78)						
RB11	12	1.002	0.999	0.998	1.003	1.001	1.004	1.004	-0.410	1.090
		(313, 42)	(105, 45)	(210, 15)						
RB12	12	1.005	1.004	0.991	1.002	1.013	1.015	1.016	0.763	0.127
		(3, 6)	(93, 1)	(192, 84)						
RB13	9	1.004	1.002	0.994	1.001	1.009	1.010	1.011	0.732	0.144
		(119, 26)	(214, 11)	(325, 61)						
RB14	19	1.007	1.006	0.986	1.001	1.021	1.022	1.024	0.908	0.048
		(78, 4)	(348, 5)	(206, 84)						
RB15	12	1.009	1.004	0.986	1.005	1.018	1.023	1.024	0.558	0.251
		(188, 10)	(96, 10)	(324, 76)						
RB16	15	1.013	1.010	0.977	1.003	1.034	1.036	1.041	0.848	0.080
		(281, 4)	(11, 5)	(152, 83)						
RB17	14	1.009	1.003	0.987	1.006	1.016	1.022	1.023	0.455	0.318
		(292, 6)	(201, 6)	(66, 81)						
RB18	7	1.009	1.001	0.990	1.008	1.011	1.018	1.018	0.169	0.528
		(120, 1)	(30, 2)	(234, 87)						
RB19	13	1.012	1.003	0.985	1.009	1.018	1.028	1.028	0.324	0.411
		(26, 15)	(277, 50)	(127, 36)						
RB20	11	1.013	1.005	0.982	1.008	1.023	1.031	1.032	0.458	0.317
		(300, 19)	(205, 15)	(78, 66)						
RB21	17	1.015	1.005	0.980	1.010	1.026	1.036	1.037	0.433	0.335
		(215, 75)	(90, 9)	(358, 12)						

[i] - N is the number of specimens. Directions of AMS principal axes are in stratigraphic coordinates.

Sedimentological context of the AMS fabric is demonstrated in Figures 14 and 15. Paleocurrent directions inferred from the Eocene AMS data tend to align in N-S azimuth (Figure 14), and accord with development process of the forearc basin [4]. Takano and Waseda [4] demonstrated that the Eocene paleo-Ishikari basin experienced differential subsidence during deposition. Such deformation may be related to longstanding strike-slip faulting around central Hokkaido [17], and tectono- / sedimentological context of the AMS fabric will be better evaluated in the light of quantitative study of basin-forming processes described in this book. For reliable interpretation of AMS data, it is necessary to assess properties of ferromagnetic minerals, such as composition, grain size and contribution to bulk magnetic susceptibility, as shown in this paper.

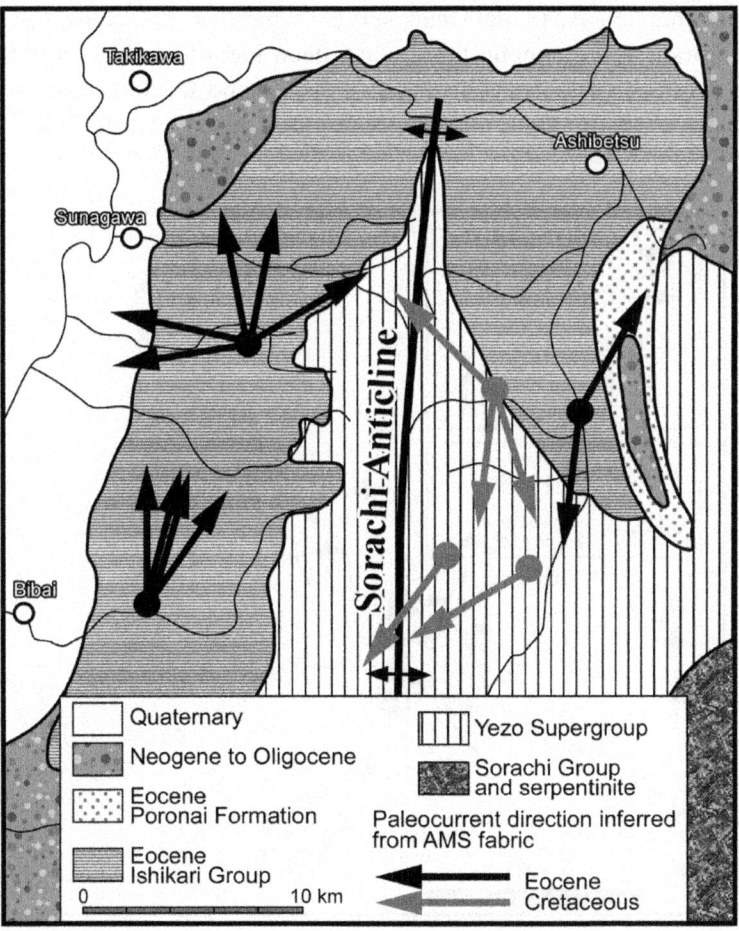

Figure 14. Paleocurrent directions inferred from AMS fabric of the Paleogene and Cretaceous samples. Geologic map is compiled from Editorial Committee of Hokkaido, Regional Geology of Japan [1] and Takano and Waseda [4].

Our field survey revealed indicators of paleocurrent directions in the Kawabata Formation along the Rubeshibe River as depicted in Figure 15. After correction for the counterclockwise rotation identified in our paleomagnetic study, most of the markers indicate a westward current direction with minor southward flow contributions. This is consistent with a tectono-sedimentary model of rapid burial of the Miocene N-S foreland basin by clastics derived from the eastern collision front presented in such research as Kawakami et al. [7]. Notably, the imbrication of the oblate AMS fabric matches visible sedimentary structures. Although the transport direction of muddy detrital material spilled out of a levee is not necessarily parallel to the turbidity current within a channel, AMS data can serve to indicate paleocurrents after the contributors to the magnetic fabric have been identified. Also note that K_1 of prolate samples (with negative T parameters) tend to align perpendicular to the paleocurrent direction, implying that elongate grains roll on the sediment surface.

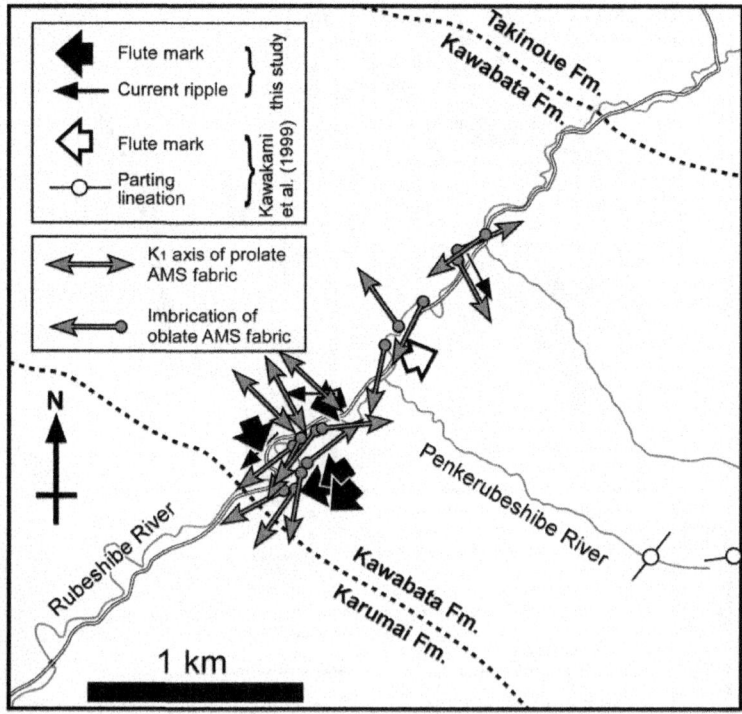

Figure 15. Paleocurrent map of the Kawabata Formation around the Rubeshibe River route. Formation boundaries are after Kawakami et al. [7].

Figure 16 delineates groups of microscopic fabrics identified in the Kawabata Formation as a function of the AMS shape parameter (T). The intensity of alignment forcing inferred from AMS data is closely related to sedimentary facies (shown on the right in the figure) determined by field observation. For example, weak hydrodynamic forcing corresponds to fine rhythmically alternating facies in channel-levee systems. Thus, the sedimentological context of muddy sediments' AMS fabric can be interpreted in the light of sandy sediments' facies analysis.

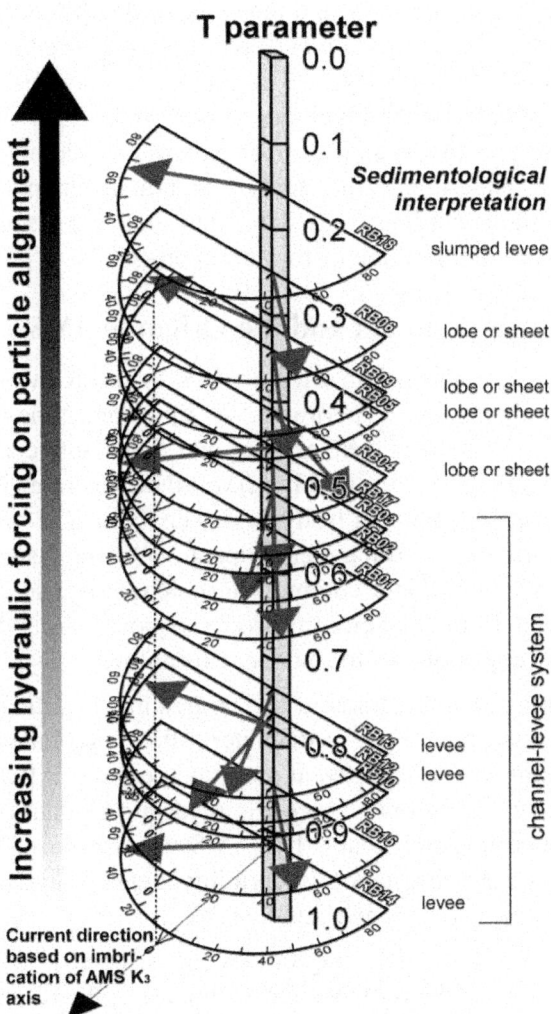

Figure 16. AMS paleocurrent indicators of the Kawabata Formation. Directions of K_1 (gray arrows) are shown as acute angles from the dotted baseline of K_3 axis imbrica-

tion. Vertical positions of the data are based on the T parameter. Samples with negative T values are excluded from the diagram because such cases have a large scatter in the K_3 directions.

Azimuths of AMS maxima in natural sediments vary significantly, reflecting the size or shape of magnetic grains and changes in current velocities (e.g., [18]). Figure 16 presents the relationship between paleocurrent proxies estimated from the imbrication of the AMS minimum axis (K_3) and the K_1 trend. Tarling and Hrouda [19] stated that the angle between K_3 and K_1 changes as a function of current velocity and the slope of the sedimentary surface. Our result suggests that the orientation between those AMS sedimentary indicators can vary, regardless of the level of hydraulic forcing, based on the shape parameter (T), which implies development of a preferred orientation. Although the AMS fabric is a diagnostic tool for patterns of sediment transportation, laboratory-based experiments that analyze natural sediments under conditions where a few of the prevailing factors are controlled, are essential to allow firm sedimentological interpretation of formation processes.

Re-Deposition Experiment and the Origin of AMS

In order to consider the origin of the AMS in the Kawabata samples, we organized a re-deposition experiment. A silty sandstone (SP1C-1) and a mudstone (SP2F-1) samples were crushed and sieved into coarse, medium and fine fractions. The fine fraction (< 63 μm) was then separated into magnetic and non-magnetic fractions with an isodynamic separator. The 'magnetic' fraction actually contained no ferromagnetic opaque minerals such as magnetite, but had abundant biotite and common hornblende. It also contained garnet, probably derived from metamorphic rocks exposed around the hinterlands during the rapid deposition of the Miocene turbidite.

A suspension of the fine fraction was poured into a vertically settled plastic tube 1 m in length and 2.5 cm in diameter, filled with water. This deposit of artificial sediment was dehydrated at room temperature. After being soaked in an adhesive resin, the samples were trimmed into standard-sized specimens for rock-magnetic measurements. The AMS was measured with an AGICO KappaBridge KLY-3 S magnetic susceptibility meter. The AMS parameters for the artificial samples are summarized in Table 4.

Figure 17 presents the magnitudes of magnetic fabrics in natural sedimentary rocks and the re-deposited sediments of the Kawabata Formation. Obviously, the magnetic separation results in remarkable decrease of both the bulk susceptibility and the degree of anisotropy (P_j). It is also noteworthy that the shape parameter (T) of the artificial sediments is almost null, suggesting a neutral magnetic fabric. The directions of the principal AMS axes (see Table 4) are not bound to the horizontal plane or to geomagnetic north. Thus, the detrital particles, free from paramagnetic minerals having shape anisotropy, like platy biotite, are deposited without any gravitational or geomagnetic forcing, creating an isotropic sediment.

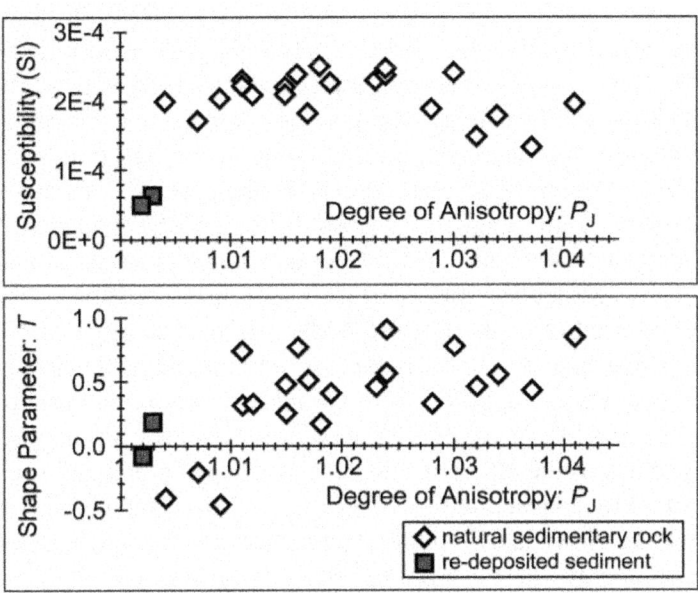

Figure 17. Magnitudes of magnetic fabrics in natural samples and re-deposited non-magnetic fine particles of the Kawabata Formation.

Table 4. AMS parameters of re-deposited non-magnetic fine fraction of the Kawabata Formation

Sample	N	κ^1	κ^2	κ^3	L (K_1/K_2)	F (K_2/K_3)	P (K_1/K_3)	P^J	T	q
		(D, I)	(D, I)	(D, I)						
SP1C-1	1	1.0009	1.0000	0.9992	1.001	1.001	1.002	1.002	-0.080	0.740
		(167, 75)	(265, 2)	(356, 15)						
SP2F-1	1	1.0014	1.0002	0.9984	1.001	1.002	1.003	1.003	0.180	0.517
		(250, 27)	(343, 6)	(84, 62)						

[i] - N is the number of specimens. Directions of principal axes of AMS are shown in *in situ* coordinates.

SUMMARY

Rock-magnetic investigation of sedimentary rocks provides insights into the basin's formation and sedimentation processes on an active margin. Cretaceous (Yezo Supergroup) ~ Eocene (Ishikari Group) strata and middle Miocene (Kawabata Formation) turbidites in central Hokkaido represent forearc and foreland settings, respectively. Progressive demagnetization successfully isolated characteristic remanent magnetization (ChRM) of the Kawabata Formation. Mean declination of the formation's ChRM exhibited significant westerly deflection, suggesting counterclockwise rotation of the study area since the middle Miocene. This differs from previous reports that indicated clockwise rotation. We attribute the difference to complicated deformation around the terminations of faults that form the N-S elongate Kawabata sedimentary basin. Anisotropy of magnetic susceptibility (AMS) principal axes were clearly determined for both the Cretaceous/Paleogene samples and Neogene samples, and regarded as a proxy of sediment influx directions. Paleocurrent directions inferred from the Eocene AMS data tend to align in N-S azimuth (Figure 14), and accord with the results of sedimentological paleoenvironment reconstruction, which suggest a northward downstream trend in fluvial to tidal estuarine systems [4]. As for the Cretaceous, further acquisition of AMS data is necessary to assess the effect of intensive syn-depositional deformation of the forearc [20]. After correcting for the tectonic rotation, most of the paleocurrent markers in the Kawabata Formation indicated a westward current direction with minor southward flow contributions, consistent with a sedimentary model that envisions burial of the Miocene N-S foreland basin by clastics derived from the eastern collision front. The intensity of alignment forcing of sedimentary particles inferred from the shape parameter (T) of the AMS data was closely related to sedimentary facies observed in the field. In investigating the origin of the AMS fabrics of turbidite deposits of the Kawabata Formation, we conducted a re-deposition experiment of fine detrital particles with no magnetic fraction including paramagnetic minerals with relatively high magnetic susceptibility, which demonstrated the significance of the alignment of paramagnetic minerals having shape anisotropy.

ACKNOWLEDGEMENTS

The authors are grateful to N. Ishikawa for the use of the rock-magnetic laboratory at Kyoto University and for thoughtful suggestions in the course of the magnetic analyses. We thank S. Oshimbe and S. Nishizaki for their help with field work. Thanks are also due to N. Yamashita and Y. Danhara for their support in mineral separation. Constructive review comments by G. Kawakami greatly helped to improve early version of the manuscript.

REFERENCES

1.　Editorial Committee of Hokkaido, Regional Geology of Japan. Regional Geology of Japan, Part 1: Hokkaido. Tokyo: Kyoritsu Shuppan; 1990.

2.　Tamaki M, Itoh Y. Tectonic implications of paleomagnetic data from upper Cretaceous sediments in the Oyubari area, central Hokkaido, Japan. Island Arc 2008; 17: 270-284.

3.　Tamaki M, Oshimbe S, Itoh Y. A large latitudinal displacement of a part of Cretaceous forearc basin in Hokkaido, Japan: paleomagnetism of the Yezo Supergroup in the Urakawa area. Journal of Geological Society of Japan 2008; 114: 207-217.

4.　Takano O, Waseda A. Sequence stratigraphic architecture of a differentially subsiding bay to fluvial basin: the Eocene Ishikari Group, Ishikari Coal Field, Hokkaido, Japan. Sedimentary Geology 2003; 160: 131-158.

5.　Takano O, Waseda A, Nishita H, Ichinoseki T, Yokoi K. Fluvial to bay-estuarine system and depositional sequences of the Eocene Ishikari Group, central Hokkaido. Journal of Sedimentological Society of Japan 1998; 47: 33-53.

6.　Miyasaka S, Hoyanagi K, Watanabe Y, Matsui M. Late Cenozoic mountain-building history in central Hokkaido deduced from the composition of conglomerate. Monograph of the Association for the Geological Collaboration in Japan 1986; 31: 285-294.

7.　Kawakami G, Yoshida K, Usuki T. Preliminary study for the Middle Miocene Kawabata Formation, Hobetsu district, central Hokkaido, Japan: special reference to the sedimentary system and the provenance. Journal of the Geological Society of Japan 1999; 105: 673-686.

8.　Otofuji Y, Kambara A, Matsuda T, Nohda S. Counterclockwise rotation of northeast Japan: paleomagnetic evidence for regional extent and timing of rotation. Earth and Planetary Science Letters 1994; 121: 503-518.

9.　Itoh Y, Tsuru T. Evolution history of the Hidaka-oki (offshore Hidaka) basin in the southern central Hokkaido, as revealed by seismic interpretation, and related tectonic events in an adjacent collision zone. Physics of the Earth and Planetary Interiors 2005; 153: 220-226.

10.　Takano O, Tateishi M, Endo M. Tectonic controls of a backarc trough-fill turbidite system; the Pliocene Tamugigawa Formation in the Niigata-Shin'etsu inverted rift basin, Northern Fossa Magna, central Japan. Sedimentary Geology 2005; 176: 247-279.

11. Bouma AH. Sedimentology of Some Flysch Deposits; A Graphic Approach to Facies Interpretation. Amsterdam: Elsevier; 1962.

12. Kawamura K, Ikehara K, Kanamatsu T, Fujioka K. Paleocurrent analysis of turbidites in Parece Vela Basin using anisotropy of magnetic susceptibility. Journal of the Geological Society of Japan 2002; 108: 207-218.

13. Day R, Fuller M, Schmidt VA. Hysteresis properties of titanomagnetites: grain-size and compositional dependence. Physics of Earth and Planetary Interiors 1977; 13: 260-267.

14. Kirschvink JL. The least-squares line and plane and the analysis of palaeomagnetic data. Geophysical Journal of the Royal Astronomical Society 1980; 62: 699-718.

15. Kodama K, Takeuchi T, Ozawa T. Clockwise tectonic rotation of Tertiary sedimentary basins in central Hokkaido, northern Japan. Geology 1993; 21: 431-434.

16. Takeuchi T, Kodama K, Ozawa T. Paleomagnetic evidence for block rotations in central Hokkaido-south Sakhalin, Northeast Asia. Earth and Planetary Science Letters 1999; 169: 7-21.

17. Tamaki M, Kusumoto S, Itoh Y. Formation and deformation processes of late Paleogene sedimentary basins in southern central Hokkaido, Japan; paleomagnetic and numerical modeling approach. Island Arc 2010; 19: 243-258.

18. Ledbetter MT, Ellwood BB. Spatial and temporal changes in bottom-water velocity and direction from analysis of particle size and alignment in deep-sea sediment. Marine Geology 1980; 38: 245-261.

19. Tarling DH, Hrouda F. The Magnetic Anisotropy of Rocks. London: Chapman & Hall; 1993.

20. Tamaki M, Tsuchida K, Itoh Y. Geochemical modeling of sedimentary rocks in the central Hokkaido, Japan: Episodic deformation and subsequent confined basin-formation along the eastern Eurasian margin since the Cretaceous. Journal of Asian Earth Sciences 2009; 34: 198-208.

Chapter 5

THEORIES ON ROCK CUTTING, GRINDING AND POLISHING MECHANISMS

Irfan Celal Engin

Afyon Kocatepe University, Engineering Faculty, Department of Mining Engineering, Afyonkarahisar, Turkey

INTRODUCTION

Tribological research studies including cutting, abrading and polishing mechanisms have firstly started with metals, metal cutting theories and formulas have been developed and then applications on rock material have started. In this part of the book, natural stone cutting, abrasion and polishing mechanisms are compiled and presented as a summary.

Processes of cutting, grinding and polishing natural stones are made as a result of grinding-abrading mechanism developed on the use of different abrasive grains (mostly diamond and SiC). Wear intensity is named as cutting, abrading or polishing according to the speed, chip size and situation of obtained surfaces.

No matter which cutting machine is used, generally cutting process of natural stones are done with the use of segments that are obtained through sintering of diamond grains and metal powders. Industrial diamond grains in these segments rubbed against the material to be cut with a certain force and material is removed, and as a result, the material is cut along this surface as the material is removed as much as segment width.

In the stage of abrading and polishing of natural stones, products called grinding stone containing SiC grains are generally used. Intenseness of material removal from natural stone surfaces can be arranged by changing grain size of this abrasive and magnitude of pressure intensity. When relatively coarser grains and higher pressures are chosen, coarse abrading process is obtained while slight abrading and polishing is obtained when slighter grains and lower pressures are chosen.

THE BASIC WEAR MECHANISMS EMERGING ALL TYPES OF MATERIALS

Wear is described in the literature as the loss of material as a result of the change in the shape of friction surfaces. Many researchers have stated that there are 4 main wear mechanisms causing the loss of material. These are adhesive wear, abrasive wear, and corrosive wear and wear resulting from surface fatigue (Archard, 1953; Moore, 1975; Suh and Saka, 1978; Williams, 1994; Summer, 1994). Similarly, many researchers have classified wear as heavy wear and light wear according to the wear magnitude.

A basic equation about wear is developed by Archard (1953). According to Archard (1953), wear on friction surfaces (w), is directly proportional to applied load (W) while inversely proportional to the strength of material (H).

$$w = K \times \frac{W}{H}$$
(1)

K which is non-dimensional in here is expressed as wear coefficient. This coefficient is changed into k=K/H including strength; this is the dimensional wear coefficient which is more widely accepted in engineering. This coefficient represents the volumetric wear (mm^3) resulting from the shift in unit distance (m) under unit load (N).

When two materials are rubbed against each other, stresses on touch point can easily reach yield point. With the shearing effect of lateral force, material transfers from the surface of soft material to the surface of hard piece and sticks on. Wear developed this way is called adhesive wear. A simple demonstration of this is presented by Archard (1953) (Figure 1):

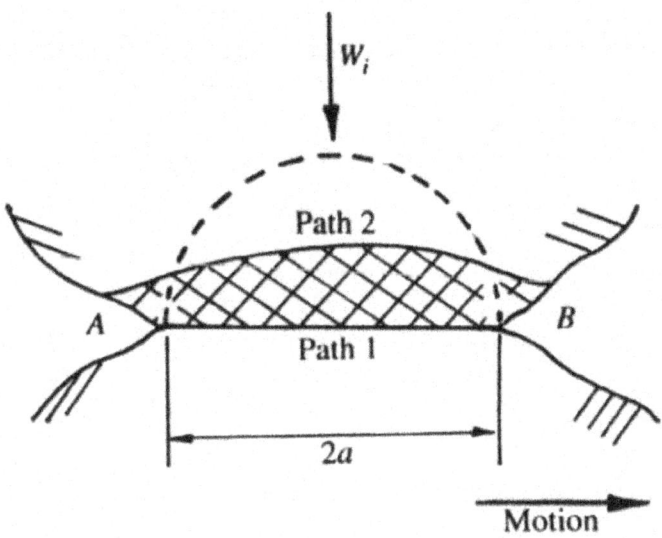

Figure 1. Material wear caused by the adhesion on friction surfaces (Archard, 1953).

Here, the diameter of contact point is shown as 2a, applied load is shown as W. It is thought that moving will be along the way shown as Path 2. For convenience, the part that will be abraded is assumed to be in the shape of a radius sphere and wear amount as a result of 2a amount of shifting. Wear per unit shifting distance is calculated as $1/3\pi a^2$, by dividing $2/3\pi a^3$ to 2a. As change in the shape is permanent, Wi load is presented as $Wi = H\,a^2$ material strength and type. At the end, total wear is shown as;

$$w = \frac{\pi}{3} \times \sum a^2 = \frac{1}{3\pi} \times \sum \frac{\pi Wi}{H} = \frac{W}{3H} \tag{2}$$

This ($W=\sum Wi$) means the total load applied by both surfaces.

If a solid material or a solid particle removes piece by scratching or rubbing, this is defined as abrasive wear. Abrasive wear comes through as long rents on surfaces in parallel with the friction direction. A simple model based on the assumption that there is not any change on grain, it only pass through soft material by rubbing it inside is presented in Figure 2. Here, normal load is shown with W, depth on the surface caused by abrasive grain is shown with h, and cone angle is shown with υ.

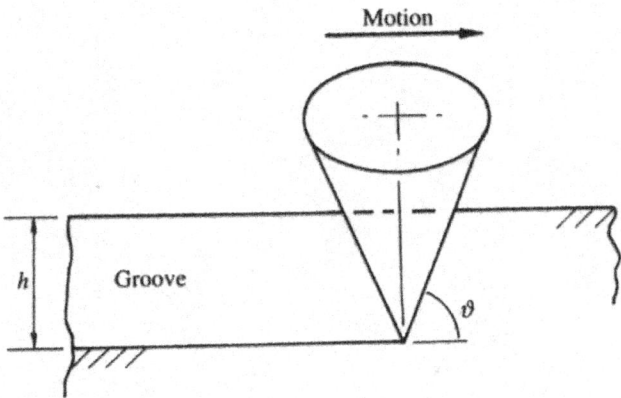

Figure 2. Movement of a cone shaped abrasive on a soft surface (Williams, 1994).

Here normal load is given as;

$$W = h^2 \times \cot\vartheta \qquad (3)$$

When depth of surface caused by the grain is given as material strength, load is defined as;

$$W = \frac{\pi}{2} \times \left(h\cot\vartheta\right)^2 \times H \qquad (4)$$

As a result, wear is given as this equation;

$$w = \frac{2\tan\vartheta}{\pi} \times \frac{W}{H} \qquad (5)$$

If abrasive grain is prismatic instead of cone, wear becomes more complex. The structure of chip created as a result of wear is based on two angles to a great extends besides affecting forces. The first is contact angle which is the surface of abrasive grain on the side of moving angle in the direction of sliding. The second is the dihedral angle (2) which is the angle between the sides of pyramid in the direction of movement (Figure 3).

Contact angle is very important for wear, because while abrasive grain cuts chips over critical contact angles (ψ_c), it only breaks through or rubs in lower angles.

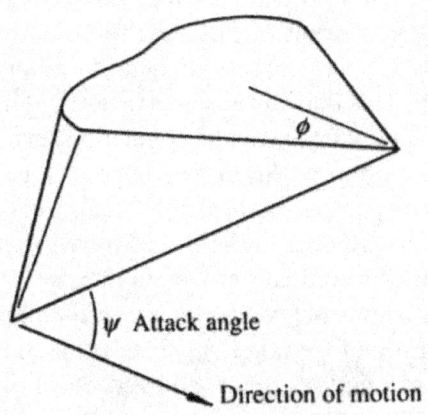

Figure 3. Geometry of prismatic abrasive grain represented with two angle ((ψ, 2φ) (Williams, 1994).

Dihedral angle also significantly affects the shape of chip. In very small 2 angles, abrasive grain breaks through the surface like a knife. When 2φ=180, it means there is a smooth surface vertical to the motion direction and this is a limiting value.

Relation between contact angle and dihedral angle developed by Kato et al. (1986) is given in Figure 4.

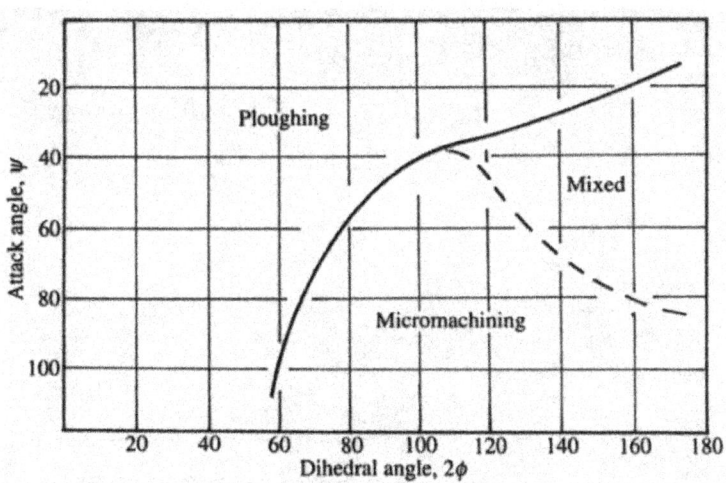

Figure 4. Relation between contact angle and dihedral angle with wear situations resulting from abrasive wear (Kato et al., 1986).

Corrosive wear occurs on surfaces that rubbed against each other with small vibrations and as a result of this, few ten micron grains are removed. Generally when irony surfaces are rubbed against one another, a reddish brown fragment is produced. This detritus is composed of solid iron oxide grains and behaves like polishing powder and make contacting surfaces smooth and shiny. This leads to the creation of a film in the shape of a protector layer on these surfaces. If wear occurs because of mechanic factors and environment involves similar atmospheric conditions, this film layer will remove and a new layer will occur as a result of re-oxidation. Grains that are formed during removal of this layer can cause abrasive wear because of their solidity. Adhesive wear can also occur as a result of friction if a part of contacting surfaces is oxidized while another part is completely non-oxidized. As a result, if this corrosion layer is continuously removing because of wear, this will have a positive effect on wear process which is named corrosive wear (Summer, 1994). Wear as a result of surface fatigue occurs generally on metal materials rolling on one another similar to the bearings. Material in contact point tightens as a result of permanent change in the shape and material embrittlement occurs. This material cracks as a result of repetitive power; they spread on the surface in time and cause breakage of material in small pieces.

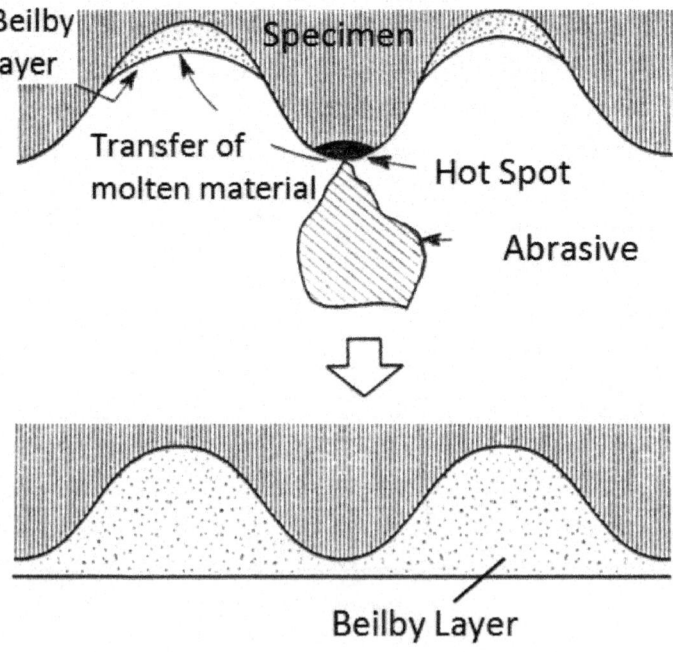

Figure 5. Beilby polishing mechanism developed by Bowden and Hughes (1937).

In the wear process resulting from surface fatigue, bigger grains remove from the surface when compared to adhesive or abrasive wear. Typical cavitations and scouring occur on these types of surfaces.

Wear mechanisms that are stated until here explain unwanted material loss on surfaces.

Abrading on the other hand, is the deliberate process of removing material from surfaces with various applications and in accordance with the purpose.

Polishing is a process of abrading and it is defined as the process of removing unevenness and visible scratches by using abrasive material (Coes, 1971). So, polished surface reflects light smoothly and in a linear way (Coes, 1971; Samuels, 1971).

According to Beilby (1921), polishing results from a smearing a material on a surface which fills the gaps on surface. Beilby (1921) stated that this material has a structure that is completely similar to amorphous and it looses crystal structure. He didn't suggest a mechanism about smearing material on these surfaces. But later on, a mechanism was developed by Bowden and Hughes (1937). These researchers determined that very high temperatures were reached in the contact points of abrasive grains, as a result of rubbing solids against each other which caused them think that heat is significant in the process of polishing (Samuels, 1971).

Unevenness on the original surface heat locally (until melting point) which is caused by friction transfers to the gaps (Figure 5). This material is transferred as a result of rapid cooling, it has an amorphous structure and constitutes Beilby layer.

On the other hand, Samuels (1971) didn't accept the existence of Beilby layer and tried to prove that Beilby layer doesn't occur with many proofs. According to Samuels (1971), there exists continuous material loss during polishing on surfaces that are physically polished. In addition, when polished surfaces are analyzed with microscope, scratches can be seen. This situation is completely opposite the existence of Beilby layer. Because Beilby stated that material fill the gaps during polishing. There shouldn't be a distinct material loss.

When physically polished surfaces are treated with acid, scratches on the surface appear. According to Samuels (1971), this situation can be explained with the deformed layer (as can be seen in Figure 6) rather than Beilby layer.

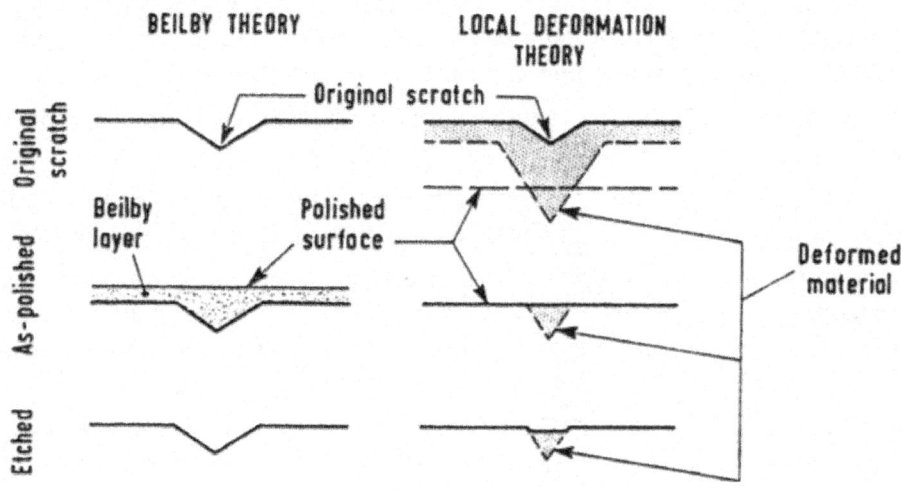

Figure 6. Comparison of Beilby and local deformation theory (Samuels, 1971)

According to Samuels (1971), one needs to explain three acceptations in order to explain physical polishing mechanism. The first is that, surfaces always have thin scratches or joint sets. The second point is that material is removed from the surface during polishing process with a constant speed. Thirdly, a layer that has permanently deformed is created. This layer is highly similar to transformed layer that is created during abrading. When grinded and polished surfaces are examined, the significant similarity between them will be seen. Scale is the only difference. Most of the researches and studies on abrading and polishing process focus on metallic materials. Studies on brittle and fragile surfaces like rock surface are very limited.

In this section, abrading and polishing processes that are based upon mechanic materials are taken into consideration. When abrading and polishing mechanisms are analyzed in these terms, it is seen that there is not a basic difference between them. By changing the abrasive material type and/or application style of gain size, abrading process can be transformed to polishing.

Explanation of abrading and polishing mechanism is possible only by revealing the type and aim of the applied process. So, mechanisms that occur at each application will be different from one another.

These applications can be lined as;

- Wear mechanism formed during the use of circular saw, grinding mills, and grinding cutting stones.
- Wear mechanism during cutting process with diamond blade saws.
- Wear mechanism during cutting process with diamond bead system
- Wear mechanism during surface polishing applications
- Wear mechanism during the use of sandpaper

Abrading processes mentioned above have significant differences in terms of basic mechanism. This is why, each one of them will be analyzed and what kind of abrading and/or polishing mechanism develops will be put forward.

WEAR MECHANISM THAT IS FORMED DURING THE USE OF CIRCULAR SAW, GRINDING MILLS, GRINDING STONES

Circular saws is the cutting tool that is used the most in cutting and sizing of natural stones that are segments containing diamonds' welded around circular metal body. Grinding mills are used for process such as cutting, graving, shaping… etc. In the abrading operation mentioned until here, the process of wear results from simple geometrical situations with a linear movement between abrasive grain and material. When grinding mills are analyzed, if abrading operation occurs on the edge of grinding mill, the situation is simple. But if abrading is on the disc, the situation is complex. In the literature, abrading mechanism that occurs here is mentioned with the word grinding. The wear that occurs here is defined as a micro-scaled grinding.

Salmon (1992) tried to make a mathematical modeling for the abrading operation on the surface of grinding mill.

Situation of grinding mill during the operation is geometrically shown in Figure 7.

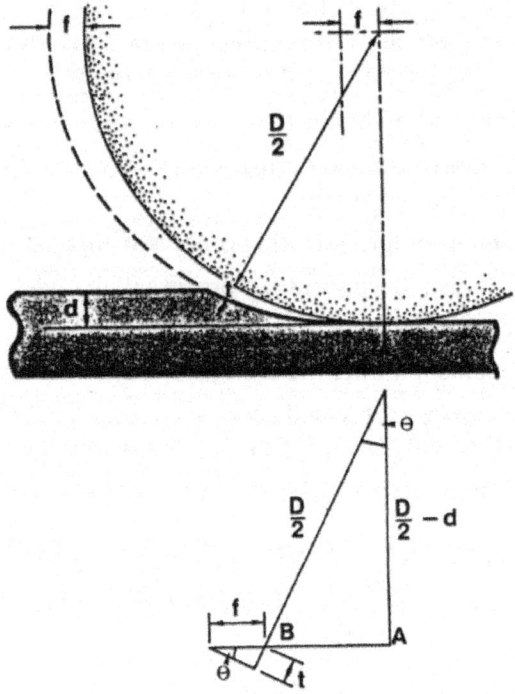

Figure 7. Geometric presentation of wear produced during grinding mill application (Salmon, 1992).

Symbols that are shown here and that will be used hereafter is shown below:

D: Diameter of grinding mill (mm)V_s: Peripheral speed of grinding mill (m/s)w: Angular speed of grinding mill (rad/s)V_w: Feedrate of material K: Number of abrasive grains along a peripheral line C: Number of active abrasive grains per unit t: Cutting depth of abrasive grains (mm)l: Length of cutting trajectory d: Cutting depth of grinding mill (mm)b': Theoretical width of each grain (mm)b: Cutting width of abrasive grain (mm)

Salmon (1992) made these approaches;

Length of cutting trajectory can be determined as below:

$$l^2 = D^2 / 4 - (D/2 - d)^2 + d^2$$

$$= D.d$$

$$l = (D.d)^{0.5} \qquad (6)$$

According to the geometrical structure in the figure;

$$AB = \left((D/2)^2 - (D/2 - d)^2\right)^{0.5}$$

Calculated as such:

$$AB = \left(d(D-d)\right)^{0.5}$$

Cutting arc is accepted to be a straight line and the mill -along a peripheral line- that contain K amount of abrasive grain, proceeds as much as f with 1/K turn. Cutting depth of abrasive grains is calculated as:

$$t = f(Sin\theta) = f(AB)/(D/2) = 2f\left(d(D-d)\right)^{0.5}/D$$

As d, is at a small value according to D,

$$t = 2f(d/D)^{0.5} \quad f = V_w/(Kw)$$

$$t = (2V_w/Kw)(d/D)^{0.5} \quad can \ be \ written \tag{7}$$

Number of abrasive grains along a peripheral line ise determined with the equation below.

$$K = \pi.D.b'.C$$

If abrasive grain is represented as such (Figure 8);

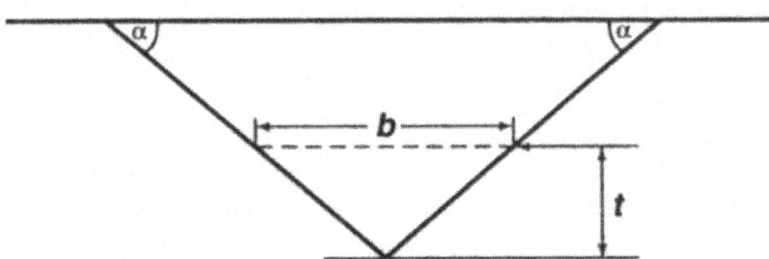

Figure 8. Frontal view of the presented grain (Salmon, 1992).

t/2 is used as average grain cutting depth. So, the ratio of grain width to cutting depth will be:

$$r = 2b'/t$$

K=π.D.b'.Cequation turns toK=π.DrtC/2. If this is put in Equation 7; t=(d/D)$^{0.5}$(4V$_w$/πDV$_s$Crt) is reached. V$_s$=π.D.ω so;

$$t^2 = (4V_w d / V_s Crl)$$

(8)

Cutting geometry when abrasive grain is considered, cutting geometry is presented in Figure 9.

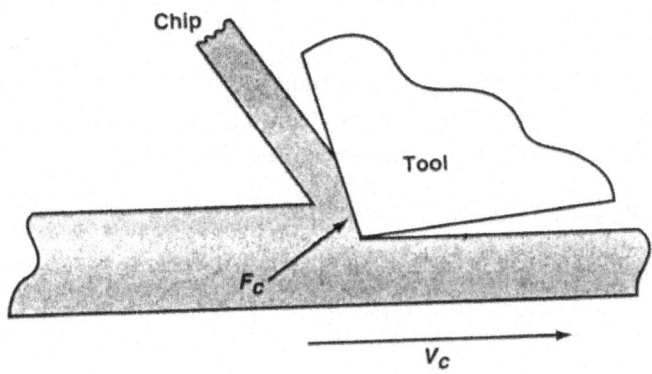

Figure 9. Cutting geometry of a abrasive grain. (Salmon, 1992).

Here;

b: Cutting width (mm)

t: Thickness of chip that is not deformed (mm)

u: Specific energy (Jmm^{-3});

For cutting operation at one point;

$$u = F_c V_c / b + V_c$$

Cutting Force $= F_c - ubt$

Cutting force in abrading is used as tangential force. So, equation is arranges as u'=F+V$_s$/V$_u$bd.

Cutting Force $= F_c$u'bdV$_u$/V$_s$

Work done in unit of time: u'Vubd

Number of abrasive grain working at unit of time: V$_s$Cb

Work done by one single grain = u'Vud / V$_s$C

Average force affecting one single grain F" is calculated by dividing the work done by one grain into the length of cutting arc.

$$F'' = u'Vud \ / \ V_sCl \qquad (9)$$

From Equation 8:

$$t^2 = 4 \ Vud \ / \ V_sCrl \qquad \Rightarrow \qquad l = 4 \ Vud \ / \ V_sCr \ t^2$$

When l is put into the place in Equation 9,

$$F'' = u't^2r \ / \ 4 \qquad (10)$$

is obtained.

According to Salmon (1992), it is possible to solve wear problems in grinding mill applications and present alternative solutions.

In terms of energy need, in order to remove material from the surface, the most efficient phase is this cutting phase. Minimum specific energy is used in this way. Here, specific energy is the energy that is needed for removing unit material form the surface and unit is joule/mm^3 or Btu/in^3.

According to Salmon (1992), energy used during abrading in which chip is shaped can occur in these ways:

- Heating on working material,
- Heat occurs at grinding unit,
- Heat that occurs at chips,
- Kinetic energy at chips,
- Radiation diffused around,
- Energy spent on producing new surface,
- Residual stress in the surface and chip's lattice structure.

Another approach for grinding mills is developed by Chen and Rowe (1996). According to Chen and Rowe (1996), when a abrasive grain on the surface of a moving chip is thought; firstly, abrasive grain combines on the material with a narrow curve. In this way, more material is removed. Secondly, productive contact point of the surface of abrading changes as long as it moves on this contact arc.

So, as can be seen in Figure 10, while abrasive grain made "ploughing" at the beginning of this arc, the rest of is can make cutting.

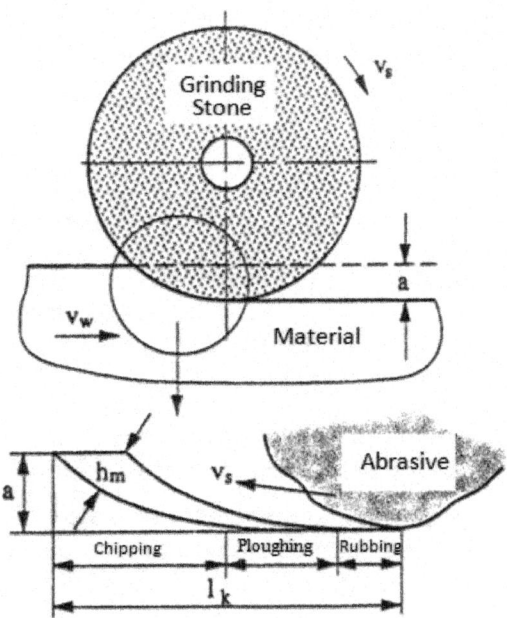

Figure 10. Phases of chip formation on the edge to grinding mill (Chen and Rowe, 1996).

In Figure 10, production of chip by grain on grinding mill during the movement of chip is seen. Cutting arc length of grain is shown with l_k, thickest chip thickness that hasn't changed shape is shown with h_m, tangential turning speed of mill is represented with V_s, progress speed of processed material is represented with V_w, cutting depth is represented with a.

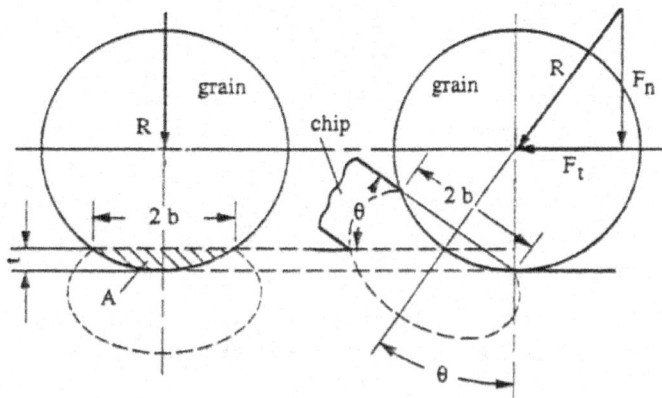

Figure 11. Behavior of circular grain in abrading (Chen and Rowe, 1996).

Movement of a circular grain on the edge of mill during abrading is shown in Figure 11. Cutting depth of grain in here is represented with t, amount of pressure with R, its horizontal component with F_t, vertical component with F_n, cross sectional area of chip that hasn't gone under any change with A, diameter of circular area which is the section of this on the surface with b, angle between power of pressure and vertical with θ. Pressure force affecting grain is represented with

$$R = \pi b \ H(C' / 3) \qquad (11)$$

Here C is the strain factor defined as the rate of average pressure affecting contact area to normal stress. Necessary specific energy is defined as;

$$e_c = F_t / A \qquad (12)$$

When $A = \dfrac{4}{3} bt$, if necessary specific energy is used for cutting,

$$e_{cc} = \frac{3R Sin\theta}{4bt} \qquad (13)$$

When is put in the equation, specific energy is calculated as below:

$$e_{cc} = \frac{3\pi}{4} \frac{b}{t} H\left(\frac{C'}{3}\right) Sin\theta \qquad (14)$$

When friction force is taken as $\mu RCos$, specific energy in friction is obtained as;

$$e_f = \frac{3\pi}{4} \mu \frac{b}{t} H\left(\frac{C'}{3}\right) Cos\theta \qquad (15)$$

total specific energy for grain is formulized as below.

$$e_g = \frac{3\pi}{4} \frac{b}{t} H\left(\frac{C'}{3}\right)(Sin\theta + \mu Cos\theta) \qquad (16)$$

In parallel with common use of grinding mills in the industry, there are many studies in the literature about grinding mills.

WEAR MECHANISM FORMED IN THE PROCESS OF CUTTING WITH DIAMOND FRAME SAW

Generally rock cutting mechanism is explained by the formation of indentation with plastic deformation and breaking mechanism of rock. When cutting depth of diamond is deep enough to produce visible cracks on a rock, breakages occur and chips are formed as a result of this. As can be seen schematically in Figure 12, there is a plastic deformation under the channel that is produced by the tangential movement of abrasive grain along the surface and there are two main crack systems named radial and lateral that are produced from this zone. Radial cracks are formed with wedge wear type when high normal force is used and when this force is removed, these cracks can continue to spread because of permanent tensile stress at the edge of crack. Lateral cracks start to be formed when he force is removed and can continue to spread with the effect of permanent tension (Konstanty, 2002).

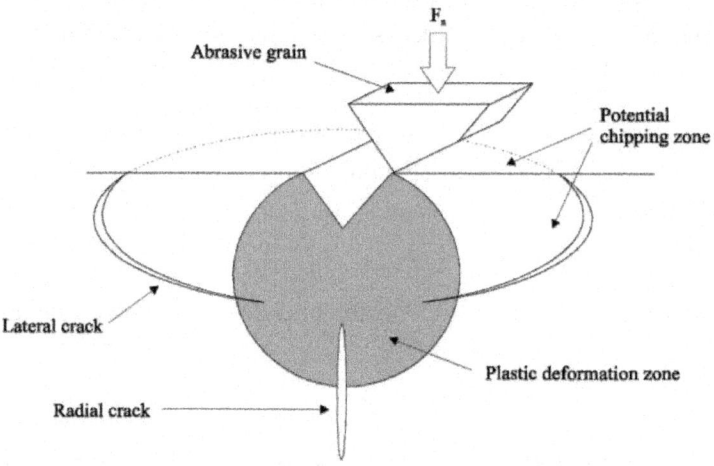

Figure 12. Schematic view of plastic deformation zone formed during cutting (Konstanty, 2002).

In the cutting process with diamond segmented blades are moved with reciprocating motion at a sinusoidal speed that is about 2 m/s. Konstanty (2002) made some researches on the cutting mechanism with diamond blade saw and defined the cutting zone during this cutting. Schematic view of cutting zone determined by Konstanty (2002) is shown in figure 13. Here, in order to make the definition, it is accepted that diamond grains in diamond zones are placed with the same protrusion height and cutting zone is the same all along the segment. But diamond grains on diamond segments have different protrusion heights. This complicates the process of defining cutting process. As can be seen in the cutting zone in figure 3.20, a pressure is produced on matrix as rock fragments accumulated in the front and behind the diamond grains cannot be removed. Magnitude of pressure resulting from the wedging of rock fragments cause wear in the contact zone that is the weakest point between diamond and matrix.

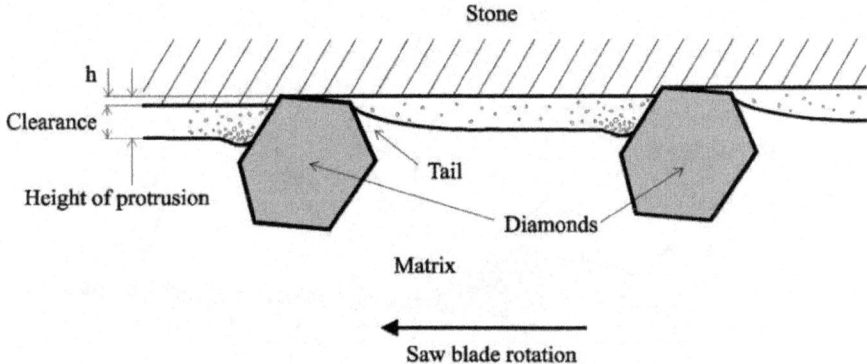

Figure 13. Schematic demonstration of cutting zone in the frame sawing system (Konstanty, 2002).

In the process of cutting with diamond blade saw, cutting of diamond segments is similar with cutting of many diamond grains. Cutting principal of diamond grain in segments are shown in Figure 14.

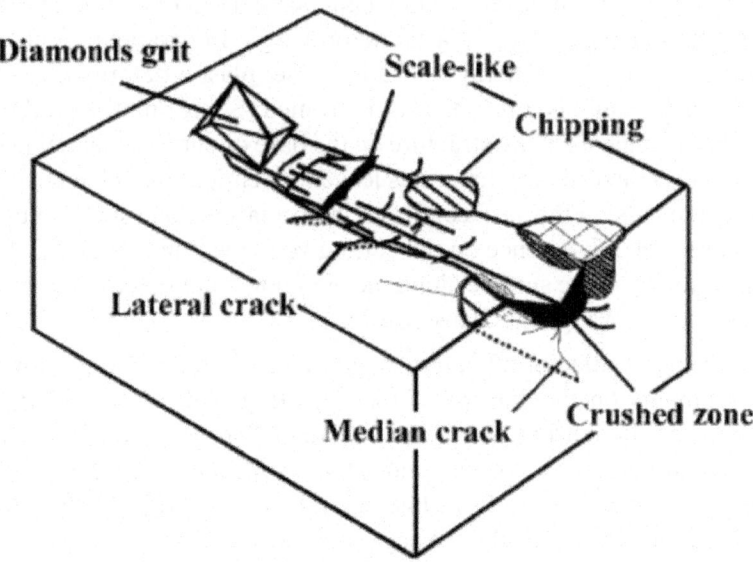

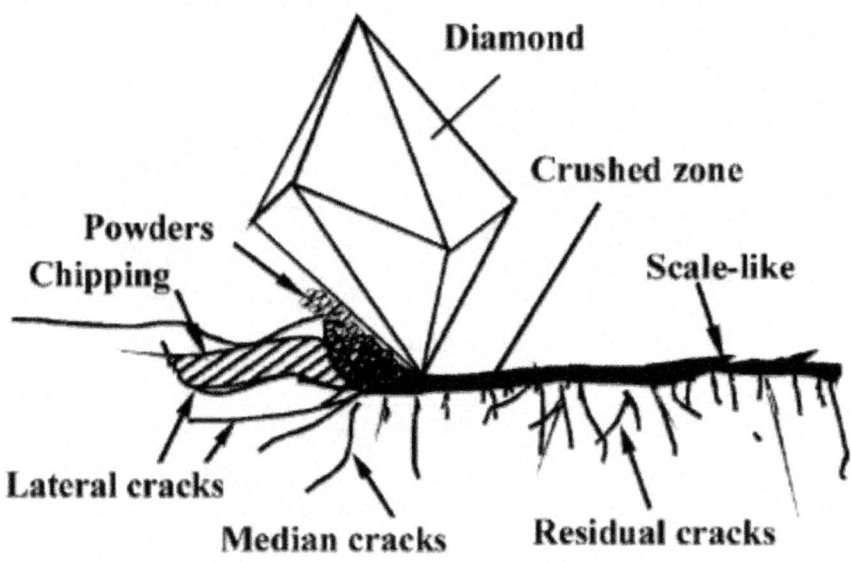

Figure 14. Cutting mechanism of marble with diamond grain (Wang and Clausen, 2002).

As can be seen in Figure 14, the main deformation of natural stone in low cutting depth is explained as plastic deformation. In parallel with the increase in cutting depth, while lateral cracks increase, plastic deformation of natural stone decreases and as a result, chip is formed. Some small lateral cracks on the surface can have a flaky structure on the base of cutting channel like a shell. Lateral cracks on different directions leave the semicircular channel behind the diamond the cuts on the surface. Plastic deformation zone stays on the base of cutting channel. Divergence on the cutting zone resulting from the increase of shearing cracks on the surface along the breaking zone seems like a continuous chip formation (Wang and Clausen, 2002).

Cutting with diamond blade saw is the continuous cutting movement of many segments on the surface of rock. Cutting with diamond segment can be defined as the cutting of a diamond cutter that cuts from many points in different cutting depths. As diamonds make chips and cuts, cracks are formed and they join and as a result of this, natural stone is broken. This situation is given in Figure 15 as stated by Wang and Clausen (2002).

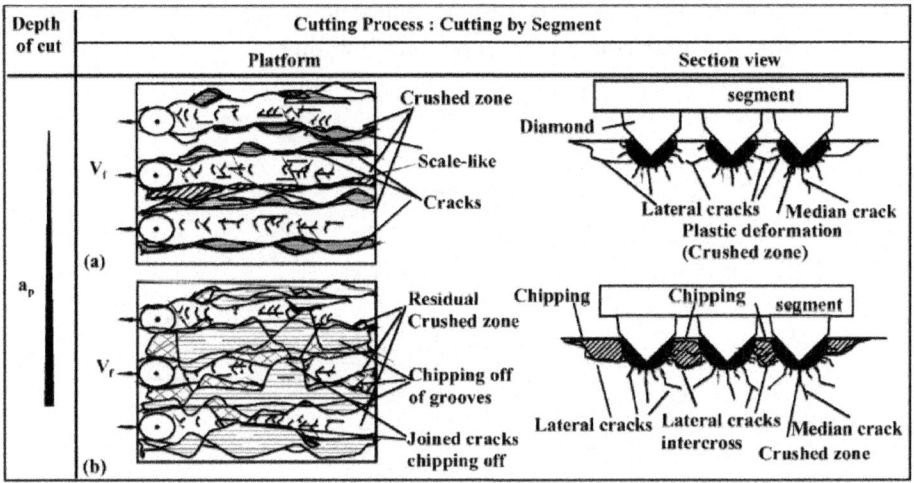

Figure 15. Cutting process of marble with diamond segments (Wang and Clausen, 2002).

Natural stone cutting mechanism with diamond blade saw system is explained as plastic deformation (breaking zone) and brittle breaking of rock. Formation of chip can be used in the explanation of cutting with frame saw system as s baseline. Plastic deformation and breaking of rock is affected from cutting conditions such as cutting depth, cooling operation, shape of cutter and aspects of rock (Konstanty, 2002; Wang and Clausen, 2002).

WEAR MECHANISM FORMED DURING CUTTING WITH DIAMOND WIRE SYSTEM

The principle behind diamond wire cutting involves pulling a spinning, continuous loop of wire mounted with diamond bonded steel beads through the stone to provide the cutting action (Figure 16). Through the combination of the spinning wire and the constant pulling force on the wire, a path is cut through the stone. In marble quarrying through diamond wire cutting, the initial step for making a vertical cut is to drill two holes, one vertical and one horizontal, which intersect at a 90° angle. The diamond wire is then threaded through these holes, mounted around the drive wheel, and the two ends are clamped together to form a continuous loop. The drive wheel may be set at any angle, from vertical to horizontal, required to facilitate cutting.

The diamond wire is comprised of a steel cable on which small beads bonded with diamond abrasive are mounted at regular intervals with spacing material placed between the beads. The beads provide the actual cutting action in this operation. They are bonded with diamond by one of the two methods: electroplating or impregnated metal powder bonding.

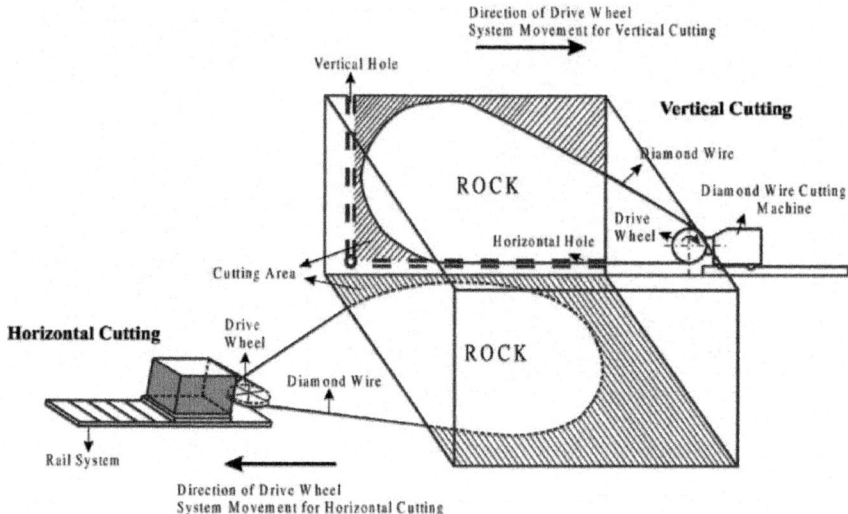

Figure 16. Situation of the diamond wire during the cutting operation (Ozcelik et al., 2002).

In the process of cutting with diamond wire system (Figure 17), diamond grains, sintered with metal powder as bead form, contact material surface similar with the circular saw and make grinding, cutting or abrading processes according to contact angle.

During cutting natural stone with diamond wire, contact angle between diamond grains and rock surface vary according to the path of steel rope (on which diamond beads are lined) in the rock rather than the diamond grains on bead.

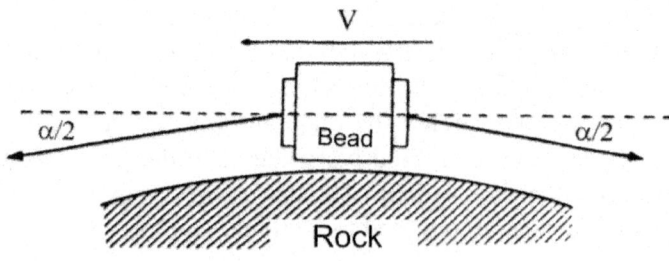

Figure 17. Schematic presentation of cutting the natural stone with diamond bead (Bortolussi et al., 1994).

WEAR MECHANISM FORMED DURING SURFACE POLISHING APPLICATIONS

This mechanism includes abrading and polishing mechanisms on grinding heads that are used for leveling and polishing of surfaces of different materials.

Although some of grinding mills are used in the literature, there are some significant differences between the grinding mechanisms of grinding mills. Abrading that is made with the help of abrasive grains on the surface of grinding heads is in fact quite similar with the abrading on sandpaper. Abrasive grains on the abrading product makes cutting, rubbing or ploughing as of their position. According to the application of abrading operation, complication of the mechanism is somewhere in between the mechanisms that are produced in sandpaper and grinding mill applications.

If grinding heads is turning around vertical axis on material surface, grains on it will make cutting, breaking through or friction during abrading. If there is the linear movement besides turning, the situation will be more complicated. Abrasive grain will be able to make these three moves during operation in different times.

In figure 18, the model that is developed by Lawn and Swain (1975) about crack movement on material surface and material removal. At the first contact point between grinding and surface, because of the applied loads high stress occurs. If the tip of the abrasive grain is perfectly sharp (namely, if radius of curvature is 0) stresses at this point will be infinite. These dense stresses relax with permanent changes in the shape and changes in density.

When applied loads reach a critical value, the middle crack shown with M start to increase because of tensile stress occur on vertical plane. In parallel with the decrease of load, middle fracture filling and when it reduced more, lateral cracks shown with L occur. These cracks are formed as there are residue elastic stresses after relaxation in contact points. Crack reaches surface with the removal of complete load and it cause wear with the breaking of material from surface.

According to Chandrasekar and Farris (1997), a few mechanisms are dominant in removing material from brittle surfaces. These are brittle break that is formed according to crack systems that is parallel and vertical to the surface and ductile cutting in the shape of chip similar to slim ribbon. The process that will occur depends on the load on abrasive grain, location and velocity of slip. Abrading process cause destruction in places close to the surface in the shape of small scaled crack, residual stress and permanent change in shape.

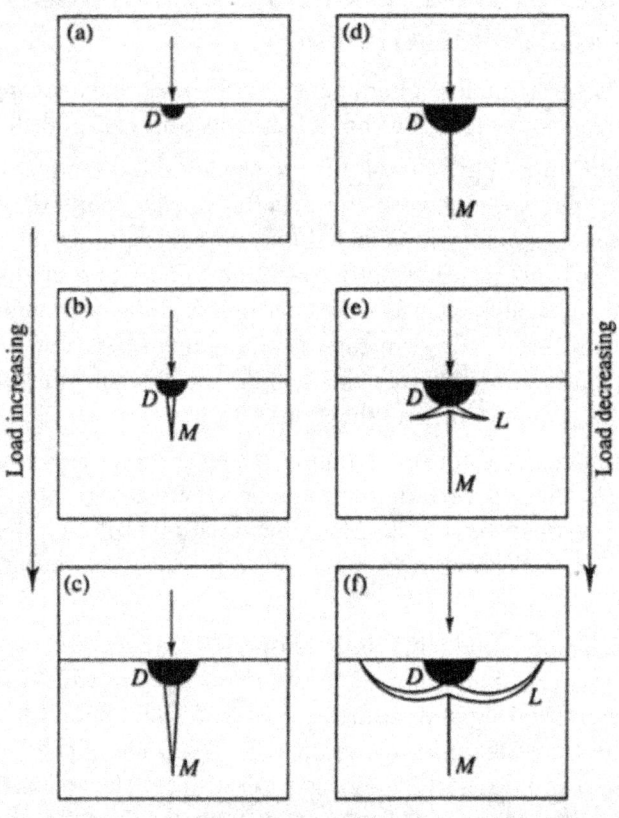

Figure 18. Formation of crack on brittle material (Lawn and Swain, 1975).

Material wear observed in the surfaces analyzed under electron microscope are in these manners; breaking of pieces by breaking of lateral cracks in parallel with the surface, big cracks resulting from breaking of grains on the surface, breakages resulting from uniting of other cracks and radial cracks and cutting movement that produce chip like metals. Formation of mechanism is proportionate to the load on abrasive grain. If load affecting abrasive grain is little, plastic micro cutting or escalloping mechanism is dominant. Surface that is formed with this process is very remarkable with it smoothness. Plastic micro-cutting movements cause creation of chip. If big loads affect discs, brittle cracks are formed on the surface. The most common types of material loss that is caused by brittle cracks are –as mentioned before- lateral crack breakings, breaking of grains on the surface and breaking of pieces from the surface in the shape of spalling. In order to understand the mechanism better, the model developed by Chandrasekar and Farris (1987) is given in Figure 19.

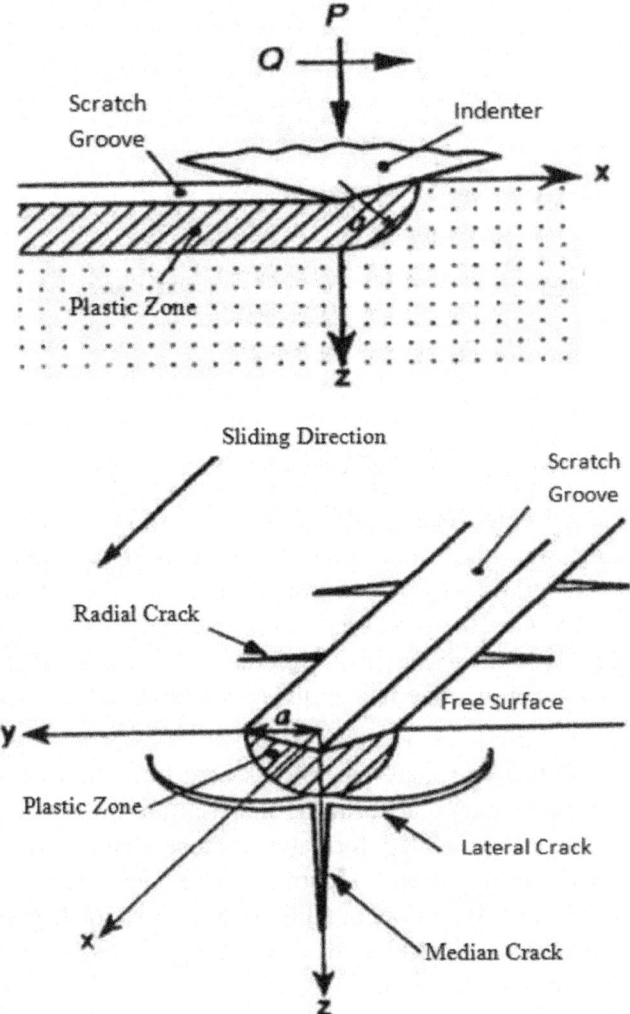

Figure 19. Slipping of grinding indenter on brittle surface and schematic demonstration of fracture after the process.

Here, it is thought that one single abrasive grain slipped over the surface and groove. Normal load is very low and groove following a permanent change in the shape without breaking. It is assumed that middle cracks are vertical to the surface and the depth is in direct proportion to the size of normal power applied on abrasive grain. Middle crack starts to unite with lateral cracks in parallel with the increase in normal power. At high loads, lateral cracks are broken and cause material wear. Again at high loads, scratches break along the middle crack and material wear occurs. A similar model is developed byRegiani et al. (2000) (Figure 20).

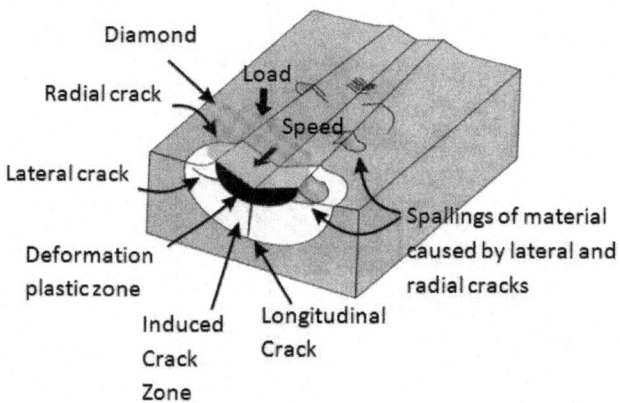

Figure 20. Cracks on the surface formed during abrasion (Regiani et al., 2000).

According to Regiani et al. (2000), basic mechanisms in material wear are; grains breaking, smashing, formation of ductile chip, and spalling. Wear of these types of materials are affected from various variables. Viscosity of used liquid, applied force on the disc, type of the disc and small scaled form of the material that is abraded are the most important of these. Small scaled form of the material has a significant effect on the development of crack that is formed as a result of abrading.

Regiani et al. (2000) stated that small scaled formation whose shape and crystal lengths of crystals that form the material are more enduring to wear than homogenous small scaled formation whose shape and whose crystal grains' lengths are similar. Again, according to Regiani et al. (2000), intrusion, gaps and crystal grain limits behave like borders for crack progress at each type of abrading process. Used liquid, size and type of disc have the secondary importance in removing material.

As a result, volumetric wear according to different operations haven't been revealed yet. Complete understanding of wear will enable the development of productive abrading processes that will create smooth surfaces.

WEAR AND POLISHING MECHANISM FORMED IN THE USE OF SANDPAPER

It is the name of grinding product formed as a result of covering abrasive grains on sandpaper, paper or on a cloth with a binding agent. Sandpaper is widely used for abrading and polishing of surfaces that are made of metal, glass, porcelain, stone, wood...etc. materials.

When glass machine profile is analyzed, it will be seen that abrading is done with abrasive grains that are lined at different positions. Effective factors in abrading process are; disc grain shape, solidity and height, its angle to the surface, applied load and form of bonding material and so the life of paper.

Contact angle of abrasive grain and surface is the most significant effective factor on wear as is given in the definition of adhesive wear.

Abrasive grains make cutting, friction and break through movements in different amounts according to the edge structure and location. Most of the studies focus on chip formation with cutting and occasions according to this.

Chip on the surface that is created by abrasive grain is given in Figure 21 in a simplified way.

In Figure 22, chip formation on brittle material is presented. Simplified chip formation model show that change of shape generally occurs in narrow space at shear plane or in an area called shear area (Figure 21). Permanent change of shape is complex in this area. But it is probably in the shape of hydrostatic component tensile type that will stay on the new surface. Namely, crack that will enable the formation of new surface is tension crack.

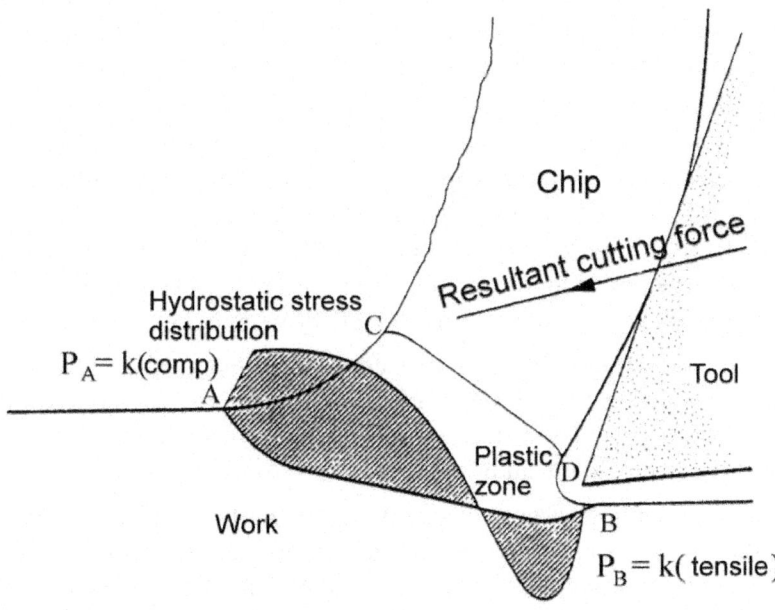

Figure 21. Chip formation model (Samuels, 1971).

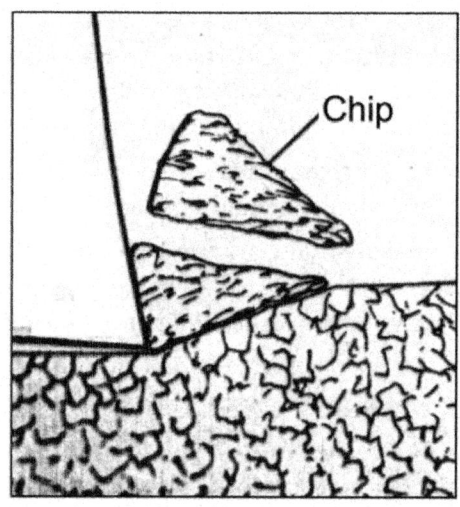

Figure 22. Discontinuous chip creation on brittle material surface (Boothroyd, 1975).

According to Boathroyd (1975), contact angle between abrasive grain and surface is very significant in the process of abrading in order to determine chip formation. on the other hand, deformation distribution of chip area is also affected from this contact angle.

As a result, there is a limiting contact angle for abrasive grains and grinding tip cuts a chip on this angle while it grooves under it. When abraded surfaces are analyzed, it was seen that very few of the scratches are formed as a result of cutting of material in the shape of chip (Samuels, 1971). Abrasive grain grooves on contact points on material surface mostly by breaking through and friction but it removes small amount of material. Very few of them create chips with grating movement and this is more effective in removing material.

Rapid material wear is ensured by applying high loads and low decreasing speed in proportion to high amounts of cutting points, namely by low wear speed of sandpaper. These factors ensure high abrading speed and prevent formation of smooth and shiny surface.

Another intended purpose of sandpaper and cloth is polishing. Mechanism in polishing process is in fact very similar to abrading. It can be said that force affecting each abrasive grain determines depth and width of scratches on the surface of basic material. So, in order to make polishing operations on abraded surfaces, very low loads should be applied on abrasive grain and abrasive grain's height should be very low. Although sandpaper is used in the first step of polishing that is made with sandpaper, polishing cloth (Figure 23) is used in the last step in order to lessen the scratch depth and form shinier surfaces.

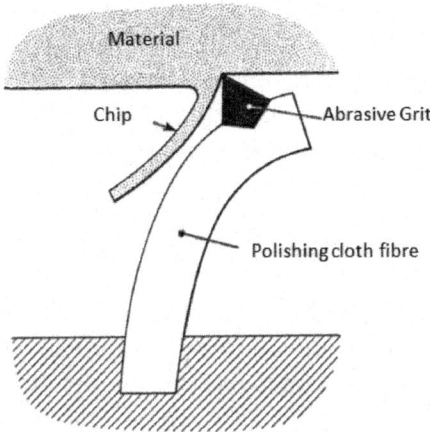

Figure 23. Mechanical polishing mechanism of abrasive grain that is clung on the polishing cloth fibre.

As can be seen in Figure 23, abrasive grains in polishing cloth is inside the fibers of cloth. Grains affect on the material surface only with elastic force and ensure the creation of narrower and shallow scratches. In this way, shinier surfaces are achieved.

REFERENCES

1. J. F. Archard, 1953Contact and Rubbing of Flat Surfaces, Journal of Applied Physics, 24, 981988

2. G. Beilby, 1921Aggresion and Flow of Solids, Macmillan, London.

3. G. Boathroyd, 1975Fundamentals of Metal Machining and Machine Tools, Scripta Book Company, Washington D. C., 64106

4. A. Bortolussi, R. Ciccu, P. P. Manca, and G. Massacci, 1994Computer Simulation of Diamond-Wire Cutting of Hard Rock and Abrassive Rock, IMM, 103August, A55A128

5. F. P. Bowden, and T. P. Hughes, 1937Proc. Roy. Soc., 575

6. S. Chandrasekar, and T. N. Farris, 1997Machining and Surface Finishing of Brittle SolidsSadhanaAcad. Proc. Engineering Sci.,, 22473481

7. X. Chen, and W. B. Rowe, 1996Analysis and Simulation of the Grinding Process. Part 2 Mechanics of Grinding, International Journal of Machine Tools and Manufacture, 368883896

8. L. Coes, 1971Abrasives, Springer-Verlag, New York, 177

9. K. Kato, K. Hokkirigawa, T. Kayabo, and Y. Endo, 1986Three Dimensional Shape Effect on Abrasive WearJournal of Tribology346351

10. J. Konstanty, 2002Theoretical Analysis of Stone Sawing With DiamondsJournal of Materials Processing Technology123146154

11. B. R. Lawn, and M. V. Swain, 1975Microfracture Beneath Point Indentations in Brittle SolidsJournal of Material Science, 10, 113122

12. D. F. Moore, 1975Principles and Application of Tribology,Mechanical Engineering Publication Limited, London, 176186

13. Y. Ozcelik, S. Kulaksiz, M. C. Cetin, 2002Assessment of the wear of diamond beads in the cutting of different rock types by the ridge regressionJournal of Materials Processing Technology, 127 (2002), 392400

14. I. Regiani, C. A. Fortulan, and B. M. Purquerio, 2000Abrasive Machining of Advanced Ceramics, IDR, 1/2000, 3765

15. S. C. Salmon, 1992Modern Grinding Process TechnologyMcGrawHill Inc.83101

16. L. E. Samuels, 1971Metallographic Polishing by Mechanical MethodsAmerican Elsevier Publishing Company, Inc., Newyork, 224 p.

17. N. P. Suh, and N. Saka, 1978Fundamentals of Tribology, Proceedings of the International Conference on the Fundamentals of Tribology, M.I.T. press, 400405

18. J. D. Summers, 1994An Introductory Guide to Industrial TribologyMechanical Engineering Publication Limited, London, 176186

19. C. Y. Wang, and R. Clausen, 2002Marble Cutting with Single Point Cutting Tool and Diamond SegmentsInternational Journal of Machine Tools & Manufacture, 4210451054

20. J. A. Williams, 1994Engineering TribologyOxford University Press, New York, 167199

Chapter 6

SIMULATION OF ASYMMETRIC DESTABILIZATION OF MINE-VOID ROCK MASSES USING A LARGE 3D PHYSICAL MODEL

X. P. Lai[1,2], P. F. Shan[1,2], J. T. Cao[1,2], F. Cui[1,2] and H. Sun[1,2]

[1] School of Energy and Mining Engineering, Xi'an University of Science and Technology, Xi'an 710054, China

[2] Key Laboratory of Western Mines and Hazard Prevention, Ministry of Education of China, Xi'an 710054, China

ABSTRACT

When mechanized sub-horizontal section top coal caving (SSTCC) is used as an underground mining method for exploiting extremely steep and thick coal seams (ESTCS), a large-scale surrounding rock caving may be violently created and have the potential to induce asymmetric destabilization from mine voids. In this study, a methodology for assessing the destabilization was developed to simulate the Weihuliang coal mine in the Urumchi coal field, China. Coal-rock mass and geological structure characterization were integrated with rock mechanics testing for assessment of the methodology and factors influencing asymmetric destabilization. The porous rock-like composite material ensured accuracy for building a 3D geological physical model of mechanized SSTCC by combining multi-mean timely track monitoring including acoustic emission, crack optical acquirement, roof separation observation, and close-field photogrammetry. An asymmetric 3D modeling analysis for destabilization characteristics was completed. Data from the simulated hydraulic support and buried pressure sensor provided effective information that was linked with stress–strain relationship of the working face in ESTCS. The results of the 3D physical model experiments combined with hybrid statistical methods were effective for predicting dynamic hazards in ESTCS.

INTRODUCTION

There are a substantial number of coal seams around the western part of China that have seam thicknesses of greater than 20.0 m and angles exceeding 45°, which are referred to as extremely steep and thick coal seam (ESTCS). In complicated geological and mining settings, coal-rock masses of mine voids that fail after coal mining may be induced by the hybrid physical-mechanics coupling effects of tectonic stress, excavation disturbed stress, and seismic forces (Alehossein and Poulsen 2010; Lai et al. 2014). Under such high stress-disturbances, asymmetric deformation of coal-rock masses in mine voids is significantly different in ESTCS (Kose and Tatar 1997; Lai et al. 2009a, b). Local strength degradation of surrounding rock may occur by asymmetric destabilization of mine voids that evolve into dynamic derivative hazards and are referred to as coal burst and dynamic collapse (Cai et al. 2005).

In the Urumchi coal field, located in the northwestern meizoseismal region of China (Dzungaria Basin), there are approximately 3.6 billion tonnes of measured and indicated coal that lie in the ESTCS category (Zhang and Lai 2008). The traditional longwall mining technique is unstable for the ESTCS with significant coal loss and poor surrounding rock structure, which not only can influence the production efficiency, but also may induce more potential hazards from mine voids similar to dynamic rock mass destabilization (Simsir and Ozfirat 2008a, b; Lauriello and Fritsch 1974; Tu et al. 2009; Kelly et al. 2001).

As a given coal mining method, mechanized SSTCC, a particular kind of top coal caving (Xie and Zhao2009), provides a powerful method to extract ESTCS in such a complicated underground circumstance. Specifically, top coal caving in ESTCS is a completely different method compared with longwall top coal caving (LTCC) in gently inclined coal seams due to limited lengths of the working faces (Vakili and Hebblewhite 2010; Zhang et al. 2011; Unver and Yasitli 2006). Generally, the length of the working face is shorter than the maximum length of 50.0 m, and the multi-narrow mine voids are above the mechanized SSTCC workings. Hence, the variability of stress and deformation in ESTCS are different from that of LTCC in gently inclined coal seams (Lai et al. 2014; Cao et al. 2011; Miao et al. 2011).

There have been numerous efforts to solve the relative hazards of dynamic rock mass destabilization with various methods (Lai et al. 2014). However, previously published literature on the destabilization characteristics of surrounding rock mainly focused on understanding the destabilization mechanisms of LTCC that are involved in slope–stress analysis, hazard–control design, and optimization and policy-making (Yasitli and Unver 2005). Jha and Karmakar (1992) investigated the factors affecting production, while Singh et al. (1992) investigated strata behavior during caving and dilution of caved top coal under in-situ conditions. Singh (1999) determined the formation and behavior of the immediate roof on 2D physical models in the laboratory, and Dian (1992) made a comparison between slice mining and top coal caving methods.

Wu (1992) tried to determine factors affecting the dilution of coal. Yasitli and Unver (2005) used the finite difference code FLAC^{3D} to optimize the crucial parameters of LTCC at the Omerler Underground Mine with in-situ conditions. However, there are a limited number of studies (e.g., Trueman et al. 2008) on mining dynamic hazard in ESTCS with effective three-dimensional physical model that can directly reveal the cause of coal-rock mass failures from mine voids after coal mining. Few studies have referred to evolution mechanism of mine voids with induced stress in meizoseismal region.

The mining dynamic hazards that have occurred in ESTCS at Urumchi Coal field are novel. Hybrid monitoring combined with 3D physical modeling is one way to theoretically predict dynamic rock mass hazards. Traditional prediction approaches for dynamic hazards were not applicable for ESTCS, and were neither analyzed nor studied in detail using a monitoring index (acoustic-seismic-wave) and their implicit relations. Particularly, the stress of deeper coal-rock mass was disturbed by collapse of the steep-broken coal-rock masses, and the magnitude of broken intensity related to the spatial–temporal-strength relation and mining disturbance. Inevitably, the indicated stress redistribution of coal-rock masses leads to the occurrence of dynamic hazards (Diaz Aguado and Gonzalez 2009).

Destabilization in the host rock mass was largely controlled by the discontinuities and structural features. In this paper, a methodology was developed for assessing destabilization potential of the host rock mass from mine voids.

The analytical method illustrated in Fig. 1 involves the development of a geomechanical model to account for asymmetric destabilization characteristics of the host coal-rock mass from mine voids. Development of a detailed geomechanical model is described and involves amalgamation of in-situ geological and mining settings, necessity of large-scale model, model material, and model device selection with hybrid monitoring. The 3D physical model provides information that can be used during mine excavation in ESTCS where asymmetric destabilization of host rock masses will be minimized and will make deep mining with effective hazard prevention possible.

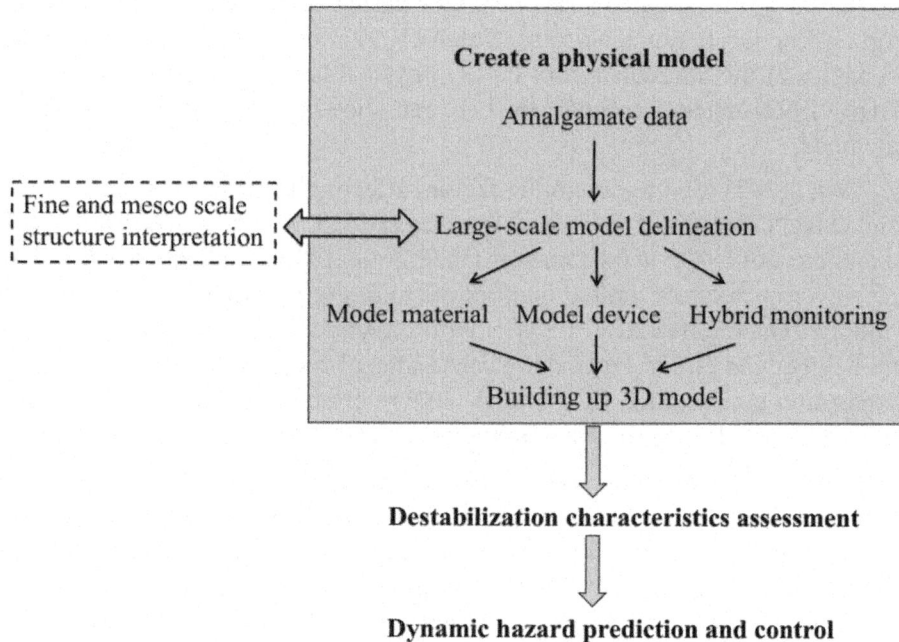

Figure. 1: Analysis of methodology for assessing asymmetric destabilization characteristics of surrounding rock from mine voids.

BACKGROUND

In-situ Geological and Mining Settings

The Weihuliang coal mine at Urumchi coal field serves as a case study to illustrate the asymmetric destabilization characteristics of rock masses from mine voids (Fig. 2). The Weihuliang coal mine is located in the pre-Cenozoic strata, approximately centered at the city of Urumchi. In-situ geological and physical settings of the case study are totally distinct (Lai et al. 2014). Fierce crushing stress is a crucial characteristic of tectonic stress and the seismic mechanism is mainly influenced by fracture of a thrust fault. Crustal movement and tectonic activity frequently occur and result in diastrophism velocity reaching up to 10 mm per year. Diastrophism is an elastico-plastic behavior, whereby plastic-shear failure occurs in the faults with strong mobility. Seismic loading would induce rock mass slide and collapse in large underground chambers when coal is extracted near past fault or well-developed joint fissure zones.

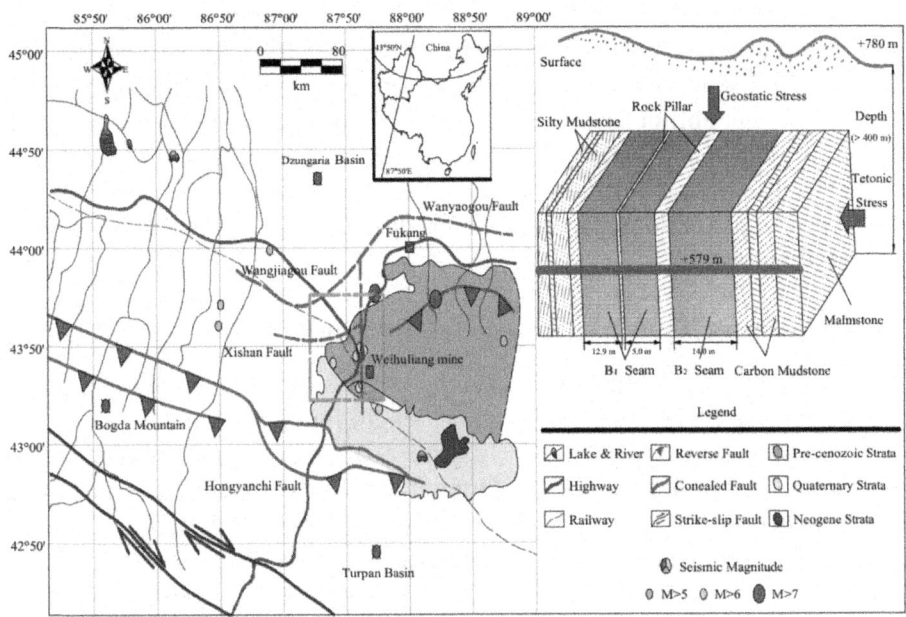

Figure. 2: Profiles of the Urumchi coal field including geological situation and sectional map with main lithostratigraphic units.

The Xishan–Wangjiagou fault group of the Urumchi coal field is located in a conversion region between a fault-fold structure of northern Tianshan Mountain and a nappe structure of Bogda Mountain. The fault-fold structure of northern Tianshan Mountain is composed of several reverse faults whose spatial span is about 30 km. The basin with a southward nappe is formed by the Xishan fault group where the depth of decollement is greater in the north and shallower in the south but generally is 10–15 km. All faults in the Xishan fault group converge to the decollement whose cover depth is about 11 km. Urumchi coal field is composed of the Badaowan syncline and Qidaowan anticline. Weihuliang coal mine is located in the Badaowan syncline with other coal mines. The total thickness of the Weihuliang coal mine is 513.77–902.90 m with 40 m of coal-bearing strata above the mine and a total thickness of approximately 117.05–175.45 m. Strike and dip angles of minable seams are 322°–335° and 46°–67°, respectively. The seam angles vary from 63° to 88° in ESTCS. B_1 and B_2 seams are extremely thick and available for practical excavation in the J_2x_1 strata. The mechanical condition of the B_{1+2} seams is relatively poor, because the incompact roof and floor do not meet prerequisite characteristics that are suitable for large-level mechanized SSTCC. However, large-scale "V" collapse grooves are formed in the surface after top coal caving and would allow precipitation to pour into the collapse grooves and induce destabilization of roof surrounding rock. Vast noxious gases assembled in the mine voids have been developed by irregular mining and continual excavation disturbance. Usually, the extra-high section height of the top coal does not hinder abrupt large-scale collapse of rock masses that may be induced by local variations of horizontal stress, methane accumulation, mine flood, or fire and coal dust with recovery process; hence, it is necessary to develop an effective methodology for assessing destabilization of rock masses with potential hazards.

Mechanism of Asymmetric Destabilization

The working face of the mechanized SSTCC could be horizontal to the direction of an ESTCS with short actual length for all staff operation. A finite-size excavation disturbed zone (EDZ) after coal mining would be regarded as a neo-active-like fault structure. Deformation of coal-rock masses is intensively concentrated on a narrow zone called an asymmetric destabilization zone under high strain rate loading. Theoretically, we have a stability analysis method for mechanical conditions and stress–strain relationships of plastic destabilization for the asymmetric destabilization of surrounding rock with abundant numerical simulations, which provides a profound understanding of asymmetric destabilization characteristics.

We pay little attention to micro-structures, which play a key role in prediction and control of the asymmetric destabilization. Because in-situ monitoring methods are limited for micro-structure evolution mechanisms on dynamic mechanical response parameters of asymmetric destabilization, we have seldom effectively uncovered the destabilization mechanism. Also, we should analyze the asymmetric destabilization characteristics with hybrid microscopic and macroscopic methods. Currently, we have developed a series of basic experimental materials and structures.

- Fine-scale (millimeter size) rock damage test was applied for quantitative analysis of mechanical properties with MTS-AE (acoustic emission) in varisized specimens.

- Multiphase medium physical simulation meso-scale (centimeter size) experiments have been used to realize spatio-temporal evolution characteristics of different media.

However, a specific large-scale experiment (meter size) should be built to scientifically realize mechanical conditions, evolution mechanisms, and macro-structure characteristics of the asymmetric destabilization. Particularly, asymmetric destabilization of coal-rock masses should be accurately predicted and used for optimization of operational facilities.

Equations on the asymmetric destabilization of coal-rock masses after coal mining are established by state vectors theory. "Damage-deformation-AE" characteristics of coal-rock masses have been revealed with different static–dynamic loading modes and were used to predict the asymmetric destabilization characteristics based on AE state vectors. The model would be regarded as a countable region that was composed by N continuous units.

In time t, $V_i(x,y,t)$ was the ith weight of $V(x,y,t)$ in the ith sub-domain, and N-dimensional vector $\vec{V}(t)$ V→(t) has been established as Eq. (1).

$$\vec{V_t} = (V_1, V_2, \cdots V_i, \cdots V_n) \qquad (1)$$

Difference value of N-dimensional vector $\vec{V}(t)$ would be another N-dimensional vector in adjacent two time $t - \Delta t$ and t as Eq. (2).

$$\Delta \vec{V}_{t-\Delta t,t} = \vec{V_t} - \vec{V}_{t-\Delta t} \qquad (2)$$

To describe the $\vec{V}(t)$ change, a new scalar value, module of N-dimensional vector $\vec{V}(t)$, M_0 was applied to state general level of the whole physical field as Eq. (3).

$$M_0 = |\vec{V}(t)| \qquad (3)$$

We defined angle (Φ) as being between $\vec{V}_{t-\Delta t}$ and $\vec{V_t}$ Another module of $|\vec{V}_{t-\Delta t,t}|$, ΔM_0 stated physical field variation in the other view. The value of M_0 and ΔM_0 explained coal-rock masses damage situation. They would increase distinctly with internal damage increased in the model, which can be described by AE total event and energy rate.

SIMULATION PREPARATION

Determining Similarity Coefficient and the Model Device

A particular model frame is required for developing a specific large-scale experiment. The model device must initially be stiff enough to ensure the stability of the model in the experiment. With current requirements in a simulation experiment, five kinds of rock masses were considered including carbon mudstone, silty mudstone, malmstone, B_1, and B_2 seam. Based on the geological section of the coal seam, the angle of the coal seam is over 64° and falls into the extremely steep and thick coal seams. We determined that, in the simulation scope, the excavation level of the Weihuliang mine had already dropped to +574 m with the mechanized sub-horizontal section top coal caving from +625 m for the whole excavation level. So, approximately 49.00 m was considered in the vertical direction. According to the geological section, the thickness of the simulation seam was 110.00 m, where the thicknesses of the B_1 and B_2 seams were, respectively, 16.90 and 14.00 m. For the actual simulation of the excavation in the trend, 72.50 m was used as the width for the simulation because it represented the advance distance (per month) of the working face. Finally, we determined that constant of geometric similarity (C_l) with comprehensive consideration of all experimental conditions using:

$$C_l = l_m / l_p = 0.04 \qquad (4)$$

In which, m represents the model parameter, and p is the parameter of in-situ rock masses. According to the properties of the simulation material and the ratio of the rock masses, the constant of density similarity (C) is represented by Eq. (5).

$$C_\rho = \rho_m / \rho_p = 0.68 \qquad (5)$$

Using the similarity principle and dimensional analysis, the constant of geometric similarity (C_l), the constant of density similarity (C), the constant of stress similarity (C), and the constant of time similarity (C_t) are related by Eq. (6).

$$C_\sigma = C_\rho C_l C_t = (Cl)^{1/2} \qquad (6)$$

The constant of stress similarity is 0.0272, and the constant of time similarity is 0.2. Considering the depositional pattern of the original strata, we established the large 3D model with 0.01 m in length direction. Due to the complex geological settings, an extensive number of faults were distributed throughout the research area. The model was unable to simulate all geological structural planes; thus, we emphasized the crucial fault that closely neighbored the B_2 seam. Furthermore, a projected coordinate system was used to locate all parts accurately in the model.

The model device not only applied geostatic stress as an analog ground stress, but also co-located loading equipment in the horizontal direction simulating in-situ tectonic stress as an auxiliary stress and as a dynamic load for the model. Ultimately, large underground structure experiment system (LUSE) was modeled as (length × width × height) 4.42 m × 2.90 m × 1.95 m in the Xi'an University of Science and Technology (XUST), China. The greatest sizes of the system is 8.80 m × 5.60 m × 12.00 m.

The large underground structure experiment system, to date is the largest device of its kind in Asia, and has several favorable aspects including loading ability, geometric size, and monitoring methods.

Physico-Mechanical Settings of Coal-Rock Masses

Core samples from geo-climatic boreholes in the experimental site were classified using the rock quality designation (RQD) system. The designation system provides a quantitative method of categorizing the engineering character of rock masses based on core hole information. A summary of the RQD for the boreholes is presented in Table 1, which shows, in general, that the RQD of rock masses of the extremely steep and thick coal seams at Weihuliang coal mine was low. Locally, mudstone and malmstone were broken and had extremely low RQD values.

Table 1: Rock quality designation (RQD) for the Weihuliang coal mine

Rock type	RQD (%)	RQD description	Integrality description
Carbon mudstone	21	Bad	Broken
Silty mudstone	38	Bad	Broken
Malmstone	29	Bad	Broken
B_1 seam	56	Medium	Relative integrality
B_2 seam	67	Medium	Relative integrality

The physico-mechanical properties of the main sedimentary rock masses used in the large physical simulation were based on evaluation of exploratory drilling and laboratory test results. In addition, the data utilized in the experiment were finished at Key Laboratory of the Ministry of Education of China for Western Mines and Hazard Prevention at Xi'an University of Science and Technology. The main properties used in the physical simulation were summarized in Table 2.

Table 2: Physico-mechanical properties of main sedimentary rock masses

Rock definition	Density (kg/m^3)	Bulk modulus (GPa)	Poisson ratio	Cohesion (MPa)	Friction angle (degree)	Compressive strength (MPa)
Carbon mudstone	2460	8.75	0.231	3.20	42.0	11.0
Silty mudstone	2510	13.50	0.217	1.20	30.0	18.4
Malmstone	2487	19.50	0.238	13.60	24.7	24.7
B_1 seam	1300	5.10	0.320	6.70	26.4	10.6
B_2 seam	1340	5.30	0.235	6.12	23.2	13.7

Determining the Model Material and Proportion

Much research work has been done on model material (Lai et al. 2012, 2013; Li et al. 2011; Pan et al. 1997). In the present work, we prepared a new kind of modeling material for establishing the specific three-dimensional model. Porous rock-like composite material (PRCM) adopted bank sand as the aggregate, with 3.9 % in the 0.50–1.00 mm size range, 29.1 % in the 0.25–0.50 mm size range, and the remaining portion in the 0.10–0.25 mm size range. We used bank sand and coal powder as the aggregate for the coal seam material being simulated. Gypsum and flour were both grouting agents and water was used as the plastic impact agent, which largely influenced the accuracy of the material proportion. The porous rock-like composite material was light material whose massive density was 1700 kg/m^3. Pebbles, clay, and mica were added separately to represent the fault and distinguish between different strata. Pebbles and clay were used to simulate the fault and improve the brittleness of the model material.

The nonlinear coupling effect of porous rock-like composite material is remarkable because of the plastic impact agent (Huang 2009; Lin 1984; Lai et al. 2013a, b). Because the porous rock-like composite material consisted of loose particles, their mechanical properties are anisotropic. Comprehensive strength of porous rock-like composite material depends on nonlinear coupling effect with water. In some cases, various ingredients would blend well with water, whereas, in other cases, the alternation effect of water on the material plays a key role in deciding strength of the modeling materials. Therefore, modeling of these effects must consider a comprehensive set of properties including static pressure (μ) and hydrodynamic pressure (τ_d). The effective stress of porous rock-like composite material (σ_a) could be divided into three stages (Lai et al. 2013a, b).

Stage I Modeling material is unsaturated. In such a situation, under an external loading, the pore water would not provide enough internal stress (σ) to create internal micro-cracks and dilatancy in the material and decrease the strength of the material. In this stage, along with increase of water, the unprofitability of static pressure would be gradually decreased and strength of the porous rock-like composite material is steadily increased while the effective stress of the material is increased, as expressed by Eq. (7).

$$\sigma_a = \sigma - \mu \qquad (7)$$

Stage II Modeling material is saturated. Under an external loading, drainage equilibrium process occurred among internal pores with the structure of the material stabilized, which would result in the skeleton and pore water bearing the external loading. The beneficial static pressure for the skeleton reaches peak value and the effective stress of the porous rock-like composite material reaches the maximum, as expressed by Eq. (8).

$$\sigma_a = \sigma + \mu \qquad (8)$$

Stage III Modeling material is oversaturated. Under such a condition, all pores are filled with flowing water that would add an extra hydrodynamic pressure and shear stress that results in incremental and tangential deformation in the skeleton. Micro-cracks are evolved into macro-cracks that may cause dynamic destabilization and damage the whole structure. Under these conditions, the mechanical properties of the porous rock-like composite material are less stable. Meanwhile, the effective pressure is sharply decreased, as expressed by Eq. (9).

$$\tau_d = R \times J \qquad (9)$$

In which, R and J are coefficient of hydrodynamic pressure and hydraulic slope angle, respectively. When flowing water streams in the porous rock-like composite material, macro-cracks would propagate remarkably forming the macro-crack plane, and the extra hydrodynamic pressure is parallel with the plane and internal static stress is perpendicular to the plane at the same time. So, R would be regarded as $br/2$. Here, b is weighted width of the macro-crack plane, and r is unit weight of the water in the experiment. Besides, J meets the Darcy's law.

The strength of the porous rock-like composite material was adjusted by changing contents of aggregate and grouting agents. Mainly, the compressive strength and bulk modulus of the porous rock-like composite material were analyzed. Standard coal-rock specimens (ϕ50 × 100 mm) with different ingredients have been made and each group had nine specimens with the same ingredients avoiding such errors. Plastic impact agent for the specimens in the experiment was water whose amount just occupied one in every ten units of the aggregate and grouting agents. Figure 3 showed stress–strain relationship of coal-rock specimens with various ingredients being adopted in the simulation.

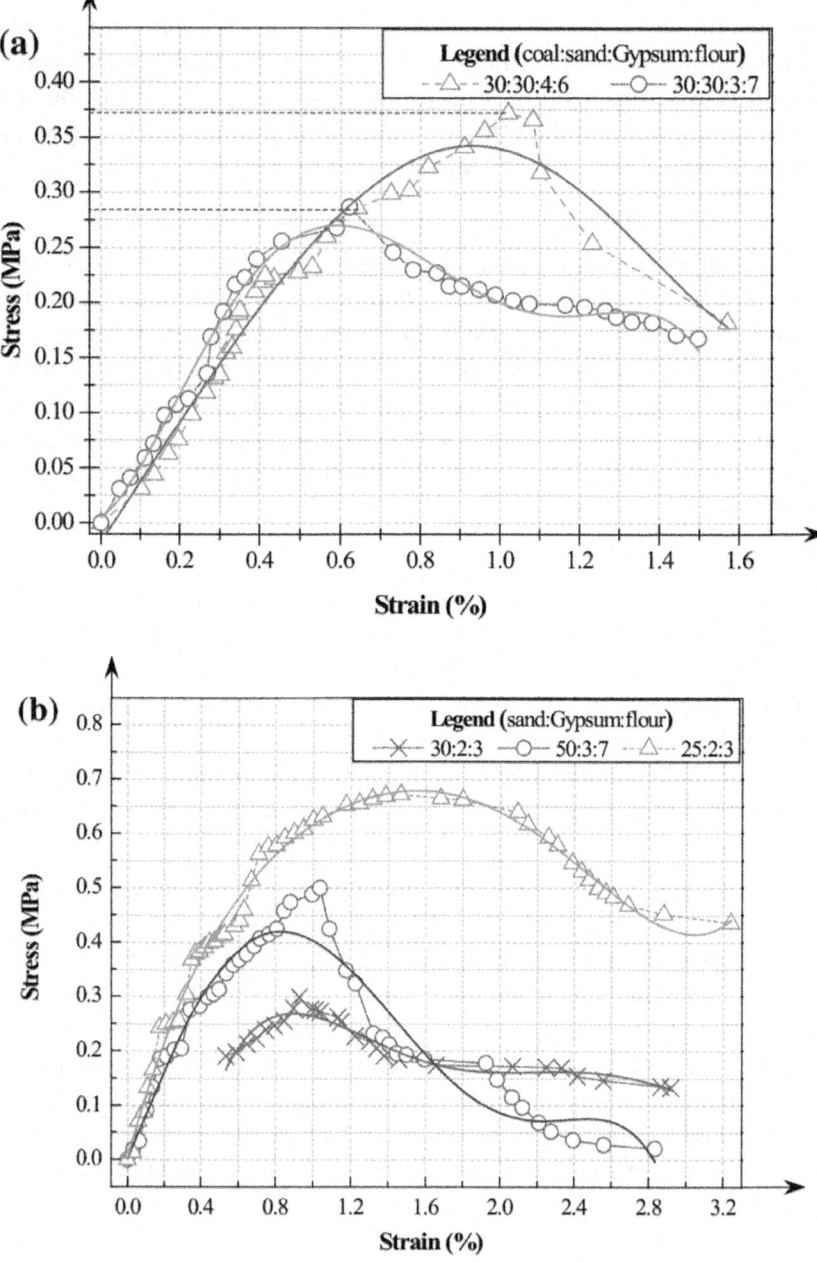

Figure. 3: Stress–strain curves of the porous rock-like composite material under uni-axial compressive loading for the experiment, **a** stress–strain relationship of the material for coal seams; **b** stress–strain relationship of the material for various rock masses.

Proportions (coal:sand:gypsum:flour) comprising the coal seam were 30:30:4:6 and 30:30:3:7, in the experiment. Maximum axial deformation of specimens with 30:30:4:6 was 1.6 %, and had an elastic deformation stage that was lower than the uni-axial stress, peak values of compressive strength and bulk modulus were 13.7 MPa and 5.3 GPa, respectively. While the proportion was altered to 30:30:3:7 with a reduction in the amount of gypsum, specimens obtained peak values of compressive strength and plastic stage. However, compressive strength and bulk modulus decreased to 10.6 MPa and 5.1 GPa, respectively. Compared with mechanical parameters of the prototype, we finally decided the model proportions of various rock masses. Tables 3 and 4 separately indicated the main mechanical parameters of the porous rock-like composite material and the state of model placement and the amount for different materials.

Table 3: Main mechanical parameters of the material in various proportions

Matching	Bulk modulus (GPa)	Compressive strength (MPa)	Field rock mass
Sand:gypsum:flour			
30:2:3	8.75	11.0	Carbon mudstone
50:3:7	13.5	18.4	Silty mudstone
25:2:3	19.5	24.7	Malmstone
Coal:sand:gypsum:flour			
30:30:3:7	5.1	10.6	B_1 seam
30:30:4:6	5.3	13.7	B_2 seam

Table 4: Specific matching ratio being adopted for the model

Rock definition	Model thickness (mm)	Total weight (kg)	Sand (kg)	Coal (kg)	Gypsum (kg)	Flour (kg)	Water (kg)
Carbon mudstone	140	437.08	374.64	0	24.92	37.52	43.10
Silty mudstone	2060	6427.20	5356.00	0	321.36	749.84	642.72
Malmstone	760	2371.20	1976	0	158.08	237.12	237.12
B_1 seam	676	2110.74	904.49	904.49	90.58	210.91	211.07
B_2 seam	560	1748.32	749.28	749.28	75.04	174.72	174.83

EXPERIMENTAL RESULTS

Model Experiment Layout

When the model was completed, we began to set all monitoring facilities and dry the model out for at least 2 months so the strength of the model were able to reach an ideal state. The mining and observation of the model complied with parameter of the constant of time similarity, and as a matter of experience and custom, we used 1 h to simulate one equivalent day. The roof collapsed method was adopted in the model and, first, caved the upper level seam at the +579 m level in Weihuliang mine to simulate the actual mine voids, as shown in Fig. 4a.

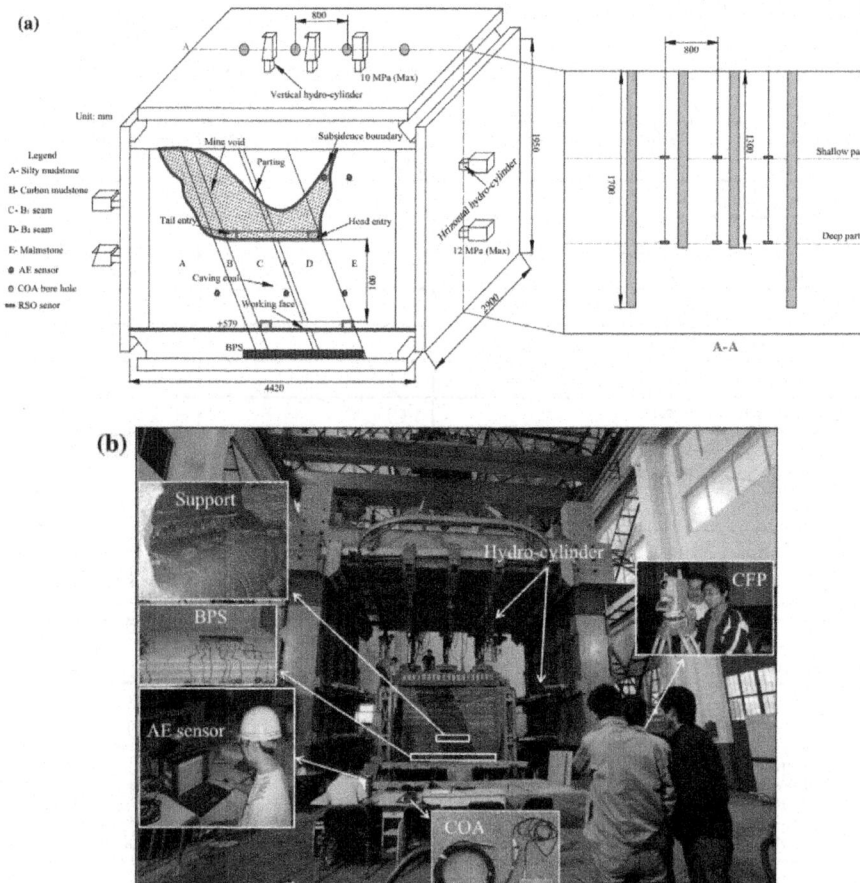

Figure. 4: Layout of model and multi-mean timely track monitoring, **a** scheme of the large 3D model experiment including general and sectional views; **b** presentation of all

monitoring facilities in actual experimental condition.

Multi-mean timely track monitoring including acoustic emission (AE) sensors, crack optical acquirement (COA), roof separation observation (RSO), and close-field photogrammetry (CFP) was applied. Data from the buried pressure sensor (BPS) was linked with the stress–strain relationship of the stope in the extremely steep and thick coal seams. Simulated hydraulic support (SHS) was simulated for parameter optimization of hydraulic supports in the extremely steep and thick coal seams. We measured the initial conditions of all facilities before coal mining. The mechanisms of asymmetric destabilization characteristics of rock mass from mine voids were clearly detected and analyzed, as shown in Fig. 4b.

Hydro-cylinders located on top and on the lateral side of the large underground structure experiment system exerted the asymmetric static–dynamic loading for the experimental purpose, which simulated the tectonic stress derived from fault-fold structures of the northern Tianshan Mountain and nappe structure of Bogda Mountain. The maximum loading of the top and lateral hydro-cylinders were 10 and 12 MPa, respectively.

AE Characteristics of Asymmetric Destabilization

Asymmetric destabilization characteristics would induce local deformation and failure of coal-rock masses, which are predisposed to mine hazards. Stress around the working face must be redistributed after coal mining by using abundant AE data on failure of the surrounding rock. We set up AE sensors in different locations to achieve information induced by rock failure in four stages, which were (I) excavation of working face, (II) first caving of top coal, (III) coal caving, and (IV) local roof caving. Figure 6 showed the mechanism of asymmetric destabilization-AE in various stages.

Initially, the model was excavated; the AE signal was less with seldom total and big events due to good condition of coal-rock masses (Fig. 5a), when the working face just advanced 0.5 m. With the working face advancing continuously, integrality of coal-rock masses was beginning to be broken and cracks were propagating. During first caving of the top coal, the AE signal waves appeared to skip. The crest value of the AE event occurred at the No. 33 monitoring serial number. The signal from the No. 5 AE sensor was weaker without severe rock mass failure than others, which indicated that coal excavation and first caving of top coal seldom affected the upper zone of the model monitored by the No. 5 AE sensor.

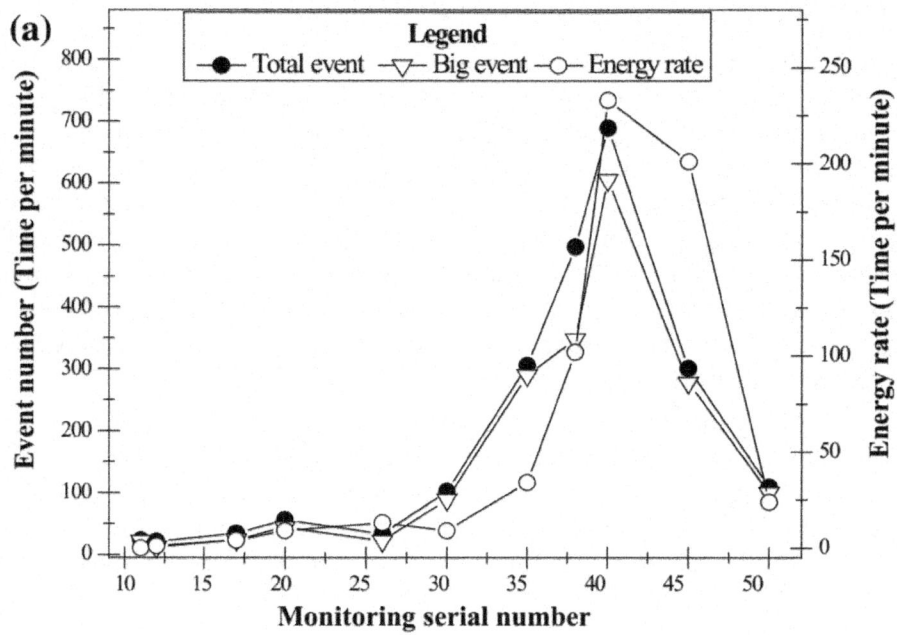

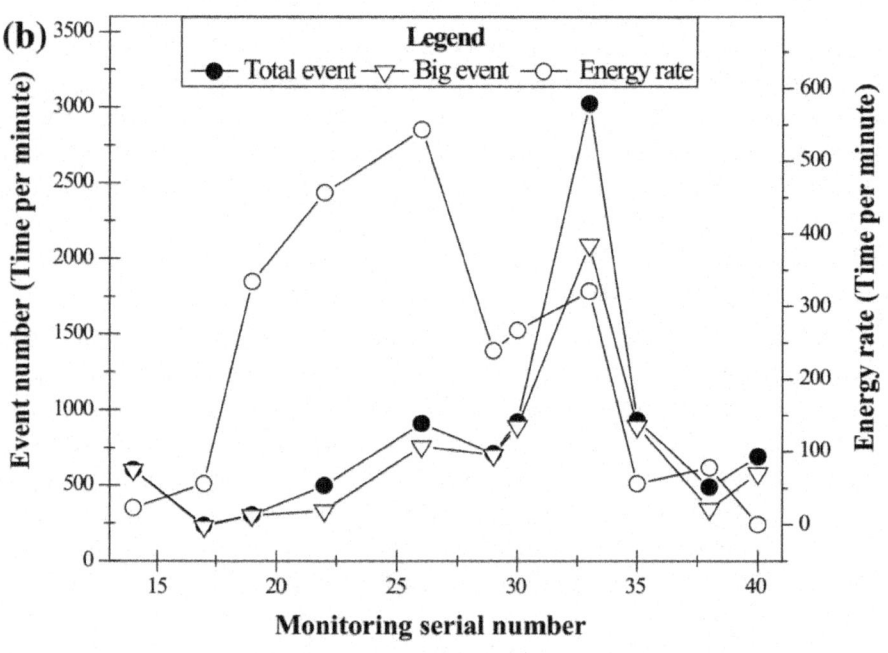

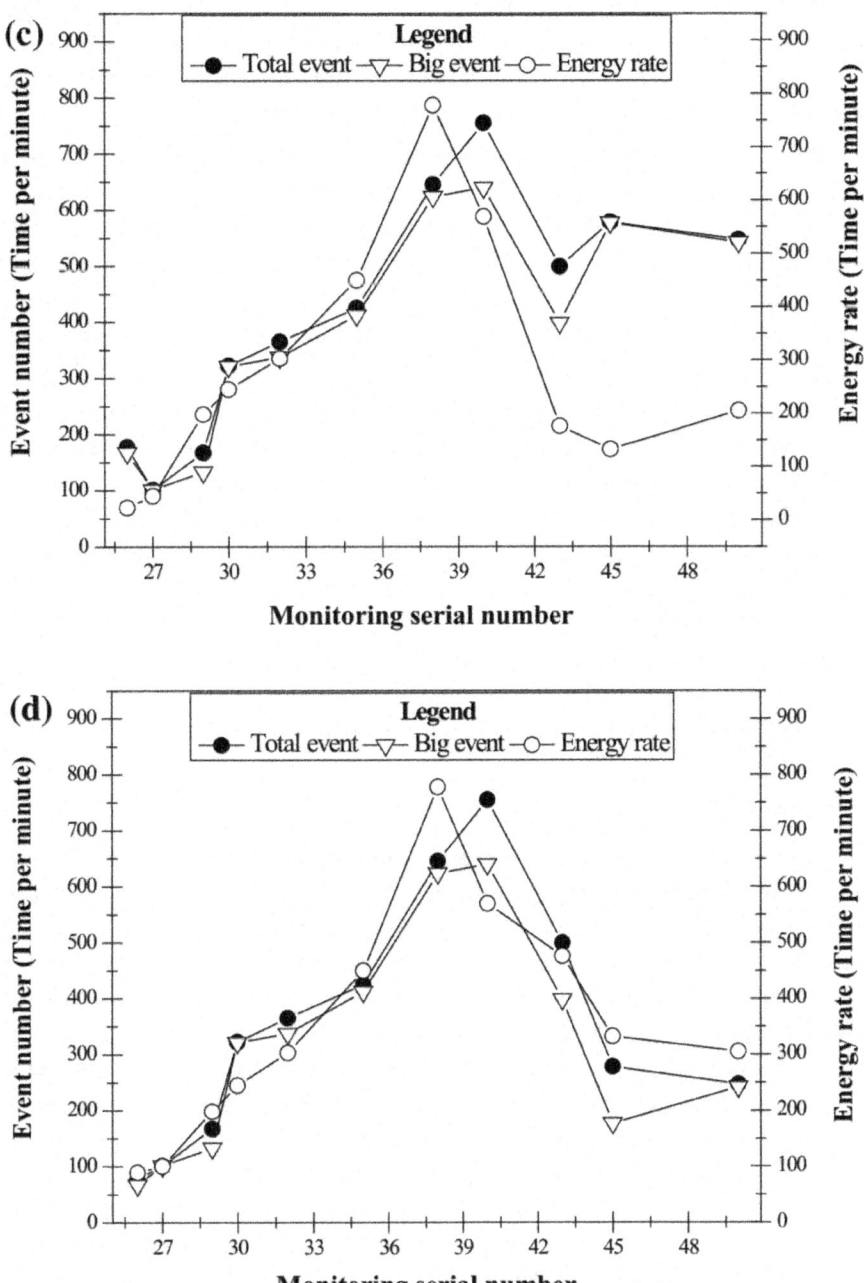

Figure. 5: AE parameters in various stages of the excavation: **a** working face; **b** first caving of top coal; **c** coal caving; **d** roof local caving stage.

Figure 5b indicated that, after first caving of the roof, several small-scale fluctuations of AE signals from the top coal emerged and cracks gradually stretched. Energy rate of coal-rock masses released abruptly up to 600, and then values of all parameters declined and tended toward a gentle state.

Figure 5c shows the parameter traits after coal caving. A large-scale roof collapsed suddenly and the AE signal surged indicating that added cracks propagated. The coal caving was a dynamic wave process where the total and big event increased to 700 and 650, respectively. The peak value of energy rate occurred at the No. 37 monitoring serial number. Thereafter, the energy rate was reduced to the point that caving no longer influenced the coal-rock masses.

The roof located upon the top coal started caving in a large scale, with Fig. 5d showing AE characteristics in the process. The AE signal rose and fluctuated repeatedly, with damage increasing to the coal-rock masses. The integrality of 3D internal part model continued degradation. According to AE trait analysis of the 3D model, the common mechanism of AE traits was summarized with parameters of the AE signal emerging as a commutative wave in time, which indicated that cracks in the coal-rock masses were propagating steadily after structural planes alternatively opened and closed. Large and small strain zones continuously alternated with time variation.

Model Asymmetric Deformation Characteristics

Internal Part of the Model

The center of the model was regarded as the origin of the coordinate system. We determined coordinate points for the No.1–No.6 boreholes for the crack optical acquirement. In preliminary stages of the experiment, the coal-rock masses were integral in all boreholes and seldom exhibited failure. After first caving of the top coal, coal caving, and roof local caving stages, the various hole walls were destroyed in different situations. Slabbing, lateral crack, oblique crack, scission, disintegration, swelling of hole walls, and oval deformation were the main failure types observed by the crack optical acquirement at the hole walls. Lateral cracks and slabbing emerged in the top coal caving process and were induced by gravity action of the overlying-strata, which resulted in asymmetric displacement of coal-rock masses. Oblique crack and scission were crucial failure types caused by internal shear stress of the 3D model. However, disintegration and swelling of hole walls were caused by stress concentration, oval deformation was caused by asymmetric static-dynamic axial loading acting on the boreholes. Failure of surrounding rock mainly occurred at 0.8–2.1 m. Rock masses of the hole tops and bottoms were seldom destroyed, as shown in Fig. 6a. Macro-cracks and joints were classified on the basis of orientation and depicted using a rose diagram of joints, as shown in Fig. 6b. We determined that six joint orientations were advantageous in the model including SE120°–130°, SW240°–250°, NW300°–310°, NW340°–350°, NE30°–40°, NE60°–70°.

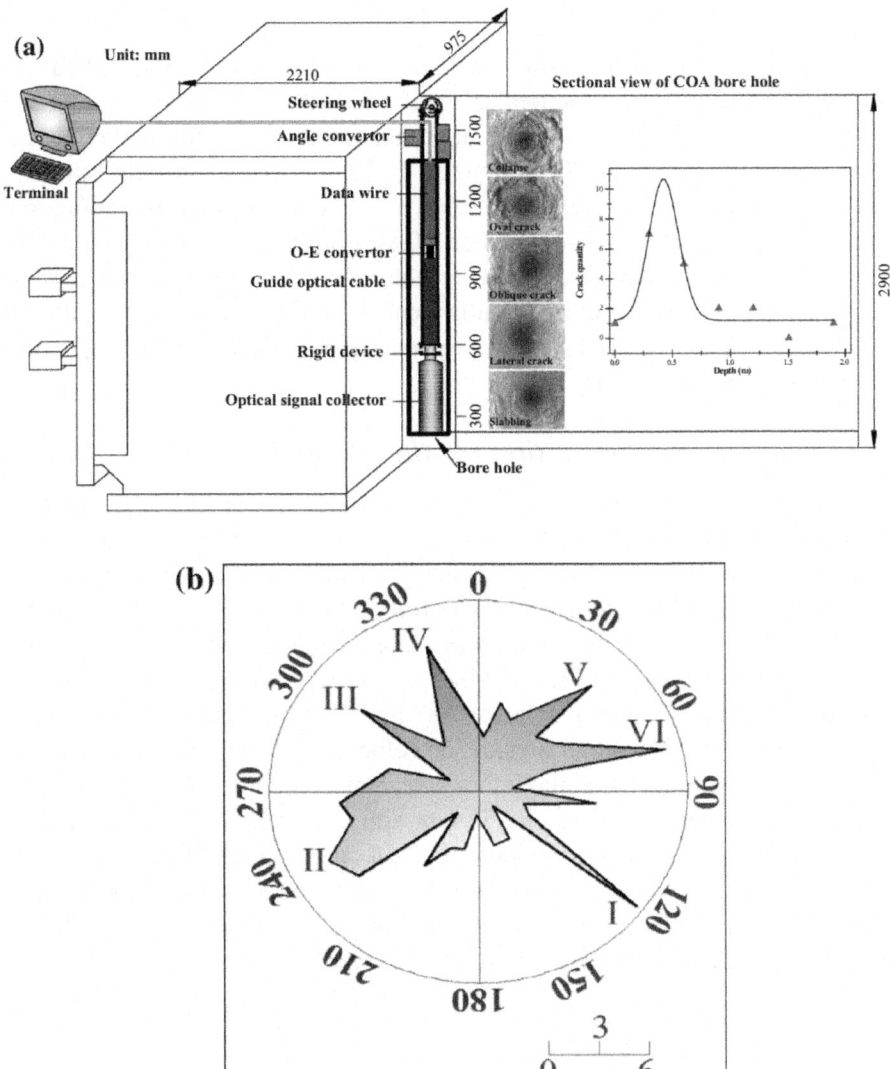

Figure. 6: Crack distribution characteristics around asymmetric destabilization rock masses, **a** shows hybrid assessment of internal deformation and crack distribution, also the specific layout of the crack optical acquirement is given. **b** Rose diagram of joints in the COA borehole.

Equation (10) indicated the quantitative statistical relations on macro-crack amounts (x) versus borehole length (y) and provides a theoretical basis for evaluation of asymmetric deformation in the internal part of the model.

$$y = 0.47206 + \frac{4.63924}{1 + e^{\frac{x-3.89602}{0.41562}}}$$

(10)

With the advent of length, crack amounts are fitted by statistical data plotting by the software. Generally speaking, fitting results originating from all bore holes are divided into three stages uniformly. First stage called fracture zone is from 0.0 to 0.5 m in the boreholes. The amount of macro-crack increases obviously in initial zone of the boreholes, and also reaches the maximum in the end. In sharp contrast, between 0.5 and 0.8 m in the boreholes (plastic zone), macro-cracks still propagate; however, the quantity of macro-cracks reduces rapidly. Asymmetric destabilization is weakened in the stage being fallen into elastic zone. When borehole length is greater than 1.0 m (elastic zone), amount of macro-crack stays steady.

Roof Separation Observation for the Model

In the experiment, we planted 19 roof separation facilities in the model that were separately located in the shallow and deep parts of the model for monitoring reparation in each stage. In Fig. 7a, the maximum subsidence and minimum subsidence of shallow parts after coal mining were 7.89 and 7.76 cm, and the average value was 7.84 cm with an average subsidence velocity of 0.163 cm per hour. After first roof caving, the maximum subsidence and minimum subsidence of the shallow part were 7.83 and 7.75 cm, and the average value was 7.77 cm with an average subsidence velocity of 0.194 cm per hour. In third stage (top coal caving), the maximum subsidence and minimum subsidence of the shallow part were 7.74 and 7.52 cm, and the average value was 7.63 cm with an average subsidence velocity of 0.254 cm per hour. When a large-scale roof caving emerged, the maximum subsidence and minimum subsidence of the shallow part were 7.48 and 7.10 cm, and the average value was 7.36 cm with an average subsidence velocity of 0.310 cm per hour.

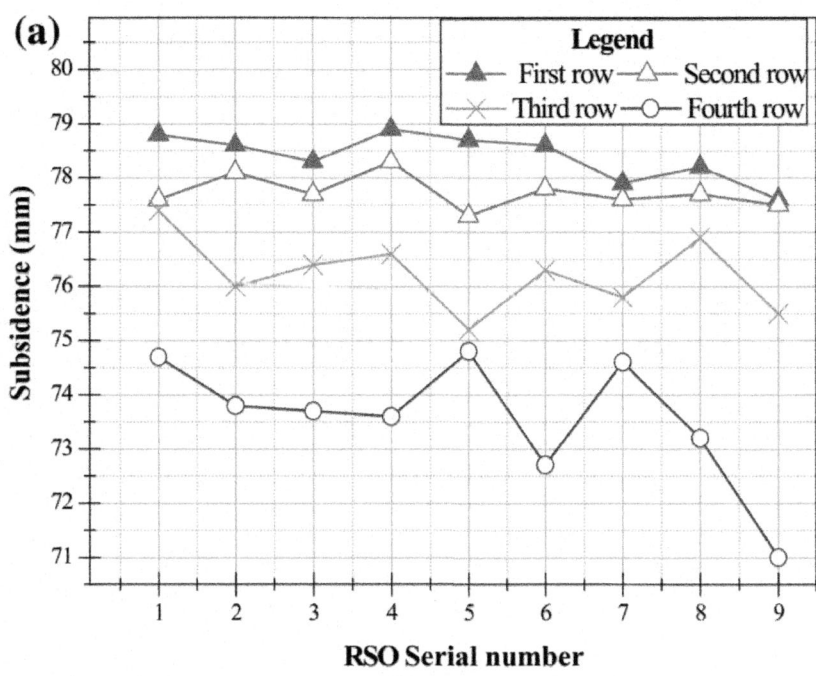

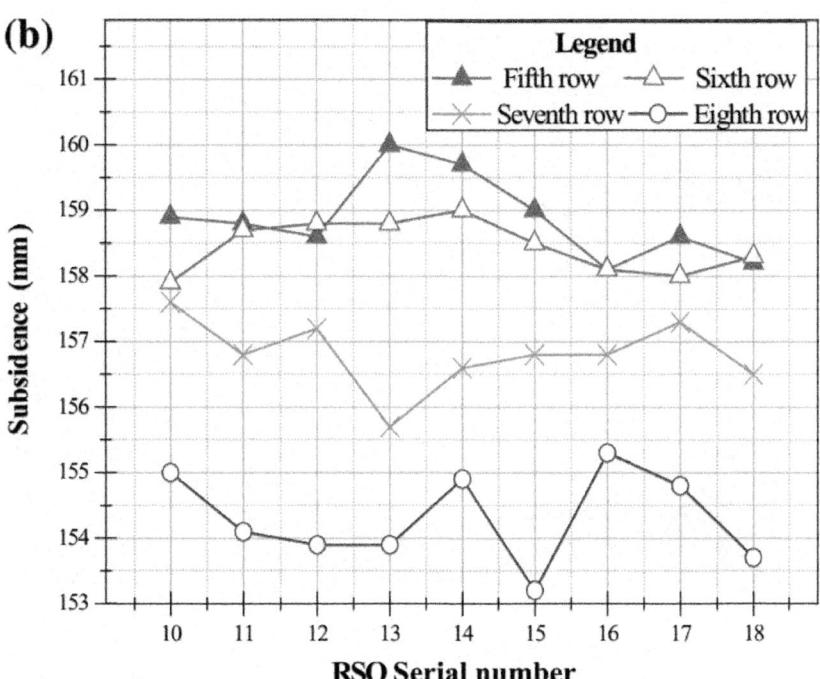

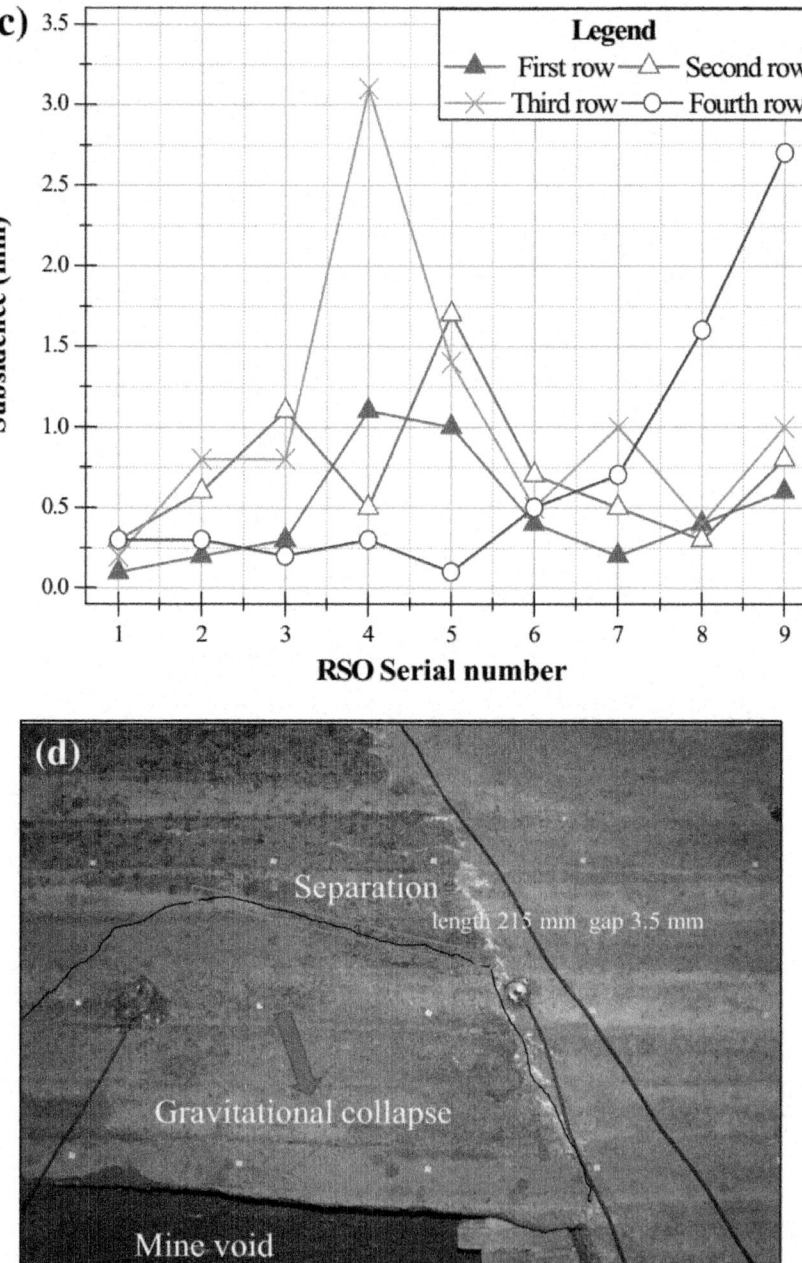

Figure. 7: Roof separation observation results for displacement of internal part of the model, **a–c** separately show roof separation characteristics. **d** Illustrates the actual separation emerged in the experiment.

The maximum subsidence and minimum subsidence of the deep part (Fig. 7b) was 15.91 and 15.81 cm, and the average value was 15.89 cm with an average subsidence velocity of 0.330 cm per hour after coal excavation. After first roof caving, the maximum subsidence and minimum subsidence of the deep part were 15.90 and 15.79 cm, and the average value was 15.87 cm with an average subsidence velocity of 0.396 cm per hour. Also, in third stage (top coal caving), the maximum subsidence and minimum subsidence of the deep part were 15.97 and 15.65 cm, and the average value was 15.73 cm with an average subsidence velocity of 0.520 cm per hour. Finally, the maximum subsidence and minimum subsidence of the deep part were 15.53 and 15.32 cm, and the average value was 15.43 cm with an average subsidence velocity of 0.640 cm per hour after the large-scale roof collapse. According to statistical analysis, accumulative maximum subsidence and minimum subsidence of strata (Fig. 7c) was 0.31 and 0.01 cm, and the average value was 0.07 cm with an average subsidence velocity of 0.014 cm per day.

From all of the above listed data, different stages had various influences on the internal situation of the model. Generally, the induced maximum subsidence and minimum subsidence were 15.53 and 15.43 cm. The crest value of separation was 0.31 cm. Under the three-dimensional loading action, separation always emerged in the model. When the working face advanced, the region of mine voids gradually developed and induced separation that evolved into macro-slabbing. Coal-rock masses near the working face would be destabilized in an oblique direction of the seams (Fig. 7d).

Surface Displacement for the Model

Surface displacement monitoring points were scattered at superficies of the model and were divided into seven rows (a–g) in which eleven points were arranged. The row spacing was 300 mm, which was prerequisite to predicting local defects of surrounding rock by observing the surface displacement of the model.

Figure 8 shows the close-field photogrammetry results with comprehensive statistics. The points in row a were so far away from the working face that the value was smaller, which indicated that initially top coal caving had less of an impact on the upper surrounding rock without large subsidence. Subsidence behavior of the points in row b were consistent with those in row a. In the process of top coal caving, subsidence of the model fluctuated violently, and the maximum value was 0.20 mm. Simultaneously, the crest value of subsidence in row c was 0.60 mm. and the average value was 0.35 mm. Subsidence of points in row d undulated sharply with the minimum values and maximum values being 0.22 and 0.58 mm, respectively. Subsidence in row e and row f also fluctuated remarkably with the maximum value being 0.59 and 0.98 mm, respectively.

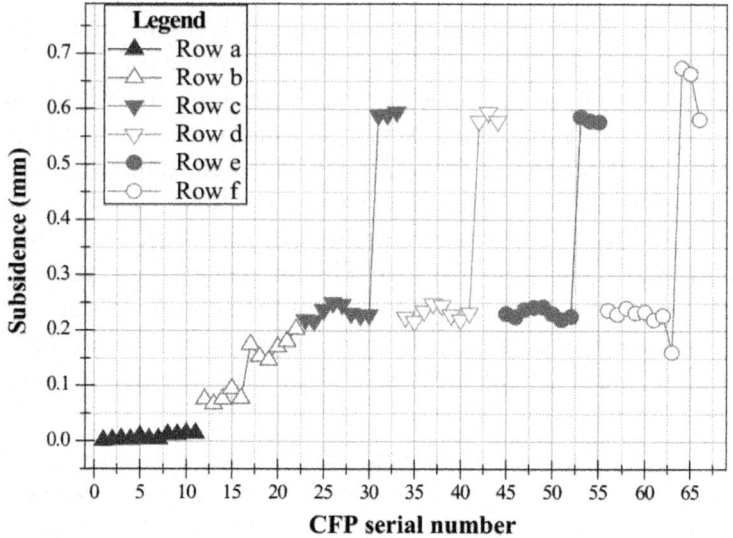

Figure. 8: Surface deformation monitoring by close-field photogrammetry under compressive loading.

According to these results, we conclude that subsidence of coal-rock masses had various characteristics in different stages of the model. During the coal caving, the subsidence values fluctuated more than in other stages. Also, quantity of caving coal affected the degree of asymmetric destabilization in the model, which means that we must control the stability of coal-rock masses by adjusting the quantity of caving coal.

Observing Support Performance

Researches on strata behavior in the extremely steep and thick coal seams indicate that parameter optimization would be needed for in-situ support (Lai et al. 2009a, b, 2014). According to various field loadings of surrounding rock, the support should be determined by comparing the theoretical analysis and numerical simulation results (Lai et al. 2009a, b). Yield loading of the in-situ supports has been set at 4500 kN by simulation theories and setting loading must be 4200 kN. The simulated hydraulic support (Fig. 9a) has been installed in the working face with static-resistance strain equipment adopted for loading observation. The strain values fluctuated in the experiment and the coal caving process had a major impact on the support (Fig. 9b). A large-scale roof caving would increase the amount of shock to the support. Based on statistical results, the setting loading of the hydraulic support was between 4000 and 4200 kN, and also the phenomena that occurred after first roof caving in the experiment were similar with the actual situation of roof failure. Particularly, the individual hydraulic support had such a low setting loading that they could not adequately maintain contact with the roof. When the roof caved in a large-scale zone, the peak value of the support loading was up to 4200 kN; however, the simulated hydraulic support with 4200 kN would ensure demand of pressure control.

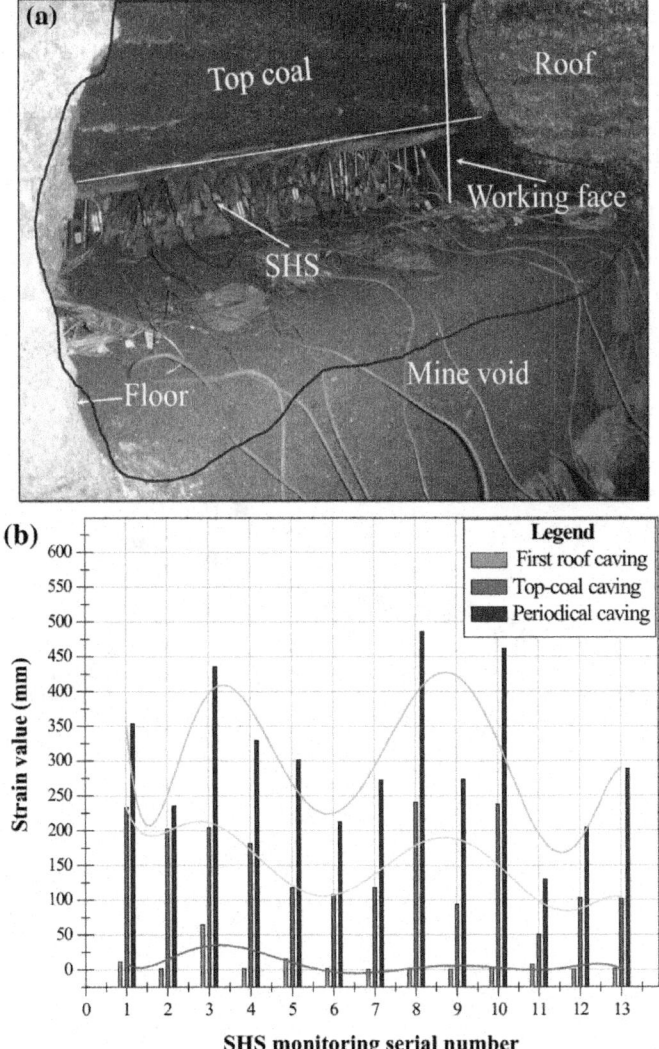

Figure. 9: Strain characteristics of the mechanized sub-horizontal section top coal caving working face by the simulated hydraulic support in first roof caving, top coal caving

and periodic caving, **a** the real arrangement of the simulated hydraulic support; **b** the deformation value of the support.

DISCUSSION

In the physical simulation experiment, asymmetric destabilization of surrounding rock around mine voids is an evolutionary process that is divided into three steps: initial failure step, separation fracture step, and three-hinge arch formation step. Figure 10 shows the asymmetric destabilization characteristics of coal-rock masses in different loading (initial failure step, separation fracture step, and three-hinge arch formation step) with the value of M_0 and ΔM_0, respectively. All statistical data comply with the basic stress–strain relationship of the actual rock masses.

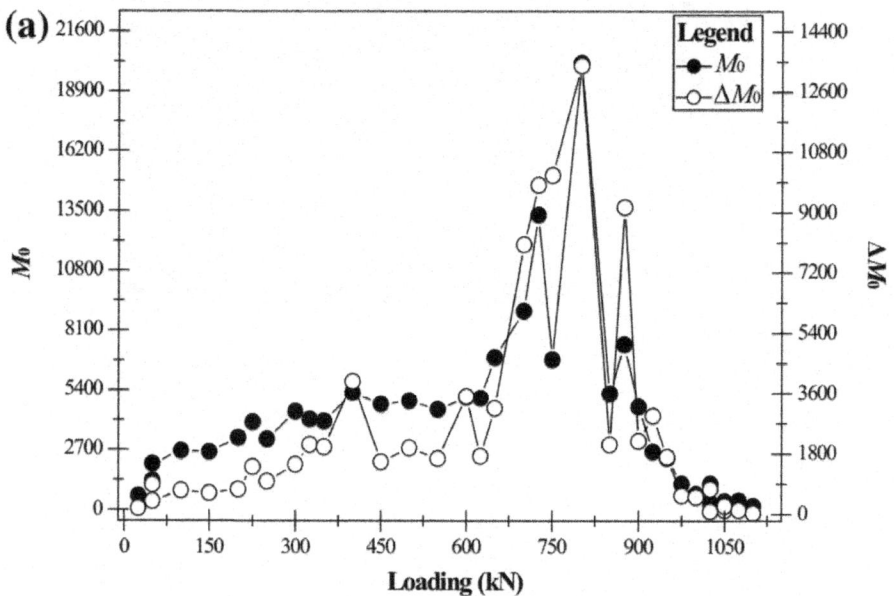

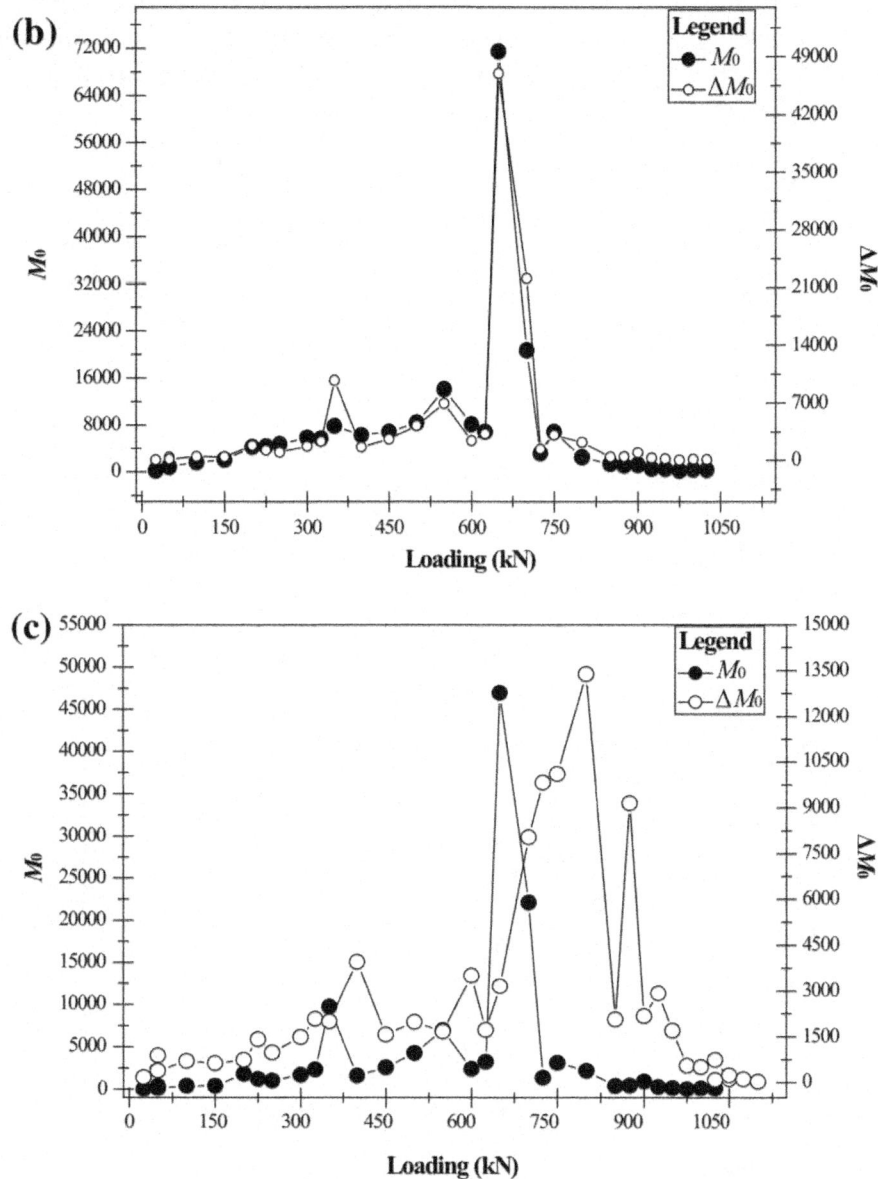

Figure. 10: Vector on damage-deformation-AE characteristics in different steps **a** initial failure step, **b** separation fracture step, and **c** three-hinge arch formation step.

Also, we revealed the mechanisms of stress–strain-AE of surrounding rocks in the experiment. Figure 11 indicates the asymmetric destabilization characteristics under horizontal and vertical dynamic loading. The results show that the abnormal area of AE trait parameters are represented with micro-crack initiation, and yielding points in the stress curve were truly initial points of coal-rock mass failure, which accelerated the destabilization of the surrounding rock.

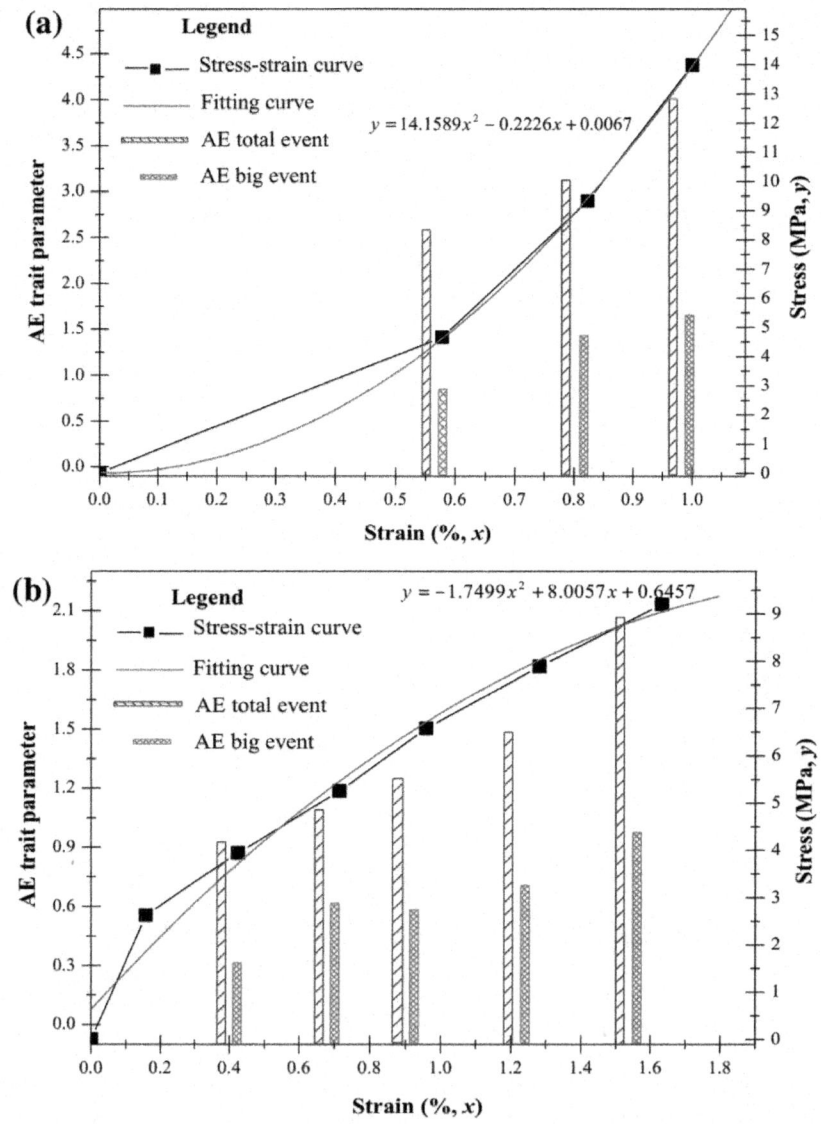

Figure. 11: Mechanism of stress–strain-AE of surrounding rock, **a** stress–strain-AE under horizontal dynamic loading; **b** stress–strain-AE under vertical dynamic loading.

The acoustic source of AE under compressive loading is represented with two disparate types:

- sources intensively distributed close to the host fracture plane of surrounding rock, and
- sources sprinkled in the whole model with obvious variances in AE time sequences due to structural traits of the surrounding rock.

In the same experimental setups and circumstance, different spatial–temporal distribution of AE may be induced by internal structure of coal-rock mass, densification of the coal-rock mass, and/or heterogeneity.

Asymmetric destabilization percentage of coal-rock masses (*ALDP*) with consideration of variability of AE trait parameters and crack interaction observed in the experiment have been defined, which can visually respond to the asymmetric destabilization degree of the large-scale experimental model. The failure extent of the surrounding rock can be quantified by *ALDP* using Eq. (11).

$$ALDP = \frac{\sum (\text{Big\_}AE)}{\sum (\text{Total\_}AE)} \times 100\% \qquad (11)$$

where, Σ (Big_AE) is accumulative amount of the AE big event, Σ (Total_AE) is accumulative amount of the AE total event. Also, Fig. 12 shows the *ALDP* degree under horizontal and vertical loading. Ultimately, *ALDP* would be employed as an effective indicator for assessing asymmetric destabilization status of the surrounding rock, which will provide valid evidence for predicting and controlling dynamic hazards in ESTCS.

A four-channel AE digital signal processing instrument was used with in-situ monitoring, to obtain a new standard with the *ALDP*, whose result can apply for assessing the asymmetric destabilization in the Weihuliang coal mine. R15 with a 1220A preamplifier as the transducer was used to detect the AE signals with a peak frequency response of 150 kHz. The preamplifiers had the ability to select the filter (high pass, low pass, or band pass), single ended or differential input, and gain.

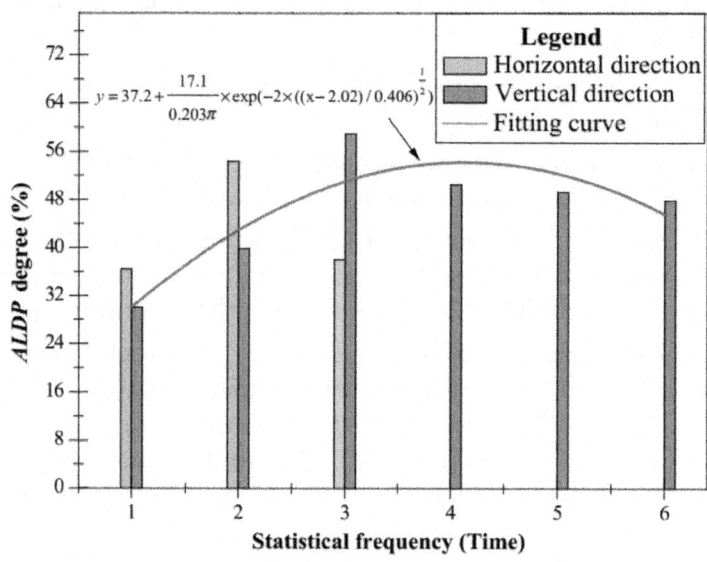

Figure. 12: *ALDP* degree under horizontal and vertical loading.

The gain in the actual operation was set at 40 dB. Here, the mine-void surrounding rocks in +574 m stope were loaded to failure with coal mining. We monitored AE signature at various stages including (I) excavation of working face, (II) first caving of top coal, (III) coal caving, and (IV) local roof caving for the asymmetric destabilization characteristics. The AE data (Table 5) indicates that all *ALDP* values in the in-situ monitoring were greater than 50 %. In the III stage, duration of the AE was longer than the duration in others. When the working face was extracted, the initial stress was disturbed by the excavation zone. However, the threshold of the asymmetric destabilization was from the final stage because of a larger energy rate and rise time of AE signals.

Table 5: Specific matching ratios adopted for the model

Serial number	Σ (Big_AE)	Σ (To-tal_AE)	Duration (μs)	Energy rate	Rise time (μs)	*ALDP*(%)	Stage
1	54	99	2678.4	64	666	54.55	Excavation of working face
2	17	31	2512.8	11	1402.2	54.84	
3	10	17	4109.4	24	89.4	58.82	
4	16	21	1227	5	26.4	76.19	
5	145	178	2574.6	57	13.2	81.46	
6	66	127	2116.2	39	634.8	51.97	

7	20	38	1113.6	11	381.6	52.63	First caving of top coal
10	158	190	3918	181	477	83.16	
11	287	301	5802.6	414	993	95.35	
12	87	129	2725.2	33	378	67.44	Coal caving
13	29	47	11,581.2	10	279.6	61.70	
14	52	82	1914	30	1.8	63.41	
15	45	49	5161.2	20	573	91.84	
16	89	121	3117	59	1510.2	73.55	
17	159	172	3769.2	229	460.8	92.44	
18	234	245	5803.2	476	843.6	95.51	
19	189	217	4052.4	184	748.2	87.10	
20	154	177	3659.4	2090	2595.2	87.01	Local roof caving
21	76	90	3034.8	7200	4436.2	84.44	
22	128	178	3392.4	1570	3577.2	71.91	
23	32	64	2973	3300	2361.2	50.00	
24	102	169	6513	1200	4575.6	60.36	

CONCLUSIONS

This study included a comprehensive site investigation of engineering, geological and geotechnical characterization of rock masses in the Weihuliang mine. The results of the investigation show that the fault-fold structure of northern Tianshan Mountain and the nappe structure of Bogda Mountain would increase tectonic stress to the mechanized sub-horizontal section top coal caving (SSTCC) face in extremely steep and thick coal seams (ESTCS), which was a crucial parameter in the simulation.

Based on field rock mass material characterization, physical–mechanical properties of rock masses, and relevant theory, we formed a new model material that could be used to accurately build a 3D geological model representative of the SSTCC. A non-systematic 3D physical modeling experiment was designed and implemented for simulating and characterizing asymmetric destabilization. The results of the 3D physical modeling experiment, combined with hybrid statistical methods show that asymmetric destabilization characteristics of coal-rock masses under different loading conditions would occur in three steps: initial failure step, separation fracture step, and three-hinge arch formation step. The acoustic sources of AE under compressive loading were mainly distributed close to the primary fracture plane of the surrounding rock. Few sources were distributed throughout the entire model with variances in AE time sequence due to structural traits of the surrounding rock. We defined an indicator (*ALDP*) that could be used to predict and control dynamic hazards in ESTCS.

ACKNOWLEDGMENTS

Financial support for this work was provided by the 973 Key National Basic Research Program of China (No. 2014CB260404, No. 2015CB251600), the Key National Natural Science Foundation of China (No.U13612030), Shaanxi Innovation Team Program (No.2013KCT-16), and the High Technology Development Program of XinJiang Municipality (No. 201432102). Support from these agencies is gratefully acknowledged.

REFERENCES

1.   Alehossein H, Poulsen BA (2010) Stress analysis of longwall top coal caving. Int J Rock Mech Min Sci 47(1):30–41

2.   Cai MF, Ji HG, Wang JA (2005) Study of the time–space–strength relation for mining seismicity at Laohutai coal mine and its prediction. Int J Rock Mech Min Sci 42(1):145–151

3.   Cao JT, Lai XP, Shan PF (2011) Hybrid analysis of dynamic destabilization to HSTCC workings in steep coal seams. In: The 2nd ISRM international young scholars' symposium on rock mechanics, 2011, pp 757–759

4.   Castro R, Trueman R, Halim A (2007) A study of isolated draw zones in block caving mines by means of a large 3D physical model. Int J Rock Mech Min Sci 44:860–870

5.   Dian C (1992) The state of the art and future of China thick seam mining technology. In: Singh TN, Dhar BB (eds) Proceedings of international symposium on thick seam mining. Central Mining Research Station, Dhanbad, pp 171–182

6.   Diaz Aguado MB, Gonzalez C (2009) Influence of the stress state in a coal bump-prone deep coalbed: A case study. Int J Rock Mech Min Sci 46:333–345

7.   Huang QX (2009) Simulation of clay aquifuge stability of water conservation mining in shallow-buried coal seam. Chin J Rock Mech Eng 28(5):988–992 (in Chinese)

8.   Jha SN, Karmakar S (1992) Thick seam mining-some experience and exaltation. In: Singh TN, Dhar BB (eds) Proceedings of international symposium on thick seam mining. Central Mining Research Station, Dhanbad, pp 191–202

9.   Kelly M, Balusu R, Hainsworth D (2001) Status of longwall research in

CSIRO. In: Morgantown: Proceedings of 20th international conference on ground control in mining, 2001, pp 16–20

10. Kose H, Tatar C (1997) Underground mining methods. DEU, Izmir

11. Lai XP, Shan PF et al (2013) Large-three-dimensional physical simulation experiments on a high and steep slope stability of open-pit mines. In: 3rd ISRM symposium on Rock Mechanics, 2013, pp 425–430

12. Lai XP, Ren FH, Wu YP (2009a) Comprehensive assessment on dynamic roof instability under fractured rock mass conditions in the excavation disturbed zone. Int J Minerals Metall Materials 16(1):12–18

13. Lai XP, Wang NB, Xu HD et al (2009b) Safety top coal caving of heavy and steep coal seams under complex environment. J. Univ Sci Tech Beijing 31(3):277–280 (in Chinese)

14. Lai X, Shan P, Ren F et al (2012) Comparative Experiment on Strength of Multi-media Composite Simulate Materials in Mine's High and Steep Slope. J Xi'an Univ Sci Tech 32(2):1–5 (in Chinese)

15. Lai XP, Shan PF, Ren FH et al (2013) Application of model material PRCM in physical simulation experiment on slope's stability in mine's high and steep. Metal Mine 441(3):1–5 (in Chinese)

16. Lai XP, Shan PF, Cao JT et al (2014) Hybrid assessment of pre-blasting weakening to horizontal section top coal caving (HSTCC) in steep and thick seams. Int J Min Sci Tech 24(1):31–37

17. Lauriello PJ, Fritsch CA (1974) Design and economic constraints of thermal rock weakening techniques. Int J Rock Mech Min Sci Geo-Mech 11(1):31–44

18. Li TB, Wang XF, Meng LB (2011) Physical Simulation Study of Similar Materials for Rockburst. Chin J Rock Mech Eng 30(supp 1):2610–2616 (in Chinese)

19. Lin YM (1984) Lab of rock mechanics. Chinese Coal Industry Press, 58 (in Chinese)

20. Miao SJ, Lai XP, Cui F (2011) Top coal flows in an excavation disturbed zone of high section top coal caving of an extremely steep and thick seam. Min Sci Tech 21(01):99–107

21. Pan YS, Zhang MT, Wang LG et al (1997) Study on rockburst by equivalent material simulation tests. Chin J Geotech Eng 19(4):49–56 (in Chinese)

22. Simsir F, Ozfirat MK (2008a) Determination of the most effective longwall equipment combination in longwall top coal caving (LTCC) method by simulation modelling. Int J Rock Mech Eng 45(6):572–582

23. Simsir F, Ozfirat MK (2008b) Determination of the most effective longwall equipment combination in longwall top coal caving (LTCC) method by simulation modeling. Int J Rock Mech Min Sci 45(6):1015–1023

24. Singh R (1999) Mining methods to overcome geotechnical problems during underground working of thick coal seams-case studies. Trans Inst Min Metall Sec A Min Indus 108:121–131

25. Singh TN, Kushwaha A, Singh R, Singh R (1992) Strata behaviour during slicing of thick seam at East Katras Colliery. In: Singh TN, Dhar BB (eds) Proceedings of international symposium on thick seam mining. Central Mining Research Station, Dhanbad, pp 237–250

26. Trueman R, Castro R, Halim A (2008) Study of multiple draw-zone interaction in block caving mines by means of a large 3D physical model. Int J Rock Mech Min Sci 45(7):1044–1051

27. Tu SH, Yong Y, Zhen Y, Ma XT, Qi W (2009) Research situation and prospect of fully mechanized mining technology in thick coal seams in China. Proc Earth Plan Sci 1(1):35–40

28. Unver B, Yasitli NE (2006) Modeling of strata movement with a special reference to caving mechanism in thick seam coal mining. Int J Coal Geol 66(04):227

29. Vakili A, Hebblewhite BK (2010) A new cavability assessment criterion for Longwall Top Coal Caving. Int J Rock Mech Eng 47(8):1317

30. Wu J (1992) The movement regularity of the roof coal around a longwall with caving and its coal recovery. In: Proceedings of international symposium on fully mechanized mining technology for high output and efficiency, 1992, pp 307–318

31. Xie YS, Zhao YS (2009) Numerical simulation of the top coal caving process using the discrete element method. Int J Rock Mech Min Sci 46(6):983–991

32. Yasitli NE, Unver B (2005) 3D numerical modeling of longwall mining with top coal caving. Int J Rock Mech Min Sci 42:219–227

33. Zhang ZF, Lai XP (2008) Segment pre-blasting of sublevel caving of steep and thick coal seam under complex conditions. J Chin Coal Soc 33(8):845 (in Chinese)

34. Zhang JG, Zhao ZQ, Gao Y (2011) Research on top coal caving technique in steep and extra-thick coal seam. Proc Earth Plan Sci 2:145–151

Chapter 7

WATER INGRESS ASSESSMENT FOR ROCK TUNNELS: A TOOL FOR RISK PLANNING

Wing Kei Kong

A-P Design, MWH Australia Pty Ltd, Level 3, 35 Boundary Street, South Brisbane, QLD 4101, Australia

INTRODUCTION

Groundwater inflow is one of the key issues impacting on the process of design and construction for tunnel projects, particularly for open face excavation methods. During tunnelling, extensive water inflow may cause unpredictable down time for the construction and may also introduce secondary effects of groundwater drawdown to the above ground, leading to ground movement or settlement impacts to sensitive buildings and utilities.

The predominant factor controlling groundwater flow towards the tunnel will, therefore, be the presence of weaknesses and jointing and the connectivity of these features to the aquifers within the rock mass. The major pathways for such flows are the geological faults that may occur at various points along the length of the tunnel route, and their associated sheared/shattered zones. Within the remaining rock mass any movement of groundwater will only occur as a result of fracture flow between interconnected discontinuities. The high strength of the rock mass and the local in situ stress state of the rock is likely to allow the presence of open jointing within the rock mass. These features, when encountered in the tunnel, may result in high groundwater inflows which need to be carefully controlled.

Knowledge in understanding and providing controls, including the designation of allowable limits of ingress, to the likely groundwater ingress that may occur during tunnelling is important. Whilst the occurrence of water is difficult to predict accurately, it is important to be prepared for large variations, both with respect to locations and volumes. Early planning to tackle the potential risks associated with water ingress is very important in order to set up appropriate mitigation measures ensuring the excavation work to be conducted in safe and controllable manners. With respect to estimating groundwater inflow into tunnels, several analytical methods can be used at the risk planning stage.

ANALYTICAL METHODS FOR ESTIMATING WATER IN-FLOW

Conventionally, there are four classical analytical methods being used to estimate water inflow into tunnels. These are:

- Goodman method;
- Heuer and Raymer method;
- Heuer Analytical method; and
- IMS method, and are discussed in the following sections.

Goodman Method

Estimations of groundwater inflow into tunnels are often carried out based on the equations proposed by Goodman et al. (1965). The relevant equation for deep tunnels states that the steady state radial groundwater inflow into a tunnel that is overlain by a column of water much larger than the tunnel radius can be approximated using the following formula:

$$q_s = \frac{2\pi k(z + h1)}{2.3 \times \log\left(\frac{r}{2z}\right)}$$

(1)

where: q_s = steady state inflow per unit length of tunnel (m³/s); k = equivalent hydraulic conductivity of rock mass (m/s); z = thickness of ground cover above the tunnel centre line (m); $h1$ = depth of standing water above ground surface, if present (m); and r = tunnel radius (m).

It is noted that the equation presented above, which is based on the paper by Goodman et al. (1965), appears slightly erroneous in the lower part, which would commonly result in a negative divisor. Later citations of this equation (Heuer 1995, 2005; Raymer 2001; McFeat-Smith et al. 1998) appear to have corrected for this fact and have amended the divisor to $\ln(2z/r)$.

Additionally, it should be noted that the Goodman function is not able to predict the water table drawdown, i.e. the transient phase. This equation is limited to steady state (long-term) inflow, supposing that the perturbed water level is known, fact that is rarely true, and does not address the short-term higher rates of heading inflow that are typically encountered during tunnel excavation. This method also takes no account of local variations in the geological conditions within each respective unit, and assumes the spread and distribution of the data already account for such features.

Heuer and Raymer Method

Based on Heuer (1995) study, an equation is proposed by Raymer (2001) for estimating tunnel water inflow from vertical recharge comprising a modified version of the Goodman equation, where a reduction factor is applied. The Goodman equation was found to overestimate tunnel inflows when reviewed against a number of tunnelling case studies. The equation proposed, using the same symbols as in Sect. 2.1, is as follows:

$$q_s = \frac{2\pi k(z + h1)}{\ln\left(\frac{2z}{r}\right)} \times \frac{1}{8}. \tag{2}$$

Similar to Goodman Method, this equation is limited to estimate steady state inflow as well as taking no account of local variations in the geological conditions, and is, therefore, unable to predict the water table draw-down.

Heuer Analytical Method

Heuer (1995, 2005) proposes a statistical method for predicting tunnel groundwater inflows based on an assessment of the frequency distribution of the rock mass hydraulic conductivity, as evaluated through water absorption packer tests within a respective geological unit. The methodology considers three possible models, based on the tunnel depth and ground conditions, in order to assess the statistical distribution of potential water inflows into the excavation. These three models comprise:

Vertical recharge. Applicable for tunnels where a recharge source of large water volume at constant head is close to the tunnel, such as beneath a large water body or within a highly permeable aquifer.

Radial flow. Comprises the standard equation for a fully penetrating well in a confined aquifer and is applicable to tunnels where water flows from all directions and the recharge source is far away.

Lateral flow. Applicable for tunnels where the predominant direction of water flow towards the tunnel is along lateral features such as weathered profiles or bedding. In this case the groundwater table is usually sufficiently close to the tunnel, and the water recharge sufficiently limited, such that tunnelling could result in notable drawdown of piezometric levels above the tunnel alignment.

Graphical presentations of typical conditions for each assessment model are presented in Fig. 1.

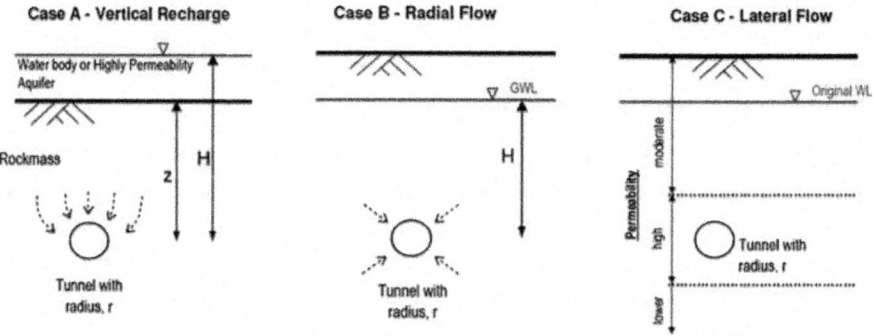

Figure. 1: Graphical presentations of Heuer analytical method.

Whilst this technique is relatively sound in theory, it is purely statistical in the sense that estimation parameters are considered and taken from Fig. 2. Similar to the above two methods, this method takes no account of local variations in the geological conditions.

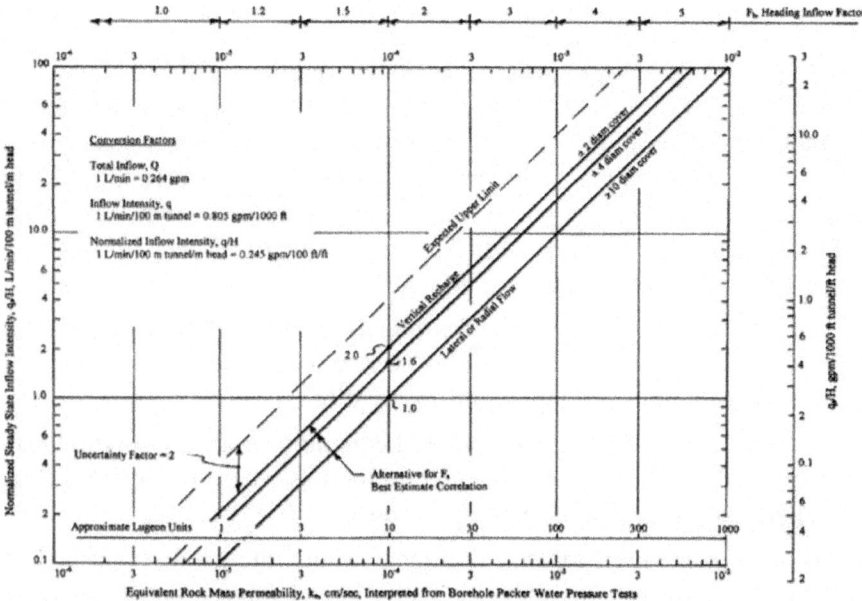

Figure. 2: Relationship between steady state inflow and equivalent permeability based on statistical data (Heuer 2005).

Based on Fig. 2, the distribution of equivalent rockmass permeability (k_e, m/s) may be divided into various ranges: $\leq 3 \times 10^{-8}$, 3×10^{-8} to 1×10^{-7}, 1×10^{-7} to 3×10^{-7}, 3×10^{-7} to 1×10^{-6}, 1×10^{-6} to 3×10^{-6}, 3×10^{-6} to 1×10^{-5}, 1×10^{-5} to 3×10^{-5}, 3×10^{-5} to 1×10^{-4}, $\geq 1 \times 10^{-4}$, and so on. However, the ranges selected for the water ingress assessment depend upon the site specific rockmass characters and the availability of field permeability testing records. Then, a histogram plot (equivalent rockmass permeability vs. frequency of test data) of field permeability data can be produced using these ranges. Based on the range of permeability the value of F_h (heading inflow factor, as shown in the upper axis of Fig. 2), and the maximum value of normalized steady state inflow intensity (q_s/H, where H = water head above the tunnel in terms of meter and q_s = total water inflow, as shown in the right axis of Fig. 2) can be determined by selecting the appropriate analytical model (e.g. vertical recharge or radial/lateral flow, as shown in Fig. 2).

For water ingress assessment, the whole length of tunnel may be divided into number of reaches based on water head (H). The q_s can be calculated as of normalized steady state inflow intensity multiplied by water head. Total water inflow rate per reach will then be calculated by summation across the permeability ranges of the product of "the percentage of distribution frequency of the range of permeability" and "the length of reach".

IMS METHOD

The IMS Method (McFeat-Smith et al. 1998) is an empirical approach based on selected Hong Kong cases mainly in granitic and volcanic strata. The methodology makes predictions of water inflow based on a number of local factors including ground conditions, which should be assessed using the IMS classification system (McFeat-Smith et al. 1985), the tunnel depth and the tunnel distance from potential groundwater/water sources. The prediction of the initial inflow (Ii) and final inflow (Fi) are proposed:

$$\text{Ii} = \text{Sf} \cdot \text{Hf} \cdot \text{df} \cdot \text{IF} \qquad (3)$$

$$\text{Fi} = (\text{Sf} \cdot \text{Hf} \cdot \text{df})^2 \cdot \text{IF} \qquad (4)$$

where, IF is the rate of inflow of different geological conditions in l/min/m.

A detailed summary of the various input parameters required for inflow assessment using notable the IMS method is provided below:

IMS classification system. The IMS classification system (McFeat-Smith et al. 1985) is based on a simple relationship between weathering grade (based on BS5930: 1999) of rock mass and fracture index (i.e. rock joint spacing), thus called "initial classification" as summarised in the left-hand side of Fig. 3.

The initial rock classes could be modified by the possibility of the rate of water inflow and sub-parallel jointing (SPJ) to the tunnel.

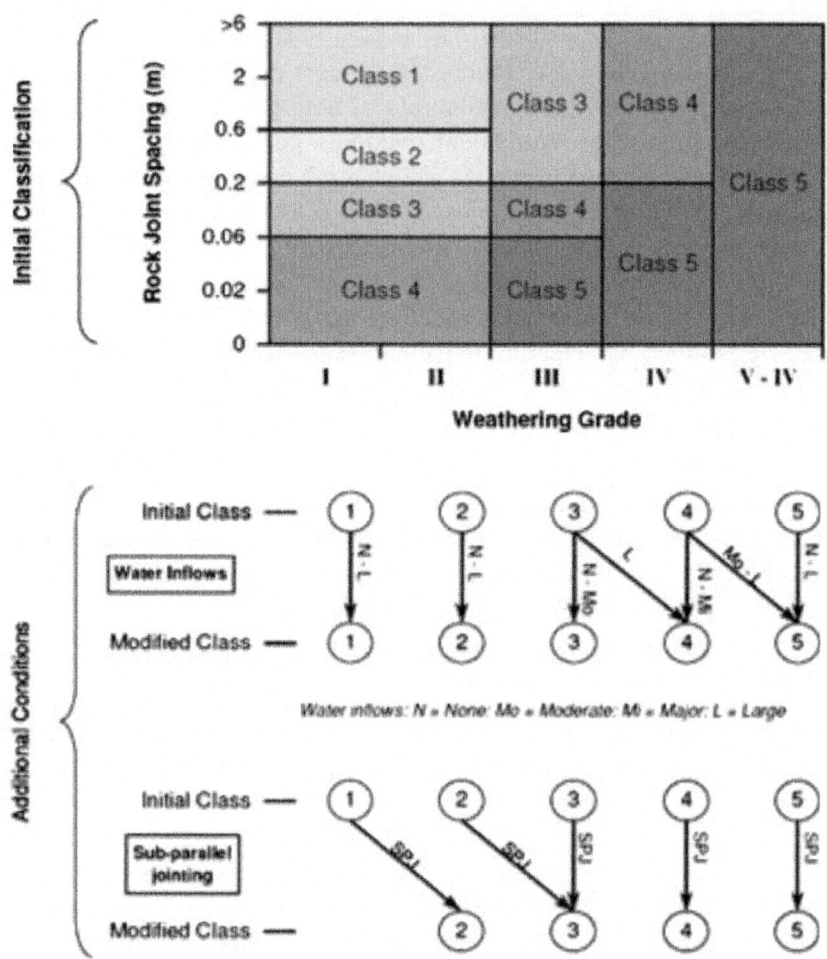

Figure. 3: Summary chart of IMS rock mass classification (McFeat-Smith et al. 1985).

As noted, the rock classes may be modified after calculation of the water inflow using initial rock classes. The rate of water inflow is defined as: None (N, <0.1 l/min/m); Moderate (Mo, 0.1 < flow rate < 2 l/min/m); Major (Mi; 2 < flow rate < 4 l/min/m); and Large (L, >4 l/min/m), as shown in the right-hand side of Fig. 3.

Water source size factor (*Sf*). This parameter describes the nature of the potential groundwater source and is summarized in Table 1:

Table 1: Values of water source size factor, Sf (McFeat-Smith et al. 1985)

Water source	Sea	Major valley/reservoir	Large valley/reservoir	Small river/reservoir	Stream	Ridge
Sf value	1.0	0.85	0.7	0.5	0.3	0.1

Head factor (Hf). The head factor is a measure of the water head above the tunnel and is summarised in Table 2:

Table 2: Values of head factor, Hf (McFeat-Smith et al. 1985)

Water head (m)	>100	100	80	50	20
Hf Value	1.0	1.0	0.8	0.5	0.2

Horizontal separation factor (df). The horizontal separation factor (as summarised in Table 3) is a measure of the plan distance between the tunnel and the water source, also "df" can be expressed as $[1 - \sqrt{(s/400)}]$:

Table 3: Horizontal separation factor, df (McFeat-Smith et al. 1985)

Separation to water source (s, in m)	0	50	100	200	300	400 or >400
df Value	1.0	0.65	0.5	0.29	0.13	0

As discussed above, the IF values for IMS rock classes (l/min/m) are summarised in Table 4:

Table 4: IF values for IMS rock classes (McFeat-Smith et al. 1985)

IMS rock class	1	2	3	4	5
IF values (l/min/m)					
High	0.6	1.4	12.2	37	3.8
Average	0.45	1.05	6.55	24	3.1
Low	0.3	0.7	0.9	11	2.4

Once the IMS rock class is determined, the IF value will be easy to obtain. Of particular note is that due to the relative tightness of zones of rock Class 5, being partly sealed by decomposed clayey/silty soils without open joints, is well illustrated. Therefore, the potential inflow rate of Class 5 is relatively smaller compared to Classes 3 and 4, as shown in Table 4.

PROCEDURES OF WATER INFLOW ASSESSMENT

Having reviewed the above analytical methods and references, the author's proposed procedures of the water inflow assessment for hard rock tunnels are outlined as follows:

Step 1. Collecting geological data, ground and groundwater information, and permeability data along and in the vicinity of the tunnel alignment in order to prepare a geological longitudinal section together with a rock quality assessment along the tunnel.

Step 2. Preparing rockmass permeability histogram chart, rock mass quality versus permeability chart, etc. In this step, all available field and laboratory permeability testing data are reviewed in relation to rock mass quality. In addition, groundwater monitoring records as well as seasonal effects should be assessed in order to determine design groundwater level for the tunnel. Once the generalized design groundwater level is determined, the tunnel is divided into a number of reaches for the estimation of water inflow. For the rock mass quality assessment, several classification systems may be used based on the Q-system (Barton et al. 1974), RMR (Bienawski 1973, 1984) and IMS classification system (McFeat-Smith et al. 1985). As discussed in Sect. 2.4, IMS method uses its classification system to estimate water inflow. The IMS rock class can be correlated to Q (Barton et al. 1974; Barton 2000) as shown in Table 5.

Table 5: Approximate correlation between Q values and IMS rock class

Classification	Class				
IMS	1	2	3	4	5
Q	>10	$4 < Q \leq 10$	$0.4 < Q \leq 4$	$0.04 < Q \leq 0.4$	$Q < 0.04$

Step 3. Estimating water inflow to the tunnel using the above methods together with the available information of geology, rock quality, rockmass permeability and water table. To have appropriate reasoning, at least two analytical methods should be used to compare the estimation results to each other in the risk planning stage. In addition, the assumptions of water inflow calculation have been based on:

- Class of rock mass quality distribution against rock mass permeability;
- statistical mean value of rock mass permeability in each rockmass quality class;
- as Goodman and Heuer and Raymer methods are adopted, the equivalent circular section area of tunnel opening may be used.

Step 4. Identifying potential high water inflow zone (criteria refer to Step 5) of the tunnel to propose additional ground investigation works and probe drill during tunnel construction. Based on the estimation results from Step 3, it may identify numbers of high water inflow sections that may impact on the tunnel driving and surface sensitivity buildings. These potential high water inflow sections should be reported in connection with proposing appropriate measures to deal with it.

Step 5. Setting up grouting and ground treatment requirements in order to control groundwater inflow to the tunnel during tunnel construction. When tunnel construction is by means of open type excavation (e.g. drill and blast method, open mode tunnel boring machine, etc.), groundwater inflow will impact on tunnel driving, and grouting works is often necessary. The allowable water inflow limits and grouting requirements for the tunnel construction depend on the ultimate function of the tunnel, and the water-draw down (or settlement) limits agreed with the owner or operator of the existing structures and utilities; and should be designed to ensure the tunnel is being excavated in safe manners and to minimize the impact on environment above the tunnel. The grouting criteria and specified high water inflow zones play a key role of risk planning for tunnel projects, and should be documented in the construction contract of the Project Performance Requirements (or Project's Particular Specification).

Step 6. Monitoring and reviewing updated tunnelling record during construction. In addition to mitigating against the effects of inflow within the tunnel, it is required to establish a comprehensive groundwater level and settlement monitoring programme during the construction phase. The monitoring programme should include continuous long-term monitoring of piezometers, settlement markers and tilt monitoring of sensitive structures, associated with contingency plan.

Case Example

For one of the tunnel projects in Hong Kong (HKG 2006), two horizontal directional drillholes (HDD) were drilled along two proposed tunnel alignments as shown in Figs. 4 and 5. The hole size of the HDD was 76 mm in diameter, and the total drill length of the holes, HDD-1 and HDD-2, were 240 and 310 m, respectively, and drilled upward at about 3°. A total of 14 and 25 permeability tests [i.e. water absorption (packer) tests] were conducted along HDD-1 and HDD-2 respectively.

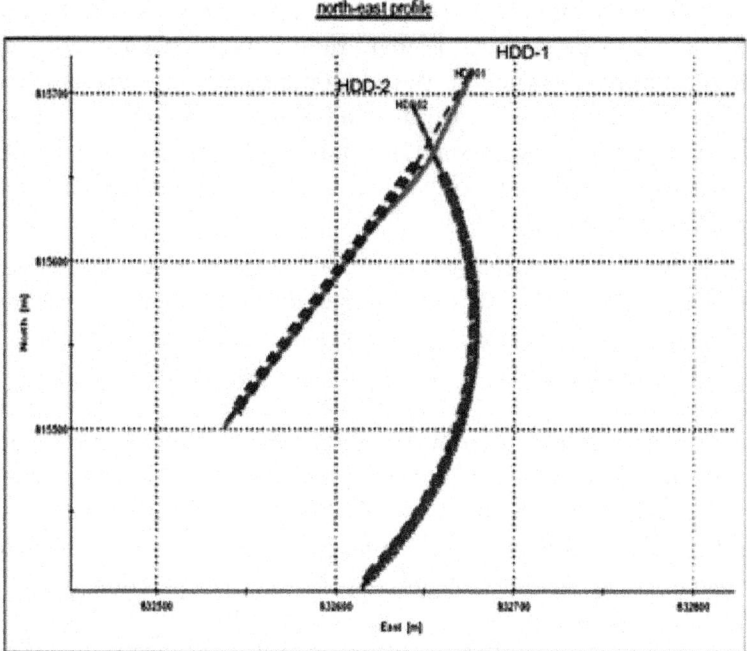

Figure. 4: Alignment of HDD holes.

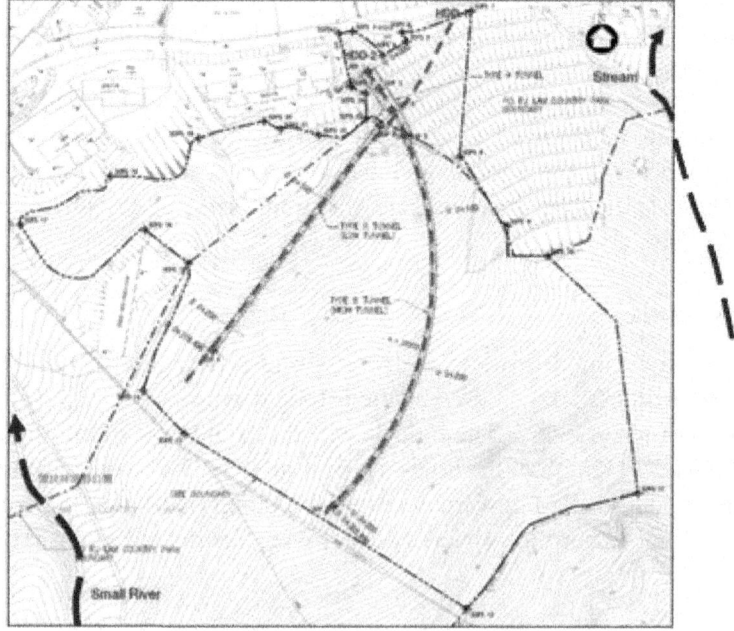

Figure. 5: Topography setting of the site.

With reference to the above procedure (Step 1): initial exercise to gather all available ground information was conducted.

- The borehole logs of HDD-1 and HDD-2 indicated the first 52 and 20 m, respectively, were completely driven within soft ground (i.e. completely to highly decomposed Tuff), with localised corestones encountered. For the rest of the drill length, moderately to slightly decomposed tuff was identified.

- The laboratory rock testing results identified that the tuff is very to extremely strong with strength ranging from 100 to 280 MPa.

- Based on the borehole log descriptions: rock mass jointing is generally rough planar, widely spaced, occasionally close to medium spaced, extremely narrow to tight, iron–manganese stained or kaolin infilled, locally with narrow soil seams.

- The groundwater monitoring records from vertical drillholes in the area indicated groundwater levels along the HDD alignments varied between 6 and 15 m.

- The site measurement recorded that the total water outflow from HDD-1 and HDD-2 was 9 and 15 l/min, respectively.

- The geological sections of the proposed tunnels (based on HDD holes information) are shown in Figs. 6 and 7.

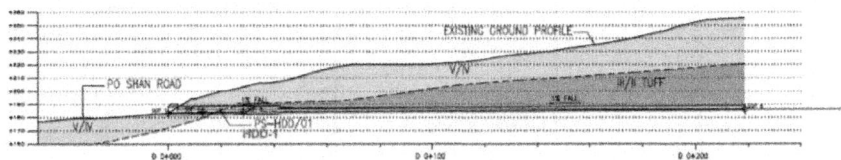

Fig. 6: Geological section of low tunnel (based on HDD-1 borehole log).

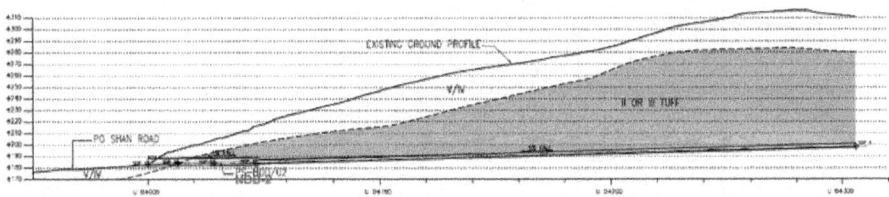

Figure. 7: Geological section of high tunnel (based on HDD-2 borehole log).

(Step 2): Rock quality assessment to horizontal holes HDD-1 and HDD-2 had also been carried out based on borehole logs and examination to the rock core. The estimated Q values (Barton et al. 1974) along HDD-1 and HDD-2 varied from 0.13 and 33.8, respectively, in hard rock (moderately weathered rock or better) sections. Further review to permeability tests against rockmass quality to the test sections, the distribution of permeability tests and rockmass quality summaries are shown in Tables 6 and 7.

Table 6: Permeability distribution summary

Permeability (m/s)	Cumulative frequency	Data range of permeability (m/s)	Frequency	Relative freq. in %	Upper value (m/s) of histogram
≤1.00E−09	0	$x \leq 1.00\text{E}{-}09$	0	0	1.00E−09
≤3.00E−09	2	$1.00\text{E}{-}09 < x \leq 3.00\text{E}{-}09$	2	6	3.00E−09
≤1.00E−08	6	$3.00\text{E}{-}09 < x \leq 1.00\text{E}{-}08$	4	12	1.00E−08
≤3.00E−08	10	$1.00\text{E}{-}08 < x \leq 3.00\text{E}{-}08$	4	12	3.00E−08
≤1.00E−07	16	$3.00\text{E}{-}08 < x \leq 1.00\text{E}{-}07$	6	18	1.00E−07
≤3.00E−07	23	$1.00\text{E}{-}07 < x \leq 3.00\text{E}{-}07$	7	21	3.00E−07
≤1.00E−06	28	$3.00\text{E}{-}07 < x \leq 1.00\text{E}{-}06$	5	15	1.00E−06
≤3.00E−06	32	$1.00\text{E}{-}06 < x \leq 3.00\text{E}{-}06$	4	12	3.00E−06
≤1.00E−05	34	$3.00\text{E}{-}06 < x \leq 1.00\text{E}{-}05$	2	6	1.00E−05
≤3.00E−05	34	$1.00\text{E}{-}05 < x \leq 3.00\text{E}{-}05$	0	0	3.00E−05
≤1.00E−04	34	$3.00\text{E}{-}05 < x \leq 1.00\text{E}{-}04$	0	0	1.00E−04
≤3.00E−04	34	$1.00\text{E}{-}04 < x \leq 3.00\text{E}{-}04$	0	0	3.00E−04
≤1.00E−03	34	$3.00\text{E}{-}04 < x \leq 1.00\text{E}{-}03$	0	0	1.00E−03
≤3.00E−03	34	$1.00\text{E}{-}03 < x \leq 3.00\text{E}{-}03$	0	0	3.00E−03
≤1.00E−02	34	$3.00\text{E}{-}03 < x \leq 1.00\text{E}{-}02$	0	0	1.00E−02
Total	34		34	100	

Table 7: Summary of permeability test results vs Q value and IMS rock class

Q value	IMS class	No. of tests	Range of rock permeability, k (m/s)		
			Mini-mum	Maximum	Median
>10	1	8	3×10^{-9}	8.8×10^{-8}	4.2×10^{-8}
$4 < Q \leq 10$	2	7	7×10^{-9}	4.1×10^{-7}	1.46×10^{-7}
$0.4 < Q \leq 4$	3	13	1×10^{-8}	2.96×10^{-6}	6.96×10^{-7}
$0.04 < Q \leq 0.4$	4	6	4.4×10^{-8}	9.96×10^{-6}	2.98×10^{-6}
$Q < 0.04$	5	0	–	–	–

(Step 3): For this project, two analytical methods: Heuer and Raymer and IMS methods, were used to compare the estimation results to each other as presented in Tables 9 and 10 of Appendix for HDD-1 and HDD-2, respectively. In the calculation, mean values of rockmass permeability and IF values were used for Heuer and Raymer method and for IMS method, respectively.

(Step 4): As shown in Tables 9 and 10, the water inflow estimation identified several localised borehole sections with water inflow rate greater than 0.2 l/min/m, along the boreholes HDD-1 and HDD-2. The total length of these high water inflow sections was about 10% of the total tunnel lengths. The result of findings is summarized in Table 8.

Table 8: Summary of the identified sections of inflow greater than 0.2 l/min/m along the HDD holes

Method	No. of section of inflow >0.2 l/min/m		Total length of sections (m)	
	HDD-1	HDD-2	HDD-1	HDD-2
Heuer and Raymer	6	7	11.3	32.72
IMS	13	9	27.1	35.5

(Step 5): Having considered the construction method by means of open type TBM and the operational function of the tunnel, it was decided to set up grouting criteria ensuring the excavation work to be conducted in safe and controllable manners and to minimize the impact on the surrounding environment. Finally three criteria were bound in Project Performance Requirements (or Project's Particular Specification) of the construction contract:

- A total discharge of greater than 10 l/min of water which persists 24 h after the completion of a 25-m length of probe hole.
- A general inflow of greater than 20 l/min which persists after a period of 24 h for the excavated section within 25 m of the current face.
- Prior to the installation of the permanent support lining, inflows of greater than 20 l/min on any 100 m length of tunnel.

(Step 6): Other than grouting criteria, a series of instrumentation including observation wells, inclinometers, building markers and settlement markers were required to be installed in the construction stage. These instrumentation schedule and requirements were also included in the Project Performance Requirements (or Project's Particular Specification) of the construction contract document.

CONCLUSIONS

Analytical predictions of groundwater inflow into a drained hard rock tunnel have to be based on several simplifying assumptions:

- homogeneous and isotropic rock mass permeability;
- steady state flow conditions are in effect;
- the tunnel has a perfectly circular cross-section that is held at constant hydraulic potential.

Whilst these assumptions clearly do not accurately represent the actual in situ conditions, especially in the case of the rock mass acting as a homogenous isotropic body, they are necessary in order to allow Darcy's Law to be applied in the calculations. Due to the assumptions, the predicted water inflows into a tunnel are only approximate. However, the water inflow assessment produces data useful for the identification of potentially problematic portions of the alignment where extremely close attention should be paid to the pre-excavation probing and grouting works.

Based on the assessment results, early risk planning should be carried out to ensure that the tunnel construction works are proceed in a safe manner, particularly for high water inflow sections of the tunnel, and thus minimizing the impact on the environment above the tunnel. The allowable water ingress rate and grouting criteria should be documented in the project's performance requirements of the construction contract.

APPENDIX

See Tables 9 and 10.

Table 9: Estimated water inflow of HDD-1 drillhole

| Drill length | | Section length (m) | Predict-ed Q value | IMS class | Ground cover z(m) | Water head, H (m) | Heuer and Raymer method | | | IMS method | | | | | | | |
From	To						Perme-ability (Lugeon)ᵃ	Water inflow (l/min/m)	Total flow of section (l/min)	Sf	Hf	df	R	Sepᵇ (m)	IF mean value (l/min/m)ᶜ	Initial flow, Ii (l/min/m)	Total flow of section (l/min)
52	54	2	0.80	3	26	20	6.96	0.09	0.18	0.3	0.2	0.396	0.024	146	6.55	0.16	0.31
54	62.54	8.54	3.33	3	29	20	6.96	0.09	0.76	0.3	0.2	0.396	0.024	146	6.55	0.16	1.33
62.54	64.4	1.86	0.25	4	31	21	29.77	0.40	0.74	0.3	0.2	0.386	0.024	151	24.00	0.58	1.08
64.4	69.52	5.12	14.25	1	31	21	0.42	0.01	0.03	0.3	0.2	0.356	0.022	166	0.45	0.01	0.05
69.52	70.71	1.19	7.50	2	31	21.5	1.46	0.02	0.02	0.3	0.2	0.356	0.023	166	1.05	0.02	0.03
70.71	73.13	2.42	0.42	3	32	22	6.96	0.10	0.24	0.3	0.2	0.356	0.023	166	6.55	0.15	0.37
73.13	79.41	6.28	13.50	1	33	23	0.42	0.01	0.04	0.3	0.2	0.346	0.024	171	0.45	0.01	0.07
79.41	83.6	4.19	0.33	4	33	23	29.77	0.43	1.81	0.3	0.2	0.327	0.023	181	24.00	0.54	2.27
83.6	92.68	9.08	4.33	2	33	23	1.46	0.02	0.19	0.3	0.2	0.318	0.022	186	1.05	0.02	0.21
92.68	97.5	4.82	1.75	3	33	23	6.96	0.10	0.49	0.3	0.2	0.309	0.021	191	6.55	0.14	0.67
97.5	98.82	1.32	0.13	4	33	23	29.77	0.43	0.57	0.3	0.2	0.309	0.021	191	24.00	0.51	0.68
98.82	100.9	2.08	0.42	3	33	23	6.96	0.10	0.21	0.3	0.2	0.300	0.021	196	6.55	0.14	0.28
100.9	102.92	2.02	3.33	3	33	23	6.96	0.10	0.20	0.3	0.2	0.300	0.021	196	6.55	0.14	0.27
102.92	104	1.08	0.13	4	33	23	29.77	0.43	0.47	0.3	0.2	0.291	0.020	201	24.00	0.48	0.52
104	106.72	2.72	0.75	3	33.5	23.5	6.96	0.10	0.28	0.3	0.2	0.275	0.019	210	6.55	0.13	0.35
106.72	108.34	1.62	0.13	4	34	24	29.77	0.45	0.73	0.3	0.2	0.267	0.019	215	24.00	0.46	0.75
108.34	110.38	2.04	7.50	2	35	25	1.46	0.02	0.05	0.3	0.2	0.267	0.033	215	1.05	0.04	0.07
110.38	126.47	16.09	14.25	1	38	28	0.42	0.01	0.12	0.5	0.3	0.275	0.039	210	0.45	0.02	0.28
126.47	129.47	3	7.00	2	39	29	1.46	0.03	0.08	0.5	0.3	0.293	0.042	200	1.05	0.04	0.13

129.47	132.5	3.03	8.33	2	40	30	1.46	0.03	0.08	0.5	0.3	0.293	0.044	200	1.05	0.05	0.14
132.5	147	14.5	33.75	1	41	31	0.42	0.01	0.12	0.5	0.3	0.311	0.048	190	0.45	0.02	0.31
147	149	2	1.90	3	43	33	6.96	0.14	0.28	0.5	0.3	0.329	0.054	180	6.55	0.36	0.71
149	151.94	2.94	33.80	1	44	34	0.42	0.01	0.03	0.5	0.3	0.329	0.056	180	0.45	0.03	0.07
151.94	153.89	1.95	1.88	3	45	35	6.96	0.15	0.29	0.5	0.4	0.329	0.058	180	6.55	0.38	0.74
153.89	162.07	8.18	10.40	1	47	36	0.42	0.01	0.07	0.5	0.4	0.348	0.063	170	0.45	0.03	0.23
162.07	165.32	3.25	10.40	1	48	38	0.42	0.01	0.03	0.5	0.4	0.348	0.066	170	0.45	0.03	0.10
165.32	171.44	6.12	5.60	2	49	39	1.46	0.03	0.21	0.5	0.4	0.368	0.072	160	1.05	0.08	0.46
171.44	179.49	8.05	10.40	1	52	40	0.42	0.01	0.08	0.5	0.4	0.368	0.074	160	0.45	0.03	0.27
179.49	182.49	3	2.92	3	52	40	6.96	0.17	0.50	0.5	0.4	0.378	0.076	155	6.55	0.49	1.48
182.49	191.88	9.39	18.75	1	52	41	0.42	0.01	0.10	0.5	0.4	0.388	0.079	150	0.45	0.04	0.34
191.88	194.53	2.65	11.67	1	53	41	0.42	0.01	0.03	0.5	0.4	0.394	0.081	147	0.45	0.04	0.10
194.53	197.74	3.21	0.52	3	53	42	6.96	0.17	0.56	0.5	0.4	0.398	0.084	145	6.55	0.55	1.76
197.74	200.74	3	17.50	1	53	42	0.42	0.01	0.03	0.5	0.4	0.408	0.086	140	0.45	0.04	0.12
200.74	201.97	1.23	0.21	4	53	42	29.77	0.74	0.91	0.5	0.4	0.408	0.086	140	24.00	2.06	2.53
201.97	203.7	1.73	0.63	3	53	43	6.96	0.18	0.31	0.5	0.4	0.408	0.088	140	6.55	0.58	0.99
203.7	205.54	1.84	4.17	2	54	44	1.46	0.04	0.07	0.5	0.4	0.408	0.090	140	1.05	0.09	0.17
205.54	220.45	14.91	33.75	1	55	45	0.42	0.01	0.17	0.5	0.5	0.415	0.093	137	0.45	0.04	0.63
220.45	221.49	1.04	1.46	3	56	46	6.96	0.19	0.20	0.5	0.5	0.415	0.095	137	6.55	0.62	0.65
221.49	227.85	6.36	18.75	1	56	46	0.42	0.01	0.07	0.5	0.5	0.415	0.095	137	0.45	0.04	0.27
227.85	230.74	2.89	3.33	3	57	47	6.96	0.19	0.56	0.5	0.5	0.419	0.098	135	6.55	0.65	1.86
230.74	239.81	9.07	31.88	1	58	48	0.42	0.01	0.11	0.5	0.5	0.430	0.103	130	0.45	0.05	0.42
							Total inflow (Heuer and Raymer)		12.00						Total inflow (IMS)		24.08

Numbers in *italics* represent zone of inflow >0.2 l/min/m

aRefer to Table 7

bSeparation between hole/tunnel section and water source (Fig. 5 for reference)

cRefer to Table 4

Table 10: Estimated water inflow of HDD-2 drillhole

Drill length		Section length (m)	Predicted Q Value	IMS class	Avg. ground cover,z (m)	Water head,H (m)	Heuer and Raymer method			IMS method							
From	To						Permeability (Lugeon)a	Water inflow (l/min/m)a	Total Flow of section (l/min)	Sf	Hf	df	R	Sepb(m)	IF mean value (l/min/m)	Initial flow, Ii (l/min/m)c	Total Flow of section (l/min)
19.9	21.9	2	0.4	4	8	2	29.77	0.05	0.09	0.3	0.0	0.378	0.002	155	24.00	0.05	0.11
21.9	35.4	13.5	0.8	3	16	9	6.96	0.04	0.59	0.3	0.1	0.378	0.010	155	6.55	0.07	0.90
35.4	45.94	10.54	1.5	3	23	15	6.96	0.07	0.73	0.3	0.2	0.378	0.017	155	6.55	0.11	1.17
45.94	50.02	4.08	3.5	3	28	18	6.96	0.08	0.33	0.3	0.2	0.378	0.020	155	6.55	0.13	0.54
50.02	58.66	8.64	2.5	3	32	23	6.96	0.10	0.88	0.3	0.2	0.378	0.026	155	6.55	0.17	1.47
58.66	60.97	2.31	12.5	1	37	27	0.42	0.01	0.02	0.3	0.3	0.378	0.031	155	0.45	0.01	0.03
60.97	62.04	1.07	0.9	3	37	28	6.96	0.12	0.13	0.3	0.3	0.378	0.032	155	6.55	*0.21*	0.22
62.04	66.81	4.77	8.0	2	38	29	1.46	0.03	0.13	0.3	0.3	0.378	0.033	155	1.05	0.03	0.16
66.81	68.55	1.74	1.7	3	40	30	6.96	0.13	0.22	0.3	0.3	0.378	0.034	155	6.55	*0.22*	0.39
68.55	73.52	4.97	6.7	2	42	32	1.46	0.03	0.14	0.3	0.3	0.378	0.036	155	1.05	0.04	0.19
73.52	81.18	7.66	21.3	1	44	34	0.42	0.01	0.07	0.3	0.3	0.378	0.039	155	0.45	0.02	0.13

81.18	87	5.82	11.7	1	49	39	0.42	0.01	0.06	0.3	0.4	0.378	0.044	155	0.45	0.02	0.12
87	89	2	4.2	2	51	39	1.46	0.03	0.07	0.3	0.4	0.378	0.044	155	1.05	0.05	0.09
89	93.73	4.73	11.7	1	54	42	0.42	0.01	0.05	0.3	0.4	0.378	0.048	155	0.45	0.02	0.10
93.73	99.01	5.28	4.2	2	57	44	1.46	0.04	0.20	0.3	0.4	0.378	0.050	155	1.05	0.05	0.28
99.01	104.32	5.31	22.5	1	58	46	0.42	0.01	0.06	0.3	0.5	0.378	0.052	155	0.45	0.02	0.12
104.32	109.35	5.03	12.5	1	59	48	0.42	0.01	0.06	0.3	0.5	0.378	0.054	155	0.45	0.02	0.12
109.35	109.86	0.51	0.6	3	59	48	6.96	0.20	0.10	0.3	0.5	0.378	0.054	155	6.55	0.36	0.18
109.86	115.77	5.91	3.3	3	60	50	6.96	0.20	1.20	0.3	0.5	0.378	0.057	155	6.55	0.37	2.19
115.77	118.06	2.29	13.3	1	63	50	0.42	0.01	0.03	0.3	0.5	0.378	0.057	155	0.45	0.03	0.06
118.06	120	1.94	13.3	1	65	52	0.42	0.01	0.02	0.3	0.5	0.378	0.059	155	0.45	0.03	0.05
120	125	5	1.5	3	66	53	6.96	0.21	1.07	0.3	0.5	0.378	0.060	155	6.55	0.39	1.97
125	130.91	5.91	2.3	3	68	55	6.96	0.22	1.30	0.3	0.6	0.378	0.062	155	6.55	0.41	2.41
130.91	137.71	6.8	21.3	1	70	57	0.42	0.01	0.09	0.3	0.6	0.378	0.065	155	0.45	0.03	0.20
137.71	140.92	3.21	11.7	1	72	59	0.42	0.01	0.05	0.3	0.6	0.378	0.067	155	0.45	0.03	0.10
140.92	143.35	2.43	3.8	3	74	61	6.96	0.24	0.59	0.3	0.6	0.378	0.069	155	6.55	0.45	1.10
143.35	144.34	0.99	0.6	3	75	62	6.96	0.25	0.24	0.3	0.6	0.378	0.070	155	6.55	0.46	0.46
144.34	146.7	2.36	13.3	1	75	62	0.42	0.01	0.03	0.3	0.6	0.378	0.070	155	0.45	0.03	0.07
146.7	149.5	2.8	1.9	3	75.5	62.5	6.96	0.25	0.69	0.3	0.6	0.378	0.071	155	6.55	0.46	1.30
149.5	153.85	4.35	11.7	1	75.5	62.5	0.42	0.01	0.06	0.3	0.6	0.378	0.071	155	0.45	0.03	0.14
153.85	157.93	4.08	2.2	3	76	63	6.96	0.25	1.02	0.3	0.6	0.378	0.071	155	6.55	0.47	1.91
157.93	175.57	17.64	21.3	1	79	64	0.42	0.02	0.27	0.3	0.6	0.352	0.068	168	0.45	0.03	0.54
175.57	184.55	8.98	6.1	2	84	70	1.46	0.06	0.52	0.3	0.7	0.346	0.073	171	1.05	0.08	0.69
184.55	187.1	2.55	0.6	3	85	70	6.96	0.27	0.70	0.3	0.7	0.339	0.071	175	6.55	0.47	1.19
187.1	193.7	6.6	4.2	2	85	70	1.46	0.06	0.38	0.3	0.7	0.331	0.070	179	1.05	0.07	0.48
193.7	196.24	2.54	0.6	3	87	72	6.96	0.28	0.71	0.3	0.7	0.329	0.071	180	6.55	0.47	1.18

196.24	270.1	73.86	22.5	1	106	90	0.42	0.02	1.53	0.5	0.9	0.293	0.132	200	0.45	0.06	4.38
270.1	278.15	8.05	11.7	1	117	103	0.42	0.02	0.19	0.5	1.0	0.302	0.151	195	0.45	0.07	0.55
278.15	293.41	15.26	22.5	1	117	103	0.42	0.02	0.36	0.5	1.0	0.329	0.165	180	0.45	0.07	1.13
293.41	296.82	3.41	11.7	1	114.5	99.5	0.42	0.02	0.08	0.5	1.0	0.333	0.166	178	0.45	0.07	0.25
296.82	310.44	13.62	33.8	1	114.5	100	0.42	0.02	0.31	0.5	1.0	0.372	0.186	158	0.45	0.08	1.14
							Total inflow (Heuer and Raymer) 15.35								Total initial inflow (IMS)		29.81

Numbers in *italics* represent zone of inflow >0.2 l/min/m

aRefer to Table 7

bSeparation between hole/tunnel section and water source (Fig. 5 for reference)

cRefer to Table 4

REFERENCES

1. Barton N (2000) TBM tunnelling in jointed and fractured rock. AA Balkema, Rotterdam, Netherlands

2. Barton N, Lien R, Lunde J (1974) Engineering classification of rock masses for the design of tunnel support. Rock Mech Rock Eng 6(4):189–236

3. Bienawski ZT (1973) Engineering Classifications of jointed rock masses. Civ Eng S Afr 15:335–344

4. Bienawski ZT (1984) Rock mechanics design in mining and tunnelling. AA Balkema, Rotterdam, Netherlands

5. British Standards Institution (1999) BS 5930:1999 Code of practice for site investigations

6. Goodman R, Moye D, Schalkwyk A, Javendel I (1965) Ground-water inflow during tunnel driving. Eng Geol Bull IAEG 2(1):39–56

7. Heuer R (1995) Estimating rock tunnel water inflow. In: Proceedings of Rapid Excavation and Tunnelling Conference 1995, pp 41–60

8. Heuer R (2005) Estimating rock tunnel water inflow—II. In: Proceedings of Rapid Excavation and Tunnelling Conference 2005, pp 394–407

9. Hong Kong Government (2006) Landslide Preventive Works at Po Shan Road, Mid-levels for Contract No. GE/2005/45. Geotechnical Engineering Office, Civil and Development Engineering Department, Hong Kong Government

10. McFeat-Smith I, Turner VD, Bracegirdle DR (1985) Tunnelling conditions in Hong Kong. Hong Kong Engineer 13(6):13–25

11. McFeat-Smith I, MacKean R, Waldmo O (1998) Water inflows in bored rock tunnels in Hong Kong: prediction, construction issues and control measures. ICE Conference on Urban Ground Engineering, Hong Kong

12. Raymer JH (2001) Predicting groundwater inflow into hard-rock tunnels: estimating the high-end of the permeability distribution. In: Proceedings of Rapid Excavation and Tunnelling Conference 2001, pp 1027–1038

CITATION

CHAPTER 1

Lixin Wu and Shanjun Liu (2009). Remote Sensing Rock Mechanics and Earthquake Thermal Infrared Anomalies, Advances in Geoscience and Remote Sensing, Gary Jedlovec (Ed.), ISBN: 978-953-307-005-6, InTech, DOI: 10.5772/8292.

CHAPTER 2

Karl Gunnar Holter, "Performance of EVA-Based Membranes for SCL in Hard Rock," doi; 10.1007/s00603-015-0844-5.

CHAPTER 3

Md. Aminul Islam and Paal Skalle, "An Experimental Investigation of Shale Mechanical Properties Through Drained and Undrained Test Mechanisms," DOI 10.1007/s00603-013-0377-8.

CHAPTER 4

Yasuto Itoh, Machiko Tamaki and Osamu Takano (2013). Rock Magnetic Properties of Sedimentary Rocks in Central Hokkaido — Insights into Sedimentary and Tectonic Processes on an Active Margin, Mechanism of Sedimentary Basin Formation - Multidisciplinary Approach on Active Plate Margins, Dr. Yasuto Itoh (Ed.), ISBN: 978-953-51-1193-1, InTech, DOI: 10.5772/56650.

CHAPTER 5

Irfan Celal Engin (2013). Theories on Rock Cutting, Grinding and Polishing Mechanisms, Tribology in Engineering, Dr. Hasim Pihtili (Ed.), ISBN: 978-953-51-1126-9, InTech, DOI: 10.5772/56046.

CHAPTER 6

X. P. Lai, P. F. Shan , J. T. Cao, F. Cui, H. Sun, "Simulation of Asymmetric Destabilization of Mine-void Rock Masses Using a Large 3D Physical Model," doi: 10.1007/s00603-015-0740-z.

CHAPTER 7

Wing Kei Kong, "Water Ingress Assessment for Rock Tunnels: A Tool for Risk Planning," doi: 10.1007/s00603-011-0163-4.

INDEX

A

acoustic emission (AE) 241
American Society for Testing and Materials (ASTM) 129
anisotropy of magnetic susceptibility (AMS) 179
arc-arc collision 174

B

bedding 112, 113, 115, 117, 118, 119, 121, 122, 126, 127, 129, 130, 132, 133, 135, 138, 139, 141, 148, 153, 156, 157, 158, 160, 163, 164, 165, 166, 167, 168
Brazilian Test 119
buried pressure sensor (BPS) 241

C

circular grain 212, 213
Circular saws 207
close-field photogrammetry (CFP) 241
crack optical acquirement (COA) 241

D

degree of capillary saturation (DCS) 59
desorption 78

detrital remanent magnetization (DRM) 178
Dihedral angle 203

E

earthquake (EQ) 1
electromagnetic (EM) 1
excavation disturbed zone (EDZ) 232
exploiting extremely steep and thick coal seams (ESTCS) 227

G

Goodman method 262
granodiorite 3, 8, 12

H

high-pressure/high-temperature (HPHT) 112
horizontal directional drillholes (HDD) 269

I

infrared radiation (IRR) 2
International Society for Rock Mechanics (ISRM) 129

L

longwall top coal caving (LTCC) 228, 259, 260

N

natural remanent magnetization (NRM) 179

P

paleomagnetic study 185, 192
Poisson's ratio 112, 127, 128, 129, 133, 134, 148, 152, 153, 156, 157, 166, 167
progressive alternating field demagnetization (PAFD) 179, 181
pseudotachylyte 23

R

Remote Sensing Rock Mechanics (RSRM) 2, 44
rock cutting mechanism 213
rockmass 265, 268, 272, 273
rock quality designation (RQD) 235
roof separation observation (RSO) 241

S

Scanning electron microscope (SEM) 97

Simulated hydraulic support (SHS) 241
solid material 201
solid particle 201
sorptivity 47, 69, 76, 78
Sorptivity 77
sprayed concrete tunnel linings (SCL) 47
stress pattern analysis by thermal emission (SPATE) 1
sub-horizontal section top coal caving (SSTCC) 227, 257

T

tectono-sedimentology 175
tensile bond 48, 49, 55, 57, 72, 73, 95, 96, 97, 98, 101, 103, 104, 106
thermal infrared (TIR) 1
thermo-elastic stress analysis (TSA) 1
Triaxial Testing 119, 122
turbidite 178, 194, 196, 197

X

X-ray diffraction analysis 116

Y

Young's modulus 112, 127, 128, 129, 130, 133, 139, 141, 148, 164, 166, 167, 169